GW01162100

Mike,
I owe you a huge debt of gratitude for your help and inspiration over the course of our friendship!
Cinnamon

Climate Change Law and Policy

Climate Change Law and Policy

EU and US Approaches

CINNAMON PIÑON CARLARNE

OXFORD
UNIVERSITY PRESS

OXFORD
UNIVERSITY PRESS

Great Clarendon Street, Oxford OX2 6DP

Oxford University Press is a department of the University of Oxford.
It furthers the University's objective of excellence in research, scholarship,
and education by publishing worldwide in

Oxford New York

Auckland Cape Town Dar es Salaam Hong Kong Karachi
Kuala Lumpur Madrid Melbourne Mexico City Nairobi
New Delhi Shanghai Taipei Toronto

With offices in

Argentina Austria Brazil Chile Czech Republic France Greece
Guatemala Hungary Italy Japan Poland Portugal Singapore
South Korea Switzerland Thailand Turkey Ukraine Vietnam

Oxford is a registered trade mark of Oxford University Press
in the UK and in certain other countries

Published in the United States
by Oxford University Press Inc., New York

© Cinnamon P. Carlarne, 2010

The moral rights of the author have been asserted

Crown copyright material is reproduced under Class Licence
Number C01P0000148 with the permission of OPSI
and the Queen's Printer for Scotland

Database right Oxford University Press (maker)

First published 2010

All rights reserved. No part of this publication may be reproduced,
stored in a retrieval system, or transmitted, in any form or by any means,
without the prior permission in writing of Oxford University Press,
or as expressly permitted by law, or under terms agreed with the appropriate
reprographics rights organization. Enquiries concerning reproduction
outside the scope of the above should be sent to the Rights Department,
Oxford University Press, at the address above

You must not circulate this book in any other binding or cover
and you must impose the same condition on any acquirer

British Library Cataloguing in Publication Data
Data available

Library of Congress Cataloging-in-Publication Data
Carlarne, Cinnamon Piñon.
Climate change law and policy : EU and US approaches / Cinnamon Piñon Carlarne.
  p. cm.
Includes bibliographical references and index.
ISBN 978–0–19–955341–9 (acid-free paper)
 1. Climatic changes—Law and legislation—United States. 2. Climatic
changes—Law and legislation—European Union countries. I. Title.
K3585.5.C37 2010
344.2404'6342—dc22   2010023801

Typeset by Newgen Imaging Systems (P) Ltd., Chennai, India
Printed in Great Britain
on acid-free paper by
CPI Antony Rowe,
Chippenham, Wiltshire

ISBN 978–0–19–955341–9

1 3 5 7 9 10 8 6 4 2

To my lovely and inquisitive daughters, Matilda and Sophia,
and my husband, John, and
my Texas family for all of their love and support.

Preface

International environmental law is a relatively new field wherein academics and practitioners are still working to discern the boundaries, principles, and norms of the discipline. Defying this youthful status, international environmental law now faces one of the greatest challenges to ever confront the international community—climate change. Global climate change threatens the integrity of the natural environment as well as the physical, economic, and social stability of the human environment. It poses environmental, ethical, and security dilemmas that defy conventional legal or political responses.

In 1992, the international community adopted the United Nations Framework Convention on Climate Change (UNFCCC) as the primary legal and political architecture for global efforts to address climate change. The UNFCCC created the foundations for international cooperation but stopped short of drafting a blueprint for collective action, prompting the adoption in 1997 of the Kyoto Protocol.

The Kyoto Protocol to the UNFCCC offered a more substantive and detailed framework for action premised on a system of targets and timetables for developed countries to reduce their greenhouse gas emissions. Since the Kyoto Protocol's entry into force in February 2005, countries all over the world have intensified their efforts to develop comprehensive national climate change regimes. Regional climate change programs have emerged in a seemingly haphazard manner within diverse and highly localized political and legal environments. These efforts have produced great successes and great failures that should be heeded as the global community strives to find effective methods for addressing climate change at the local, national, and international levels.

With the recent near-collapse of international negotiations at the 2009 Copenhagen Climate Change Conference, the urgency of reviewing the disparate legal and political strategies key actors are using to address climate change becomes all the more important. However, there continues to be a dearth of research analyzing and comparing the diverse legal and political strategies that key states and regional actors are using to address the threats posed by climate change. This book endeavors to help fill this gap by exploring how the United States and the European Union converge and diverge in their approaches to climate change.

While this book focuses on the US and the EU, it does not presuppose their preeminence in global climate negotiations. The US is no longer the prevailing hegemonic power and the EU possesses greater symbolic than substantive authority in international politics. Both the US and the EU have ceded great power to the rapidly developing economies during the first decade of the new millennium. A new political order is emerging wherein power is shared among developed and developing nations in new and, as yet, undetermined ways. Within the more specific context of international climate change politics, power is in flux and no single nation maintains the ability to coerce the global community down a specific pathway. Despite ongoing changes in global climate politics, however, the EU and US continue to occupy important symbolic and substantive roles. As key political actors and global polluters, the actions and omissions of the US and the EU influence not only the parameters of global climate negotiations but also the material ability of the global community to reduce net greenhouse gas emissions.

In exploring the actions of the US and the EU, this book adopts a comparative approach. However, in order to more accurately reflect the profound interplay between legal and regulatory institutions and social, cultural, economic, and political processes, it also explores the relationship among the social and legal processes that influence US and EU climate change policymaking.

The aims of the manuscript are threefold. First, it aims to contribute to the development of more effective and equitable climate change laws and policies. Second, it seeks to contribute to the emerging field of comparative environmental law. Third, it endeavors to improve understanding of the socio-legal factors that influence climate policy in different contexts. Underlying these goals is a desire to provide an accessible analysis to a wide audience of readers, including academics, practitioners, and interested members of civil society.

As we approach the 20th anniversary of the UNFCCC, we must look back to move forward. Exploring the climate policymaking processes and the root causes of the successes and failures of regional initiatives offers an important starting point for improving the viability of climate change law and policymaking efforts at a critical moment in the history of the global climate regime.

Acknowledgments

This book would not have been possible without the support of a number of people and organizations, to whom I am extremely grateful. First, I am indebted to my friends and colleagues at Wadham College, the Centre for Socio-Legal Studies, and the Faculty of Law at the University of Oxford for their encouragement and advice. Particular debts of gratitude are owed to Michael Depledge, Laura Hoyano, Jeffrey Hackney, Liz Fisher and Bettina Lange. Second, I am grateful to my colleagues at the University of South Carolina School of Law for their support and insight. Third, special thanks are offered to all of my invaluable student research assistants. Finally, I am deeply thankful to my family for their enduring patience and unwavering support throughout the writing process.

Contents

List of abbreviations xvii

PART I: THE POLITICS OF INTERNATIONAL CLIMATE CHANGE

1. A Brief History of Climate Politics and the Roles of the US and EU Therein ... 3

PART II: CLIMATE CHANGE LAW AND POLICY IN THE UNITED STATES

2. Law and Policy in the United States ... 21
3. Sub-Federal Laws and Policies in the United States ... 61
4. Litigation, Regulation, and International Law as Law and Policy Drivers in the United States ... 98

PART III: CLIMATE CHANGE LAW AND POLICY IN THE EUROPEAN UNION

5. Law and Policy in the European Union ... 143
6. Member State Laws and Policies ... 192

PART IV: US AND EU CLIMATE CHANGE LAWS AND POLICIES COMPARED

7. US and EU Laws and Policies Compared ... 237
8. Socio-Legal Factors Influencing US and EU Law and Policymaking ... 309

PART V: THE FUTURE OF INTERNATIONAL CLIMATE CHANGE POLITICS

9. Conclusions and the Way Forward ... 347

Bibliography ... 363
Index ... 375

Detailed Contents

List of abbreviations	xvii

PART I: THE POLITICS OF INTERNATIONAL CLIMATE CHANGE

1. A Brief History of Climate Politics and the Roles of the US and EU Therein	3
Concise History of the International Climate Regime	4
The Role of the US and the EU in International Climate Change Politics	11

PART II: CLIMATE CHANGE LAW AND POLICY IN THE UNITED STATES

2. Law and Policy in the United States	21
American Federalism: Federal Governance	21
The Federal Government and the States 23	
Environmental Law and Policy in the USA	25
The Environmental Decade 26	
Shifting Tides 29	
Federal–State Relations in US Environmental Law 31	
US Federal Climate Change Policy	34
President Bush's National Climate Change Strategy: An Overview 37	
Recent Political Activity: Congressional Engagement and President Barack Obama 49	
Conclusion	58
3. Sub-Federal Laws and Policies in the United States	61
Introduction	61
Overview of State Climate Change Initiatives	63
Regional Climate Change Initiatives 64	
California—Climate Policy Driver 77	
State Climate Change Laws and Policies: A Review 88	
Local Climate Change Laws and Policies 89	
State and Local Climate Change Policies in Review	95

4. **Litigation, Regulation, and International Law as Law and Policy Drivers in the United States** — 98

   Climate Change Litigation in the United States — 98
      Private Suit Against the Federal Government: *Friends of the Earth, Inc et al v Spinelli et al*   99
      State Suit against Private Industry: *State of Connecticut et al v American Electric Power Co et al*   101
      Climate Change Litigation Reaches the Supreme Court: *Massachusetts et al v EPA*   104
      Fuel Economy Regulatory Challenge: *Center for Biological Diversity v National Highway Traffic Safety Administration*   110
      Other On-going Climate Change Litigation   112
   Regulatory Actions for Climate Change — 114
      The Endangered Species Act   114
      The Clean Water Act   119
      The Clean Air Act   122
   Reverse Momentum — 123
   International Mechanisms for Domestic Climate Change Action — 125
      The Inter-American Commission for Human Rights   125
      The World Heritage Convention   129
      Complementary International Activity   133
   Overview of Sub-Federal Climate Change Initiatives — 137
   Conclusion — 140

PART III: CLIMATE CHANGE LAW AND POLICY IN THE EUROPEAN UNION

5. **Law and Policy in the European Union** — 143

   Introduction to the European Union — 143
   European Union Decision-Making Institutions — 144
      The Council   144
      The Commission   145
      The European Parliament   146
      The Court of Justice of the European Union   146
      Governance Overview   147
      The Legislative Process   150
   European Union Competences in Environmental Law — 152
      Community Programme of Action on the Environment 1973   152
      Single European Act of 1986   154
      Maastricht Treaty of 1992, Amsterdam Treaty of 1999, Nice Treaty of 2001   154
      The Consolidated EU Treaty   155
      The Treaty of Lisbon   156
      European Environment Agency 1993   157

Climate Change Law and Policy in the European Union 158
  EU-15 159
  Beyond the EU-15 160
  Allocation of Responsibilities within the European Union 160
  European Union Climate Change Policy Framework 162
Progress and Challenges in European Union Climate Change Policy 182
  Prioritizing Climate Change 186
  Miles to Go: Revision and Recession 187
  Basic Differences 188
Conclusion 189

6. **Member State Laws and Policies** 192
Introduction 192
Germany 194
  German Governance and Environmental Politics 194
  Rule of Law in Germany 196
  Foundations of German Environmental Law 197
  Introduction to German Climate Change Law 198
  Components of German Climate Strategy 203
  Progress and Challenges 213
The United Kingdom 215
  UK Climate Policy in Context 216
  Evolution of the UK Climate Strategy 221
  Progress and Challenges 224
Beyond the EU-15: The Case of Poland 225
  Polish Climate Policy 227
Conclusion 232

PART IV: US AND EU CLIMATE CHANGE LAWS AND POLICIES COMPARED

7. **US and EU Laws and Policies Compared** 237
Introduction 237
EU and US Climate Change Policy Pathways Compared 239
  Early Negotiations 239
  Behind the Scenes 241
From Kyoto to Copenhagen: Transatlantic Divergences 244
  Measures for Progress and Commitment to GHG
    Emission Reductions 245
  Policy Mix 246
  Source of Leadership 253
  Policy–Evolution 257
  Commitment to International Negotiations, Past and Present 259
  Summary of Divergences 263

From Kyoto to Copenhagen: Transatlantic Convergences 264
    Systemic Pushes and Pulls   264
    Economic Centrality   276
    Security: The New Driver   283
    Shared Trouble Spot: Transport Sector   292
    Other Macro-Level Convergences   301
Conclusion 308

8. **Socio-Legal Factors Influencing US and EU Law and Policymaking** 309
Introduction 309
Giving Context to Transatlantic Policy Choices 310
    Governance   311
    Polity   317
    Scientific Uncertainty and Risk   325
    Media and Civil Society   330
    Modes of Capitalism   334
    Ethics   336
Conclusion 341

PART V: THE FUTURE OF INTERNATIONAL CLIMATE CHANGE POLITICS

9. **Conclusions and the Way Forward** 347
The Road Travelled 347
    The Escalation of Climate Change Politics in 2009   348
    The 2009 Copenhagen Climate Change Conference   352
    Reassessing the Roles of the European Union and the United States   359
The Path Ahead 361

Bibliography 363
Index 375

List of abbreviations

CO2	carbon dioxide
G-77	The Group of 77
ACES	American Clean Energy and Security Act
AEP	American Electric Power
AOSIS	Alliance of Small Island States
ARRA	American Recovery and Reinvestment Act
AWG-KP	Ad Hoc Working Group on Further Commitments for Annex I Parties under the Kyoto Protocol
AWG-LCA	Ad Hoc Working Group on Long-Term Cooperative Action under the Convention
BACT	best available control technologies
BMU	German Federal Ministry for the Environment, Nature Conservation and Nuclear Safety
CAA	US Clean Air Act
CAFE	corporate average fuel economy standards
CARB	California Air Resources Board
CBD	Convention on Biological Diversity
CBDR	common but differentiated responsibilities
CCC	UK Committee on Climate Change
CCP	Cities for Climate Protection Campaign
CCPM	common coordinated policies and measures
CCSP	US Climate Change Science Program
CCSTI	Cabinet-level Committee on Climate Change Science and Technology Integration
CCTP	Climate Change Technology Plan
CDM	clean development mechanism
CDU	German Christian Democratic Union
CEQ	White House Council on Environmental Quality
CFSP	EU Common Foreign and Security Policy
CHP	combined heat and power
CIEL	Center for International Environmental Law
COP	conference of the parties
CSIS	Center for Strategic and International Studies
CSU	German Christian Socialist Union
CWA	US Clean Water Act
DECC	UK Department of Energy and Climate Change
DEI	Diversified Energy Initiative
DG	directorates-general

List of abbreviations

DOD	US Department of Defense
DOI	US Department of the Interior
DOT	US Department of Transportation
EAB	US Environmental Protection Agency Environmental Appeals Board
EAP	Environmental Action Programme
EC	European Community
ECCP	European Climate Change Programme
ECR	Eastern Climate Registry
EEA	European Environment Agency
EEC	European Economic Community
EEG	German Renewable Energy Sources Act
EESC	European Economic and Social Committee
EIA	Environmental Impact Assessment
EP	European Parliament
EPA	US Environmental Protection Agency
EPCA	US Energy Policy and Conservation Act of 1975
ESA	US Endangered Species Act
ET	emissions trading
EU	European Union
EU-ETS	EU Emissions Trading System
Ex-Im	US Export-Import Bank
FWS	US Fish and Wildlife Service
GDP	Gross Domestic Product
GHG	Greenhouse Gas
GMO	genetically modified organism
IACHR	Inter-American Commission on Human Rights
ICJ	International Court of Justice
ICLEI	International Council for Local Environmental Initiatives
IMA	Interministerial Working Group
IPCC	Intergovernmental Panel on Climate Change
IRENA	International Renewable Energies Agency
JI	joint implementation
MEA	multilateral environmental agreement
MEP	Member of the European Parliament
MGGRA	Midwestern Greenhouse Gas Reduction Accord
MGR	Midwest Greenhouse Registry
NAP	national allocation plan
NCPP	German National Climate Protection Programme
NEG/ECP	New England Governors' Conference Climate Change Action Plan 2001
NEPA	US National Environmental Policy Act
NGO	non-governmental organization

List of abbreviations

NHTSA	US National Highway Traffic Safety Administration
NRDC	Natural Resources Defense Council
OCC	UK Office for Climate Change
OECD	Organisation for Economic Co-operation and Development
OMB	US Office of Management and Budget
OPIC	US Overseas Private Investment Corporation
PPI	Powering the Plains Initiative
PSD	Prevention of Significant Deterioration
QELRC	Quantified Emission Limitation and Reduction Commitment
R&D	research and development
RGGI	Regional Greenhouse Gas Initiative
ROC	renewable obligation commitment
SCCI	Southwest Climate Change Initiative
SEA	Single European Act (1986)
SIPs	state implementation plans
SPD	German Social Democratic Party
TEU	Treaty on European Union
TSD	technical support document
TVA	US Tennessee Valley Authority
UCS	Union of Concerned Scientists
UK	United Kingdom
UKCCP	United Kingdom Climate Change Programme
UN	United Nations
UNEP	United Nations Environment Programme
UNFCCC	United Nations Framework Convention on Climate Change (1992)
USCM	US Conference of Mayors
WBGU	German Advisory Council on Global Change
WCG	West Coast Governors' Global Warming Initiative
WCI	Western Climate Initiative
WGA	Western Governors' Association
WMO	World Meteorological Organization
WTO	World Trade Organization

PART I

THE POLITICS OF INTERNATIONAL CLIMATE CHANGE

1

A Brief History of Climate Politics and the Roles of the US and EU Therein

As the twentieth anniversary of the United Nations Framework Convention on Climate Change (UNFCCC) and the end of the first compliance period for the Kyoto Protocol to the UNFCCC nears, the scientific and political atmosphere surrounding climate change reflects a sense of urgency unparalleled in the history of international environmental law. Discussions over the future of the international climate regime pervade environmental, economic, and security forums and rank high on political agendas worldwide. News on climate change science, climate change politics, and climate change consequences and solutions appear with increasing frequency in local, regional, and global media outlets. While once a fringe issue primarily confined to scientific deliberation and heralded largely by environmentalists, global climate change now dominates mainstream debate and has become part of popular culture in much of the world. From science, to religion, to economics, to medicine, to politics, to development studies, to law, there is virtually no academic or political forum devoid of discussions on climate change.

The proliferation of 'climate discourse' reflects a widespread and growing recognition of the threats posed by climate change. Containing the causes of climate change requires much more than mere awareness, however; it demands the modification of primary economic, social, and legal structures, while containing and adapting to the consequences of climate change requires global cooperation to a degree never before witnessed in international environmental law. The extent of mitigation and adaptation efforts needed to maintain adequate standards of living is still being debated and the complexities of this debate are only marginally reflected in conventional thought and discourse. Complicating this process is disciplinary and jurisdictional fragmentation.

There is a growing body of academic literature exploring questions of climate change at the margins between the physical and social sciences. A smaller body of literature analyzes and compares approaches to climate change mitigation and adaptation worldwide. However, there is a dearth of literature dealing with the intersection between socio-legal studies and comparative law analysis. Thus, our understanding of how and why localities, States, and regional entities are converging and diverging in response to climate change remains patchy. For this reason, this book focuses primarily on examining climate change law and policymaking at multiple levels in the European Union and the United States to shed light on the processes by which state and federal entities endeavor to address climate change and the consequences of these endeavors for larger global efforts to structure an effective global climate regime.

With these points in mind, this introductory chapter sets the scene by briefly reviewing the history of climate change regime building before laying the foundations for the ensuing comparative analysis of United States (US) and European Union (EU) climate change law and policymaking.

Concise History of the International Climate Regime

The evolution of climate change from a peripheral topic to one of the most highly contentious themes of political debate can be traced through the development of the international climate change legal regime, as embodied by the UNFCCC[1] and the Kyoto Protocol[2] to the UNFCCC.

The UNFCCC and the Kyoto Protocol represent international efforts to address the causes and consequences of global climate change. Both the UNFCCC and the Kyoto Protocol promote the goal of 'stabilizing atmospheric concentrations of greenhouse gas at a level that would prevent dangerous anthropogenic interference in the climate system'.[3] Together, the UNFCCC and the Kyoto Protocol create the backbone of the international climate change regime.

The breadth and objectives of the UNFCCC and, especially, the Kyoto Protocol, represent a new era in international environmental law—one that reflects the dual realities that environmental change can no longer be regarded as tangential to social and economic well-being and that efforts to prevent and adapt to environmental change will require concerted international cooperation and transformations in how we think about and interact with the natural environment. The Kyoto Protocol, more than any existing multilateral environmental agreement, impacts local and national economies, lifestyle choices, political beliefs, and ethical perspectives.

Early concern for the health of the global atmosphere dates back to the beginning of the global environmental era in the late-1960s. In 1965, the US President's Science Advisory Committee identified the process of global climate change and suggested the possibility that climate change might be par-

---

[1] United Nations Framework Convention on Climate Change (adopted 9 May 1992, entered into force 21 March 1994) 31 ILM 849 (UNFCCC) art 2.
[2] Kyoto Protocol to the United Nations Framework Convention on Climate Change (adopted 11 December 1997, entered into force 16 February 2005) 37 ILM 32 (Kyoto Protocol).
[3] UNFCCC (n 1) preamble.

tially attributed to human activities and could have important consequences.[4] Following this early warning on the problem of global atmospheric warming, in the 1970s, Drs Sherwood Rowland and Mario Molina highlighted the threat that chlorofluorocarbons (CFCs) posed to the Earth's ozone layer.[5] This revelation initiated a period of intense research, culminating in the enactment of the 1985 Vienna Convention for the Protection of the Ozone Layer[6] and the 1987 Montreal Protocol on Substances that Deplete the Ozone Layer.[7]

Ozone depletion highlighted the global scale of many emerging environmental problems and represented the first time that scientists and policymakers worldwide worked cooperatively to identify and address a problem of atmospheric pollution. The international regime for curbing ozone depletion is largely heralded as a success with the Montreal Protocol being 'recognized as a landmark accord in the most effective international environmental regime to date'.[8] The time-span between when scientists first identified the risks that CFCs pose to the ozone layer and the time that the Vienna Convention and the Montreal Protocol were implemented was remarkably short. Ozone depletion was quickly identified as a global problem and, despite contentious negotiations, the international community was able to devise and implement an effective legal regime. The epistemic communities formed to analyze ozone depletion, the model of international negotiations pursued to formulate the legal regime, and the resulting international ozone regime offered promising models for subsequent problems of managing the global environmental commons. In the wake of international ozone negotiations, there was great hope that the international community could follow a similar path to develop an international climate change regime.

Earlier warnings notwithstanding, the climate change debate emerged in the midst of international efforts to address ozone depletion. During the early 1970s, scientists became increasingly concerned about the impact of anthropogenic releases of greenhouse gases on the Earth's atmosphere. By the late 1970s, scientists had identified links between CO2 emissions and suspected global temperature increases, spurring a new era in global climate research. It was not until the late 1980s, however, with the establishment of the Intergovernmental Panel on Climate Change (IPCC) in 1988, that international cooperation on global climate change began in earnest. By the mid-1980s, it was apparent that climate change—although similar in many ways to the problem of ozone

---

[4] Shardul Agrawala, 'Context and Early Origins of the Intergovernmental Panel on Climate Change' (1998) 39 *Climactic Change* 605, 606.
[5] Richard Elliot Benedick, *Ozone Diplomacy* (2nd edn, Harvard University Press, Cambridge, 1998) 11.
[6] Vienna Convention for the Protection of the Ozone Layer (adopted 22 March 1985, entered into force 22 September 1988) 1513 UNTS 293.
[7] The Montreal Protocol on Substances That Deplete the Ozone Layer (adopted 16 September 1987, entered into force 1 January 1989) 1522 UNTS 293.
[8] Laura Thoms, 'A Comparative Analysis of International Regimes on Ozone and Climate Change with Implications for Regime Design' (2003) 41 *Colum J Transnatl L* 795, 797.

depletion—posed vastly more complex questions in terms of scientific understanding, economic implications, cultural change, and political solutions.

Due to this complexity, international negotiations on climate change were divisive almost from the very beginning. Leading up to the negotiations for the UNFCCC, State perspectives ranged from small island States calling for dramatic, immediate, and binding emissions reduction obligations, to developing country members of the G77/China warning against the imposition of any obligations that would hinder economic growth, to the US and the EU wavering between the relative priorities of environmental protection and economic supremacy.[9]

Despite contentious international relations, at the United Nations Conferences on Environment and Development in Rio de Janeiro in 1992, following intense negotiations, the global community adopted the UNFCCC. The UNFCCC is a framework convention that sets out the broad, overarching goals of the international community in relation to climate change. The Convention requires Annex I countries[10] to adopt national policies that would enable them to limit their greenhouse gas emissions to 1990 levels and to 'adopt national policies and take corresponding measures on the mitigation of climate change, by limiting...anthropogenic emissions of greenhouse gases and protecting and enhancing its greenhouse gas sinks and reservoirs'.[11] The UNFCCC did not, however, create a system of detailed and legally binding obligations for its member states. The final form of the UNFCCC reflected the global community's ability to agree upon the need to 'stabiliz[e] atmospheric concentrations of greenhouse gas at a level that would prevent dangerous anthropogenic interference in the climate system',[12] but it also revealed a continuing inability to agree upon what that level would be or how to best achieve the requisite emissions reductions.

UNFCCC negotiations were fraught with problems from beginning to end. Developed countries disagreed about levels and types of obligations; developed and developing countries argued over proper roles and responsibilities; developing countries quarrelled amongst themselves over the proper level of participation and response; and overhanging the entire debate were continuing questions of whether human-induced climate change was, in fact, even a scientifically verifiable reality.

Both the US and the EU played key roles in the negotiations for the UNFCCC. The climate negotiations, however, saw the US and the EU reversing roles from those they had adopted only a few years before during the ozone negotiations. In the negotiations for the Vienna Convention, the US led negotiations for a

---

[9] Dana R Fisher, *National Governance and the Global Climate Change Regime* (Rowman & Littlefield Publishers Inc, Lanham, 2004).
[10] The Annex I Countries include the 40 industrialized countries that bear primary responsibility for both creating and addressing the problem of global climate change.
[11] UNFCCC (n 1) art 4.2.   [12] UNFCCC (n 1) preamble.

strong international regime whereas the EU initially opposed stringent legal limitations. In contrast, with the onset of climate negotiations, the US expressed tempered skepticism over the science and politics behind the treaty negotiations while the EU emerged as an international leader in promoting aggressive action to address climate change. Thus, while the EU delegation affirmed the threats posed by climate change and the wisdom of early action, the US delegation questioned the validity of the science on human-induced global warming and the wisdom behind adopting legally binding greenhouse gas reductions that would result in economic strain and social upheaval. The positions that the US and the EU adopted during UNFCCC negotiations signaled their shifting roles in global environmental politics and foreshadowed the difficulties to come in the negotiation of the Kyoto Protocol.

Negotiations for a Protocol to supplement the Convention with more robust emission reduction commitments began in 1995 at the first meeting of the Conference of the Parties to the UNFCCC (COP-1). The primary outcome of COP-1 was the drafting of the Berlin Mandate, which specified that, by 1997, the Parties to the Convention would draft a Protocol that included additional commitments for industrialized countries for the post-2000 period but imposed no new commitments on developing countries.[13] The Berlin Mandate reflected widespread support for supplementing the UNFCCC with a more detailed, binding framework for achieving requisite levels of greenhouse gas emission reductions. The stipulation in the Berlin Mandate that the Protocol limit legally binding emissions reductions obligations to developed countries established the parameters for the debate, both past and present.

The decision to exclude developing countries from binding greenhouse gas reduction commitments created a permanent rift in global climate change negotiations. In particular, the inclusion of this condition in the Kyoto Protocol divided the US and the EU and underscored the US decision not to participate in the Protocol.

During the negotiations for the UNFCCC and the early stages of the negotiations towards the drafting of the Kyoto Protocol, the atmosphere in the US towards climate change was one of scepticism, but it was not one of open hostility. President George H.W. Bush signed the UNFCCC on behalf of the US and the US was one of the first countries to ratify the UNFCCC.[14] In 1995, however, the tide took a dramatic turn towards open hostility when the Republicans assumed control of Congress. Although the executive branch was now controlled by a Democrat—President Bill Clinton—the general tide had turned against

---

[13] S Agrawala and S Andresen, 'US Climate Policy: Evolution and Future Prospects' (2001) 12[2–3] *Energy and Environment* 117.

[14] President Bush brought the UNFCCC to the US Senate for ratification on 8 September 1992. The Senate ratified the treaty less than one month later on 7 October 1992, with the requisite two-thirds majority vote. President Bush then signed the instrument of ratification one week later on 13 October 1992. Thus, it took less than two months for the US to ratify the UNFCCC.

adopting new environmental obligations at either the domestic or the international level.

Tensions between the US executive and legislative branches ran high during the mid-1990s. With tensions simmering at home, Vice-President Al Gore arrived in Berlin to represent the US at COP-1. During the COP-1 negotiations, Al Gore's perceived failure to represent the interests of the US on the issue of developing country participation in a legally binding regime created a political rift between Congress and the Clinton Administration—so much so that one commentator has described Al Gore's decision to sign the Berlin Mandate as 'haunt[ing] [the Democrats] since 1995'.[15] This gulf was deepened even further when, at the COP-2 in Geneva, the Clinton Administration supported legally binding obligations applicable only to industrialized countries.

In response to the US executive branch's apparent disregard for Congressional priorities, on 25 July 1997, the US Senate passed the Byrd–Hagel Resolution by a margin of 95–0.[16] As will be explored further in Chapter 8, this Resolution expressed the view of the Senate that the US should not be a signatory to *any* protocol that exempted developing countries from legally binding obligations. The passage of the Byrd–Hagel Resolution virtually precluded the possibility that the US would ratify the subsequently drafted Kyoto Protocol.

Despite clear Congressional opposition to almost any possible outcome, during the final negotiations for the Kyoto Protocol in 1997, Al Gore continued to pursue an aggressive climate change agenda on behalf of the US. In fact, while the US negotiation team went to the Kyoto negotiations with the idea of agreeing to a 1–2% reduction obligation, Al Gore ultimately committed the US to a 7% reduction of greenhouse gas emissions below 1990 levels by 2008–2012. The Vice President's actions were greeted by Congressional claims that he had 'caved in'[17] and that he had 'destroyed the last bit of confidence that the Senate and climate change opposition had in the administration'.[18]

Entrenched political tensions within the US made it unfeasible that Congress would ratify the Kyoto Protocol, regardless of its final form. Nevertheless, international negotiations continued, led by an aggressive EU delegation. While the US bickered and stalled, the EU took advantage of the growing gap in leadership to craft a reputation as the global leader on climate change policy and to push ahead with Protocol negotiations. Despite US inconsistency and European efforts to lead, the US had a heavy hand in shaping the final version of the Kyoto Protocol, as adopted in 1997.

---

[15] Fisher (n 9) 124.
[16] Eg John Vogler and Charlotte Bretherton, 'The European Union as a Protagonist to the United States on Climate Change' (2006) 7[1] *Intl Studies Perspectives* 17, emphasizing that 'the question of the future commitments of the developing economies remains at the heart of debates about the climate regime'. [17] Fisher (n 9) 129.
[18] Ibid 131. Congress responded to the executive branch's perceived failures by passing a rider known as the Knollenberg rider, which 'made it illegal for government agencies and federally funded organizations to spend their time working on Kyoto Protocol-related work'. Ibid 132.

The US profoundly influenced the substantive content of the Kyoto Protocol. Its subsequent decision not to ratify the Treaty, however, left the international community holding a potentially lifeless Treaty. The Kyoto Protocol required 55 ratifications, accounting for at least 55% of total CO2 emissions emanating from industrialized States in 1990, to come into effect. This provision made US participation critical. The US was responsible for 36.1% of the 1990 CO2 emissions of industrialized countries. In 2001, President George W. Bush clarified that the US had no intention of ratifying the Kyoto Protocol. President Bush's decision pushed the Protocol to the brink of collapse, but it also created an opening for the EU to more effectively exercise its role as global climate change leader. Once it became clear that the US had truly abandoned the Treaty, the global community's outrage allowed the EU to move out from the shadow of US influence to take on a more meaningful role in regime-building efforts.

The EU only accounted for 24.2% of CO2 emissions in 1990, meaning that even with EU ratification the Kyoto Protocol would still fall short of achieving the requisite 55% of 1990 emissions by 30.8%. In order to salvage the Protocol, the EU would have to take a leap of faith and ratify the Protocol absent participation by its leading economic competitors and then try to convince Japan (accounting for 8.5% of 1990 emissions) and Russia (accounting for 17.4% of 1990 emissions) to ratify the Treaty, as well.[19] The EU's decision to ratify the Protocol on 31 May 2002 represented a critical turning point in the history of the Kyoto Protocol and in the dynamics of international environmental leadership. The EU's subsequent success in convincing Japan (4 June 2002), Russia (18 November 2004)[20]—and, eventually, Australia (12 December 2007)—to ratify the Protocol enabled the Kyoto Protocol to come into force on 16 February 2005, ensuring the continuing viability of the international climate regime.

Due in large part to EU efforts, the Kyoto Protocol supplemented the UNFCCC to create the regulatory backbone of the still-evolving international climate change regime. The Kyoto Protocol augments the UNFCCC by establishing legally binding obligations that require industrialized countries to incrementally reduce human-induced greenhouse gas emissions to an average of 5.2% below 1990 emission levels by 2012.[21]

The legally binding obligations that it creates for industrialized countries, ie, Annex I countries,[22] are the hallmark of the Kyoto Protocol. Kyoto commitments are expressed in terms of emissions allowances, or Assigned Amount Units, that are equal to a nation's allowable greenhouse gas emissions. In order to facilitate

---

[19] Miranda Schreurs and Yves Tiberghien, 'Multi-Level Reinforcement: Explaining European Union Leadership in Climate Change Mitigation' (2007) 7[4] *Global Environmental Politics* 19.
[20] UNFCCC Secretariat, Press Release, 'Russian Decision on Ratification—Major Step in Entry into Force of Kyoto Protocol' (7 October 2004) <http://unfccc.int/files/press/releases/application/pdf/pr040930.pdf> accessed 28 September 2009.
[21] Kyoto Protocol (n 2) arts 4.1, 10, 12.
[22] Annex I parties must return to 1990 levels by 2012. Ibid.

compliance with Treaty obligations and reduce the cost of emission reductions, the Kyoto Protocol embodies three key economic flexibility mechanisms. These flexibility mechanisms—(1) Joint Implementation (JI),[23] (2) the Clean Development Mechanism (CDM),[24] and (3) Emissions Trading (ET)[25]—enable parties to trade and transact Emission Reduction Units and Certified Emission Reductions in order to more efficiently meet their Assigned Amount Units. As discussed in more detail in Chapter 7, while the US played a key role in negotiating the inclusion of economic flexibility mechanisms in the Kyoto Protocol, they are now widely employed by Annex I parties and, in particular, the EU and its member states.

The legal obligations that the Kyoto Protocol creates and the implementation measures that it embodies represent an unheralded international commitment to protecting the global commons. The Kyoto Protocol is unlike any existing multilateral environmental agreement in the scale of commitment that it requires from member states. The Kyoto Protocol represents the first time that developed nations have jointly agreed to reduce emissions from such a wide range of gases and across such a cross-section of the economy, and the first time that a multilateral environmental agreement has created the framework for an elaborate global market in emissions trading.[26] The Kyoto Protocol defies traditional understandings of the parameters of environmental problems and requires parties to modify primary social, political, and economic structures, including transformations within domestic consumption, energy, transportation, manufacturing, agricultural, and investment sectors. The impacts of emission reduction commitments under the Kyoto Protocol resonate through all levels of the public and private sectors.

In sum, the Kyoto Protocol challenges the global community to create a sustainable and equitable[27] international regulatory framework for combating global climate change over the long term. The Kyoto Protocol requires States to balance scientific, social, and economic considerations to develop laws and policies that ensure the long-term survivability of a common resource—the Earth's

---

[23] Enabling Annex I parties to jointly implement GHG reduction projects and to trade in emission reduction units (ERUs) arising from such projects. Ibid art 3.

[24] Allowing Annex I parties to fund emission reduction projects in the territories of non-Annex I parties, thus enabling developed States to meet article 3 commitments and developing State Parties to be involved in the emission reduction process without suffering the costs associated with emission caps. Ibid art 12.

[25] Allowing Annex I parties to trade in emissions or ERUs. Ibid art 17.

[26] A Petsonk, 'The Kyoto Protocol and the WTO: Integrating Greenhouse Gas Emissions Allowance Trading Into the Global Marketplace' (1999) 10 *Duke Envtl L & Policy Forum* 185.

[27] At the 2001 meeting of Kyoto's Conference of the Parties, the Parties drafted the Marrakesh Accord, which outlines the present context of the climate change regime. The Accord emphasizes that 'economic and social development and poverty eradication are the first and overriding priorities of the developing countries' and that efforts to combat climate change must take place against this economic backdrop. 'The Marrakesh Accords & the Marrakesh Declaration' UNFCCC Conference of the Parties (Marrakesh 29 October 2001–10 November 2001) UN Doc FCCC/CP/2001/13/Add.1 [Vol. I (D), (H)].

atmosphere—without jeopardizing the short- or long-term welfare of the global citizenry. Yet, for all of its complexity and ambition, the Kyoto Protocol is incomplete and insufficiently aggressive. Even if Annex I parties meet their Kyoto Protocol obligations, the global community will have won only one small victory in a steep, uphill battle to address global climate change. The UNFCCC and the Kyoto Protocol create the foundations for a successful global climate change regime, but they require major modification and expansion to create an effective and sustainable legal regime. This goal cannot be achieved without truly global cooperation—global cooperation, which at a minimum, includes meaningful participation on the part of the largest global emitters regardless of economic status and takes into consideration the needs of the most vulnerable States.

Global regime building must be led by powerful industrialized countries, such as the US and EU member states, who have the historic responsibility and the social and economic capacity to respond to threats posed by global climate change, but must also include the rapidly developing economies who increasingly account for high percentages of total global greenhouse gas emissions. The persistent refusal of the US federal government to be a full and active player in international efforts to halt climate change impairs efforts to form a global consensus and to incentivize participation and cooperation on the part of the rapidly developing economies. Only in February 2007, when global leaders negotiated a non-binding agreement that commits both developed and developing countries alike to legally binding emission reduction obligations, did the US show even a glimmer of willingness to re-engage with global climate change regime building efforts. Translating this non-binding agreement into a post-Kyoto legally binding Convention or Protocol will test the authenticity and strength of EU climate leadership and US commitment to global cooperation and re-engaging in global climate change regime building.

The Role of the US and the EU in International Climate Change Politics

Since no single nation can efficiently and effectively deal with global challenges such as climate change, counterterrorism, non-proliferation, pandemics and natural disasters on its own, we commit ourselves to strengthening our cooperation to address these challenges.
EU–US Summit Declaration, Vienna, July 2006.

The dynamics of global politics are changing. Superpowers are no longer so super and power is no longer so centralized. Yet, power imbalances remain and global climate change regime-building continues to be disproportionately affected by a small handful of countries, both developed and developing. The role of developing States is becoming increasingly important as a consequence of shifts in economic prowess and carbon footprints. China and India, in particular, emerge as influential and essential components of the international climate regime.[28] The financial crisis of 2008 offered a daunting geopolitical setback to US and EU economic primacy that revealed the extent to which China and India have made gains in amassing global political influence, as further demonstrated by the political dynamics that emerged at the Copenhagen Climate Change Conference in December 2009.[29] Despite economic and environmental shifts, however, the US and the EU continue to exert a high level of influence in international climate negotiations. As two of the most politically and economically powerful entities in global politics, the actions and omissions of the EU and the US profoundly influence the contours and effectiveness of the international climate change regime. Political and environmental leadership on the part of the EU and the US continues to be central to ongoing efforts to develop an effective and equitable international climate change regime.

The US and the EU are the two largest economies in the world, together accounting for almost half of the entire world economy.[30] The EU and the US also share the world's largest bilateral trading and investment partnership[31] with transatlantic flows of trade and investment amounting to approximately US$1 billion a day and combined US–EU global trade accounting for almost 40% of world trade.[32] Independently, both the US and the EU possess great global economic and political power. When working together on issues of global governance their combined authority is unparalleled.

The history of US–EU relations is one of prickly camaraderie. In international relations, the US and the EU are both über-allies and über-competitors. Since World War II, the US and the countries that now form the EU have regularly acted together as a formidable force on international security issues—eg, during the Cold War, the first Gulf War, the war in Afghanistan, and other UN-sanctioned actions. This relationship has strained in recent years due to perceptions of US unilateralism,[33] but it persists. In the all-important realm of economics and international trade, the US and the EU have worked jointly to liberalize international trade and further global development while simultaneously

---

[28] Eg Roger C Altman, 'The Great Crash, 2008: A Geopolitical Setback for the West' (2009) 88[1] *Foreign Affairs* 2; Daniel W Drezner, 'The New World Order' (2007) 86[2] *Foreign Affairs* 34.   [29] As further discussed in Ch 9.

[30] EU Delegation to the US, 'EU–US Facts & Figures' (2009) <http://www.eurunion.org/eu/index.php?option= com_content&task=view&id=1746&Itemid=9> accessed 28 September 2009.

[31] Ernst-Ulrich Petersmann and Mark A Pollack (eds), *Transatlantic Economic Disputes The EU, the US, and the WTO* (OUP, Oxford, 2004) 66.

[32] EU–US Facts & Figures (n 30); 'EU–USA Summit: An EU27 surplus in trade in goods with the USA of 63 bn euro in 2008, Surplus of 11 bn in trade in services in 2007' (Eurostat News Release, STAT/09/47 2009) <http://europa.eu/rapid/pressReleasesAction.do?reference=STAT/09/47&format=HTML&aged=0&language=EN&guiLanguage=en> accessed 3 April 2010.

[33] Eg Robert Kagan, 'Power and Weakness: Why the United States and Europe see the world differently' (2002) 113 *Policy Rev* 1; 'Conclusions on Iraq' (Council of Ministers' Conclusions, 2003) <http://www.eu-un.europa.eu/articles/fr/article_3013_fr.htm> accessed 3 April 2010.

competing ruthlessly for economic primacy. Disputes between the US and the EU, for example, have dominated the General Agreement on Tariffs and Trade (GATT) and World Trade Organization (WTO) dispute settlement forums.[34] Competition and conflict pervade US–EU economic affairs and international trade relations and yet the US and the EU continue to maintain an unprecedented global trade and investment partnership.

Transatlantic environmental relations are similarly characterized by a tumultuous partnership. The EU and the US share common environmental challenges, many of which are by-products of industrialization and many of which are global in scale.[35] They also share common goals of erecting systems of environmental law and policy that protect human health and well-being, with varying capacity to incorporate non-human concerns. In the realm of domestic and international environmental lawmaking, the US and the EU have taken turns bearing the mantle of environmental leader. In recent years, however, the US and the EU have frequently diverged on questions of science, regulation, and priority-setting on environmental issues with the effect of undermining transatlantic environmental cooperation that dates back to the early days of the environmental movement.[36]

As early as 1974, for example, the US and the EU established a bilateral framework for cooperating on environmental issues and for holding annual environmental consultations between high level governmental officials. Early bilateral consultations offered a convenient forum for sharing information and facilitating transatlantic cooperation on a wide range of environmental issues. The last consultation, however, took place almost ten years ago in May 2000. The bilateral consultation process was superseded, in part, by the 1995 New Transatlantic Agenda which expanded cooperative efforts to a wider variety of political and economic matters, including:

1) promoting peace and stability, democracy, and development around the world; 2) responding to global challenges (including preservation of the environment; 3) contributing to the expansion of world trade and fostering closer ties; and 4) building bridges across the Atlantic.[37]

The cessation of bilateral consultations, however also reflected shifting priorities. Transatlantic economic relations increasingly overshadowed environmental cooperation during the late 1990s and 2000s as the US and the EU worked to minimize barriers to bilateral trade and investment and solidify global economic

---

[34] André Sapir, 'Old and New Issues in EC–US Trade Disputes' (Conference on Transatlantic Perspectives on US–EU Economic Relations: Convergence, Conflict & Cooperation, 2002).
[35] Eg point source water pollution, hazardous waste, toxic air pollution, food and agricultural safety.
[36] Miranda A Schreurs, Henrik Selin, and Stacy D VanDeveer (eds), *Transatlantic Environment and Energy Politics* (Ashgate, 2009).
[37] European Commission, 'Bilateral relations—USA: Basis for Co-operation' (Statement on International Issues 2007) <http://ec.europa.eu/environment/international_issues/relations_usa_en.htm> accessed 28 September 2009.

supremacy following the end of the Cold War. During this period, environmental relations were further strained by diverging perspectives on climate change, global biodiversity protection, food safety, and general notions of risk and precaution.

Following the 2001 EU–US Summit in Göteborg, Sweden, the US and the EU took steps to renew transatlantic environmental cooperation by initiating a process for bilateral dialogue on climate change. In so doing, the US and the EU prefaced their dialogue on points of agreement and disagreement. The US and the EU joined in agreeing that climate change poses a pressing global problem requiring intensified bilateral and global cooperation, but articulated patent disagreement over the appropriateness of the Kyoto Protocol as a vehicle for addressing climate change. The US and the EU framed their new dialogue around a mutually agreed goal of promoting 'sustainable development for present and future generations'.[38]

The first summit of the EU–US High-Level Representatives on Climate Change took place in Washington, DC in April 2002. The meeting revealed embedded differences in approach to addressing climate change, with particular reference to the Kyoto Protocol, but suggested potential areas for cooperating on research and development. In summarizing the meeting, then EU Environment Commissioner Margot Wallström stated:

We need to revitalize cooperation between the EU and the US on certain environmental issues. This visit has been a useful launch pad to move our cooperation forward. Obviously, we do not agree on everything and we have different approaches to tackling environmental problems. The Kyoto Protocol is one notable example. But we do need to work together on climate change and we have now identified some areas for joint co-operation.[39]

Following the modest success of this initial meeting, formal bilateral climate consultations collapsed and were not revived until 2005. High level meetings between EU and US leaders in April 2005 and June 2006 initiated a new series of transatlantic meetings, this time more broadly focused on the 'serious and linked challenges in tackling climate change, promoting clean energy and achieving sustainable development globally'.[40]

The first meeting of the new EU–US High Level Dialogue on Climate Change, Clean Energy and Sustainable Development was held in Helsinki, Finland, on

---

[38] Delegation of the European Commission to the USA, 'EU–US Summit: A Guide: Göteborg, Sweden, June 14, 2001' (Conclusions from EU–US Summit 2001) <http://www.eurunion.org/partner/summit/Summit0106/Statement.htm> accessed 28 September 2009.

[39] European Commission, 'Commissionner Wallström comments on transatlantic environment cooperation' (Summary of press conference 2002) <http://www.eu-un.europa.eu/articles/en/article_1329_en.htm> accessed 28 September 2009.

[40] The US Mission to the EU, 'EU, US to Continue Climate, Energy and Sustainable Development Dialogue (Joint EU–US press release 2006) <http://useu.usmission.gov/energy/oct2506_high_level_dialogue.html> accessed 3 April 2010.

24–25 October 2006 while the second meeting was held in Washington on 7 March 2008. The meetings brought together high level officials from the US and the EU to review domestic climate and energy policy developments, emissions technologies, areas for bilateral climate and clean energy cooperation as well as measures to stem biodiversity loss and illegal logging.[41] The onset of these meetings in the latter half of President George W. Bush's presidency reflects an easing in US posture towards multilateral cooperation on climate change. Yet, the substance of the meetings reflects continuing hesitancy on the part of the US to engage in negotiations leading to binding emissions reduction commitments. The meetings focused primarily on market-based responses to climate change, energy efficiency, and clean technology development leaving substantive debate on regulatory responses and international commitments for other forums. The low profile and modest scale of these meetings reflected enduring tensions between the EU and the US on the question of how to respond to climate change.

EU and US disagreement over the appropriate political responses to human-induced climate change dates back to the commencement of UNFCCC negotiations. Since that time, at the federal level, the EU has adopted a leadership role in pushing forward international and regional climate change regime-building while the US has challenged the legitimacy and viability of the international climate change regime and failed to implement domestic climate-based legislation. The nuances of EU and US approaches to climate change will be closely examined in the remainder of the book. Brief mention of the contours of EU and US climate programs is made here, however, to lay the foundations for later analyses.

Climate change regimes in both the US and the EU contain complex mixtures of regulatory, market, voluntary, and research-based strategies. The EU, however, has adopted an approach to climate change that is based on mandatory greenhouse gas emission reductions; it is based upon an enforceable regulatory framework and accompanied by numerous policies and 'soft' law measures at the regional and member state level. The EU approach supports the internationalization of climate change measures and the adoption of an increasingly stringent, emission reduction-based international law regime.

The US federal approach to climate change, in contrast, has carefully avoided mandatory emission reduction obligations and focused instead on using a variety of 'soft' measures to encourage—rather than mandate—greenhouse gas emission reductions in an economically sound, market-driven manner. Unlike the EU, which has embraced the legitimacy of binding emission reduction commitments under the Kyoto Protocol, the US has questioned the legitimacy of the Kyoto Protocol and disregarded fundamental tenants of the international climate regime.

---

[41] The US Mission to the EU, 'EU, US Advance Climate Change, Clean Energy and Sustainable Development Dialogue' (Joint EU–US press release, 2008) <http://useu.usmission.gov/dossiers/climate_change/mar0708_us_euadvanceclimatechange.html> accessed 3 April 2010.

Fundamental differences aside, both EU and US climate change regimes have fallen short on substantive and procedural grounds. Neither the EU nor the US has developed a comprehensive system of implementable climate policies conducive to the long-term decoupling of fossil fuels and social and economic stability. As a result, neither of the most powerful and affluent members of the global community offer a comprehensive or easily replicable policy roadmap for achieving equitable and efficient emissions reductions. Despite obvious short-comings, a critical examination of US and EU approaches to climate change reveals important lessons.

Primary among these lessons are the relevance of US and EU actions and omissions to global climate change regime building efforts and the necessity of a regulatory framework to achieving sustainable emissions reductions. First, variations in the language, substance, and goals of EU and US climate programs transcend bilateral relations to impact the greater process of global climate governance. Second, while the US approach to climate change lacks credibility and offers little in the way of transferrable policy frameworks, the EU approach creates the regulatory foundations for building an effective climate change regime and offers the beginnings of a roadmap in climate policy successes and failures.

Drawing upon these two observations, the remainder of this book examines US and EU climate change law and policymaking at multiple levels of governance to reveal key convergences and divergences and underlying socio-legal drivers and to suggest how transatlantic policy choices impact the viability of the larger international climate regime. The theoretical framework for this book is based on a conventional convergence and divergence approach to comparative law, but it seeks to broaden upon traditional comparative law methodologies by more closely examining the relationship between law, society, and culture in the climate change law and policymaking process.[42]

Comparative analysis of US and EU environmental policy is still an emerging area of legal inquiry,[43] and much of the existing literature focuses on trends in environmental policymaking and governance, generally, and on trade and sustainable development policies, specifically. Comparative environmental analysis of US and EU climate change policies remains limited despite a growing need for this type of study.[44] Further, many of the climate change policy analyses that exist focus on the policies of the US, the EU, and their sub-federal entities in isolation, leaving critical gaps in our understanding of transatlantic law and

---

[42] Mathias Reimann, 'The Progress and Failure of Comparative Law in the Second Half of the Twentieth Century' (2002) 50 *Am J Comp L* 671, 685.
[43] Eg Norman J Vig and Michael G Faure (eds), *Green Giants: Environmental Policies of the United States and the European Union* (MIT Press, Cambridge, 2004).
[44] Eg Miranda A Schreurs, 'The Climate Change Divide: The European Union, the United States, and the Future of the Kyoto Protocol' in Norman J Vig and Michael G Faure (eds), *Green Giants: Environmental Policies of the United States and the European Union* (MIT Press, Cambridge, 2004).

policymaking. Throughout, and in particular in Chapter 7, this book seeks to begin filling these gaps and instigating continuing dialogue in this regard.

Recognizing, however, that climate change policies reflect a profound interplay among legal and regulatory institutions and social, cultural, and political processes, the book expands upon classical comparative law approaches by situating the legal analysis of climate change in its cultural setting and by engaging with climate change research coming out of other social sciences disciplines. By engaging with interdisciplinary materials and incorporating a socio-legal perspective,[45] in Chapter 8, this study looks beyond the composition of US and EU climate change laws and policies to explore how and why they are formulated and implemented at different levels of government, and to suggest what this can teach us about domestic and international climate change policymaking moving forward post-2012.

As the end of the first Kyoto Protocol compliance period nears, global climate change regime building stands at a critical crossroads. The ability of the global community to create an enduring international climate change regime hinges in no small part on US and EU participation. The US and the EU hold themselves out to be global, political, economic, and environmental leaders and the policy positions they adopt affect policy opportunities and preferences worldwide. This is particularly true in the climate context where industrialized countries bear great responsibility for past harms and possess greater present capacity to support mitigation and adaptation measures. Even as the US and the EU increasingly express a shared recognition that climate change is a global political priority, the extent to which Europe and America accept global leadership roles continues to differ. At a 2006 US–EU summit, for example, European and American leaders identified four priority areas for transatlantic cooperation, including: promoting peace, human rights, and democracy worldwide; confronting global challenges, including security and non-proliferation; fostering prosperity and opportunity; and advancing strategic cooperation on energy security, climate change, and sustainable development.[46] The inclusion of climate change as one of four priority areas for bilateral cooperation suggests that climate change has risen to the top of political agendas worldwide, even in the most resistant of political contexts. Yet, despite this rhetorical agreement, a careful analysis of policy preferences and legal frameworks reveals that EU policies and programs far exceed anything proposed or adopted by the US federal government.

The analysis does not end there, however. Looking beyond the federal level in the EU and the US, a more complex picture of political pushes and pulls emerges. The pushes and pulls of climate change politics in the US and the EU are

---

[45] Although there is no generally accepted definition of 'socio-legal studies', the term is used here to refer to inter-disciplinary analysis of law within its wider social context.

[46] President George W. Bush, 'US–EU Summit Declaration: Promoting Peace, Human Rights and Democracy Worldwide' (Press Release 2006) <http://georgewbush-whitehouse.archives.gov/news/releases/2006/06/20060621-2.html> accessed 3 April 2010.

characteristic of modern environmental law and policymaking. Tensions between environmental protection and economic development and between national sovereignty, international law, and the plethora of governmental and civil society views that define environmental politics challenge efforts to develop domestic consensus on climate change. Patterns of interaction between federal, state, and local climate change law and policymaking in the EU and the US reflect larger trends in international climate change policy, whereby intense dialogue over the appropriate legal and political responses to climate change take place at multiple levels of governance. The interplay between governmental institutions at multiple levels and among social, political, and economic factors shapes climate change law and policymaking in the US and the EU and provides insight into the nuances of climate policy formulation, implementation, and effectiveness.

US and EU climate policymaking processes offer critical lessons in success and failure to state and global efforts to formulate effective climate policies in the post-2012 period. The remainder of this book examines US and EU climate change law and policymaking in order to further understanding of how and why transatlantic policy approaches vary and to suggest how these variations affect efforts to build an equitable and effective global climate change regime.

PART II

CLIMATE CHANGE LAW AND POLICY IN THE UNITED STATES

2

Law and Policy in the United States

Compared to many European Union member states, the United States has a relatively short legal history. As compared with many other fields of law, environmental law is a young legal field. For these reasons, environmental law presented the US with the opportunity to put its mark on an evolving field of law. And, in the nascent days of environmental law the US seized the opportunity to lead and shape the discipline. Thirty plus years after the dawning of the era of environmental law, however, the US faces a critical challenge in maintaining its status as an innovator and trendsetter in the field of environmental law and policy—global climate change.

American Federalism: Federal Governance

American climate change laws and policies must be viewed within the context of the US system of federalism. In the US, federal power is constitutionally divided between the executive, legislative, and judicial branches. The US Constitution defines the remit of each branch of government and aims to balance power between the three branches in order to create a system of political checks and balances. The following discussion offers a brief review of the American system of governance.

The legislative branch, ie, Congress—consisting of the House of Representatives and the Senate—is the primary lawmaking body. It is responsible for drafting and enacting the rules of the land, including primary environmental legislation. The executive branch, led by the President, is then responsible for carrying out the rules and/or regulations that are necessary to implement the laws. Finally, the judicial branch, consisting of a tiered system of federal courts, settles disagreements about the meaning and scope of laws and regulations and decides how laws will be applied and whether they violate the US Constitution. The lines between legislative, executive, and judicial powers blur at times. Overall, however, the American system of governance functions within a carefully prescribed set of parameters—especially as compared to the system of governance in the EU, which is functional rather than formal.

The US Constitution and Congressional legislation provide the foundations for primary law, creating overarching norms for public and private action at the federal, state, and local level. The Constitution and Congressional legislation define the parameters for state and local legislation, which cannot conflict with or subvert federal legislation in most domains. Federal case law is also integral to legal governance, especially in sectors such as the environment, where definitions, knowledge, and capacity are constantly evolving.

Primary rules are supplemented by a body of secondary rules that are increasingly important in environmental law—federal regulations. Regulations refer to rules created by administrative agencies, such as the Environmental Protection Agency (EPA), the Department of the Interior (DOI), and the Fish and Wildlife Service. Administrative law functions in much the same way as protocols and annexes function within international framework agreements—it fills in the details of the overarching laws, creating meaningful and enforceable obligations.

Administrative agencies and administrative law are fundamental components of US environmental law. Administrative agencies, such as the EPA, are hybrid governmental organizations that do not fit clearly within the executive, legislative, or judicial branches. They are tasked with creating, implementing, enforcing, and interpreting regulations to give meaning to federal law. In this way, administrative agencies take on tasks inherent to all three branches of government, creating what many people refer to as the fourth branch of government—or the 'headless' branch of government. To constrain and legitimize administrative agencies, these agencies are made accountable to each of the three branches.[1]

Despite legislative, executive, and judicial constraints, administrative agencies exercise a significant amount of power and influence over federal governance, especially in the context of environmental law—and potentially in the area of climate change law. They create the specific mechanisms for implementing much federal law; they utilize a full range of investigative and enforcement tools to implement regulations; they possess self-contained adjudicative systems that can conduct hearings, render decisions between public and private entities, and issue administrative penalties. For example, the EPA has a set of Administrative Law Judges that hear disputes between the Agency and individuals, businesses, governmental entities, and other organizations that are subject to EPA regulations. Once the Administrative Law Judge has made a decision, either party can appeal the decision to the Environmental Appeals Board. In this way, many agencies create a tiered system of adjudication that mimics the federal judiciary.

---

[1] First, administrative agencies are created by legislative mandate; Congress enacts legislation that creates the agencies and defines their mission and jurisdiction. Congress also controls financial appropriations, which effectively influence agency agendas. Second, administrative agencies are generally housed within the umbrella of the executive branch, and the President often has the right to appoint agency heads, giving the executive branch considerable control over the agenda of the agencies. Third, regulations and administrative decisions are reviewable by the federal judicial branch.

The majority of modern US environmental governance—especially lawmaking and enforcement—takes place at the administrative level. Much of the evolution of environmental law since the 1980s originated in decisions made by federal agencies, whether through regulatory enactments, agency enforcement proceedings, or judicial challenges to agency actions and omissions. As will be discussed in Chapter 4, one of the most significant decisions on the question of federal responsibility for climate change involved a judicial challenge to the authority and responsibility of the EPA to regulate greenhouse gas emissions.

The Federal Government and the States

Federal–state relationships are at the crux of the American system of governance. In particular, as will be discussed in Chapters 3 and 4, climate change policymaking is intricately bound to the complex relationship between federal and state entities.

The Constitution defines the division of power between the federal government and the states. It creates a political system that divides power among the three branches of the federal government and between the federal and state governments—much like power is divided between the EU and its member states—but with significantly different relationships between the institutions.

The Constitution expressly grants the federal government specifically enumerated powers while reserving remaining powers for the states. In many areas, including environmental policy, the federal and state governments share legislative jurisdiction. While jurisdiction and power are shared, in the US federalist system—unlike in the EU system of governance—final authority in international and domestic affairs is vested in the federal government.

Article I of the Constitution limits federal legislative authority to a set of enumerated powers,[2] while the Tenth Amendment reserves residual authority for the state governments. Despite these enumerated limits, federal powers have been greatly expanded over the years through the 'necessary and proper clause'[3] which gives the federal government the implied powers to act in ways that are necessary and proper to carry into effect its enumerated powers. The Constitution also provides for federal spending powers. In addition, the regulatory authority over interstate commerce provided for in the Constitution[4] has significantly expanded federal environmental regulatory powers. Finally, the Supremacy Clause provides that federal law pre-empts conflicting state law,[5] giving the federal government a powerful trump card. In these ways, the authority and influence of the federal government greatly exceed its enumerated powers.

Over time, the relationship between federal and state powers has been explored and modified primarily by the US Supreme Court. By and large, federal authority

---

[2] US Const art I.   [3] Ibid art II, § 2, cl 2.   [4] Ibid art I, § 8, cl 1–7.
[5] Ibid art VI, cl 2.

has been significantly extended, allowing the federal government 'to impose uniform solutions to individualized social problems that differ significantly in their scope and effects across different states'.[6] During the 1990s, the Rehnquist Supreme Court began a concerted campaign to reign in the powers of the federal government to 'rediscover[ed] limits on powers vested in the federal government by the U.S. Constitution'. Despite notable restrictions put in place during this period,[7] the federal government maintains broad authority to govern beyond its strictly enumerated powers. This is especially true in the context of environmental law.

The federal government has shaped and controlled much of US environmental law and policy since the 1970s.[8] Federal jurisdiction over environmental policy is grounded in the arguments that federal environmental policy: (1) is more efficient than state regulation at achieving specific goals; (2) is capable of regulating pollution across state borders—relying on authority under the Commerce Clause; (3) prevents states from reducing social welfare in response to competition for industry—ie, prevents a 'race to the bottom' in environmental standards; (4) more properly takes environmental interests into account than state political processes; and (5) codifies moral rights.[9] These arguments focus on the benefits of harmonization, uniformity, and economies of scale. Chapter 5 will explore how the European Union, in many ways, takes a reverse approach to power-sharing in environmental law.

In the US, the federal government has taken decisive control in the context of environmental law. While it shares authority over environmental issues with the states, since the 1970s, the federal government has consistently dictated the environmental agenda. Further, unlike the European Union, where the principle of subsidiarity dictates that actions should be taken as close to the local level as possible, in the US, environmental law follows the doctrine of pre-emption. The doctrine of pre-emption 'limit[s] local involvement in environmental law and policy' and 'uphold[s] the supremacy of the higher level of government'.[10] Thus, in the US, the federal government has broad authority to govern in the area of environmental law. The federal government generously shares its implementation authority with the states. However, as will be demonstrated in the context of climate change, it shares lawmaking and agenda-setting authority much less generously.

---

[6] Jared Bayer, 'Re-balancing State and Federal Power: Toward A Political Principle of Subsidiarity in the United States' (2004) 53 *Am U L Rev* 1421, 1424; Paul S Weiland, 'Federal and State Preemption of Environmental Law: A Critical Analysis' (2000) 24 *Harv Envtl L Rev* 237.

[7] Eg, *United States v Lopez* 514 US 549 (1995); *United States v Morrison* 529 US 598 (2000).

[8] Cinnamon Gilbreath, 'Federalism in the Context of Yucca Mountain: Nevada v Department of Energy' (2000) 27 *Ecology LQ* 577, 592; Joshua D Sarnoff, 'The Continuing Imperative (But Only from a National Perspective) For Federal Environmental Protection' (1997) 7 *Duke Envtl L & Poly F* 225, 227; Robert V Percival, 'Environmental Federalism: Historical Roots and Contemporary Models' (1995) 54 *Md L Rev* 1141, 1157–78.

[9] Gilbreath (n 8) 592, citing Sarnoff (n 8) 230.   [10] Weiland (n 6) 283.

Here, it is worth briefly reviewing the evolution of environmental law and policy in the US in order to understand the legal and political backdrop of the climate change debate.

Environmental Law and Policy in the USA

During the first 100 years of America's existence, federal policies promoted colonization, westward movement, and taming—rather than protecting—the great American wilderness. Similarly, American states adopted policies that promoted rapid industrialization and economic development, creating the conditions for a 'race to the bottom'[11] for environmental standards. Conservation and preservation were not the fashion of the day.

This early era of rapid expansion and taming of the environment was called into question in the mid-1800s with the advent of writings and activism by the likes of Henry David Thoreau and George Perkins Marsh and by Congress's decision in 1864 to pass legislation giving Yosemite Valley to the state of California to be designated as a park. Subsequently, in 1872, Congress passed legislation making Yellowstone the country's and the world's first official national park, initiating an era of American trendsetting in environmental protection. The same year, however, Congress passed the now notorious Mining Law. In direct contrast to the preservationist tone of the designation of Yosemite as a state park and the creation of Yellowstone National Park, the Mining Law permitted companies and individuals to buy the mining rights for public lands believed to contain minerals for US$5 or less per acre—creating a push-and-pull trend that still continues today.[12]

Towards mid-twentieth century, visible conflicts between environmental change and human health permanently altered the tone of the debate. In 1948, an atmospheric inversion in Donora, Pennsylvania, caused death and illness on an unprecedented scale. Between 1936 and 1969, in Ohio, the heavily polluted Cuyahoga River caught fire multiple times. The same year as the dramatic 1969 fire in the Cuyahoga River, a massive oil spill off the Santa Barbara coast captured the imagination of the US public and caused outrage and concern from coast to coast. The publication of Rachel Carson's book *Silent Spring*[13] in 1962 further

---

[11] Eg, Kristen H Engel and Scott R Saleska, 'Facts are Stubborn Things: An Empirical Reality Check in the Theoretical Debate over the Race-to-the Bottom in State Environmental Standard Setting' (1998) 8 *Cornell J L & Pub Poly* 55; Kristen H Engel, 'State Environmental Standard-Setting: Is There a "Race" and is it "to the Bottom"?' (1997) 48 *Hastings L J* 271.

[12] Eg, in 1890, Congress passed legislation establishing Sequoia National Park and Yosemite and General Grant National Park. One year later, in 1891, Congress passed the Forest Reserve Act, granting the President the right to create 'forest reserves'. In contrast, national policies also promoted mineral exploration and exploitation—spurred on by the discovery of Spindletop on 10 January 1901 and the Lakeview Gusher in 1910. In addition to encouraging mineral exploitation, the national government also promoted massive damming projects, eg, Congress authorized putting a dam in the Hetch Hetchy Valley in Yosemite National Park in 1913.

[13] Rachel Carson, *Silent Spring* (Houghton Mifflin, Boston, 1962).

heightened public concern for the state of the environment. By the end of the 1960s, the scene was set for a remarkable social and legal revolution.

The Environmental Decade

The 1970s heralded a decade of environmental mobilization and environmental policymaking at the domestic and international level. In the US, the era was characterized by a shift in political and legislative power from state governments to the federal government and increased coordination between public, private, and governmental institutions to address environmental problems. During this era, environmental protection was a relatively bi-partisan issue—it was treated as commonly desirable and for the common good. Taking advantage of the popular and political consensus, between 1970 and 1979, Congress enacted eight major environmental laws. Prior to the lawmaking barrage of the 1970s, the federal government played a relatively limited role in environmental policymaking.

The changing role of the federal government in the 1970s can be attributed to the intersection of a variety of factors. First, US society was more affluent and more educated than at any point in its history and, as a result, had the time and the resources to focus on improving its quality of life,[14] which was increasingly seen as intricately tied to environmental factors. Consequently, the public demanded that the federal government take action to preserve and improve the environment in ways that would contribute to the public's quality of life.[15] Today, there is a similar phenomenon taking place around the issue of climate change in many parts of the world, including Europe and in many parts of the US—as Chapters 3 and 4 will demonstrate.

Second, due to the increased visibility of environmental problems,[16] there was a sense of urgency associated with environmental protection.[17] The public pressured the government to take quick and effective action to solve potentially devastating environmental problems—much as the European public now pushes its government(s) to address the threat of global climate change. The public tenor of urgency was prompted by deeply held suspicions about state and local governments' ability and willingness to promote environmental protection. That is, the public attributed environmental failure to state and local government inaction and feared that, in the absence of federal intervention, sub-federal governments would continue to minimize environmental

---

[14] Norman J Vig and Michael E Kraft (eds), *Environmental Policy: New Directions for the Twenty-First Century* (6th edn, CQ Press, Washington DC, 2006) 12.   [15] Ibid.
[16] This is best exemplified by the Union Oil Company Disaster in 1969, where a 'blowout' at an offshore oil and gas well resulted in an ecological disaster along the Southern California coast. The event was widely publicized through the increasingly important medium(s) of television and news media. The ecological consequences of the spill caused public outrage and spurred the public to put political pressure on government officials to respond. As a result, politicians were clamoring to be the first and the best to mollify the public—from Senator Muskie, to President Nixon, to Governor Reagan, to consumer activist Ralph Nader, the political response was overwhelming. KC Clarke and Jeffrey J Hemphill, 'The Santa Barbara Oil Spill: A Retrospective' (2002) 64 *Yearbook Association Pacific Coast Geographers* 157–62.   [17] Vig and Kraft (n 14) 13.

regulations in order to attract industry and promote economic development. Interestingly, as will be discussed in Chapter 3, a reverse phenomenon is currently taking place in the US with regards to climate change, with the public now turning to state and local governments as agents of positive change.

Third, the public put inordinate faith in the capacity of technology to address existing environmental problems. Namely, citizens faulted big business for contributing to environmental problems but believed that, given the right incentives, these businesses could access technologies that would allow them to improve their environmental performance. This belief holds true today, as President George W. Bush demonstrated during his term by putting explicit confidence in technological solutions to climate change.

Fourth and finally, this faith in technological change was coupled with an unprecedented belief in the ability of the federal government to craft effective environmental laws and policies. In the context of climate change, there is still considerable belief in the capacity of the federal government to craft effective laws, but less faith in its willingness to do so. In the 1970s, however, the federal government was viewed as the clear herald for positive change.

The 1970s were thus characterized by a sense of environmental urgency, scepticism towards traditional state roles, and faith in the power of technology and the ability of the federal government. In addition, the 1970s heralded a shift in the role of public participation; the proliferation and increased profile of non-governmental organizations (NGOs) and the creation of new avenues for public participation in decision-making and regulatory oversight shifted power away from traditional state actors toward the federal government with more room for input by individuals, as well as interest groups. This shift shapes modern US environmental law.

During the 1970s, US environmentalism became one of the principal social movements of the twentieth century.[18] Social, economic, and political change converged to initiate an unparalleled era of environmental lawmaking. Between the early 1970s and the early 1980s, Congress enacted virtually all of the federal environmental laws that continue to shape environmental decision making at the federal and state level in the US today.

Key environmental legislation passed during the period includes, but is not limited to:

- The National Environmental Policy Act (1969)[19]
- The Clean Air Act (1970)[20]

---

[18] Setting aside over 100 million acres of parks, wildlife refuges, and wilderness areas. Cf. ANWR debate.

[19] National Environmental Policy Act of 1969, 42 USC §§ 4321–4370 (2009) (showing a holistic approach focusing on decision-making processes, and requiring the federal government to examine its own major actions for environmental consequences).

[20] Reflecting the non-partisan nature of environmental issues at the time, the Clean Air Act passed in the House of Representatives by a vote of 374–1. Clean Air Act of 1970, 42 USC §§ 7401–7671 (1972) (prior to 1990 amendment). (Calling for sharp reduction in major pollutants

- The Federal Water Pollution Control Act (aka Clean Water Act) (1972)[21]
- The Coastal Zone Management Act (1972)[22]
- The Ocean Dumping Act (1972)
- The Marine Mammal Protection Act (1972)
- The Toxic Substances Control Act (1972)[23]
- The Endangered Species Act (1973)[24]
- The Safe Drinking Water Act (1974)
- The Hazardous Waste Transportation Act (1974)
- The Resource Conservation and Recovery Act (1976)
- The Federal Land Policy Management Act (1976)
- The Whale Conservation and Protective Study Act (1976)
- The Soil and Water Conservation Act (1977)
- The Surface Mining Control and Reclamation Act (1977)
- The National Energy Act (1978), the Endangered American Wilderness Act (1978), the Antarctic Conservation Act (1978)
- The Comprehensive Environmental Response Conservation and Liability Act (aka Superfund Legislation) (1980)[25]

In addition to forming much of the substance of current US environmental law and policy, the federal environmental laws of the 1970s were groundbreaking at the international level. They set the standards for environmental protection in the US and provided paradigms for environmental lawmaking worldwide.

---

almost immediately, focusing requirements on big business and on NSPS, forbidding consideration of economic or tech feasibility, allowing federal officials to delegate to states responsibility to carry out its provisions, providing for public participation, and allowing citizen suits.)

[21] Passed despite legislative veto by President Nixon.
[22] Coastal Zone Management Act of 1972, 16 USC §§ 1451–1464 (2009) (supporting broad land use planning in coastal counties).
[23] In 1972, Congress also passed amendments to the Federal Insecticide Fungicide and Rodenticide Act (FIFRA) that shifted regulatory responsibility for the Act from the USDA to the EPA and required that the EPA certify that any pesticide developed after 1972 does not have 'unreasonable adverse effects on public health or the environment' before the pesticide can become commercially available. Federal Insecticide Fungicide and Rodenticide Act of 1972, 7 USC § 136 (1972) (amended).
[24] Endangered Species Act of 1973, 16 USC §§ 1531–1544 (1973) (amended) (including national uniform requirements and provisions, stating federal agency decisions could not jeopardize endangered or threatened species or damage their critical habitats, and making it illegal to take endangered species).
[25] Further important environmental laws passed in the 1980s include: the Coastal Barriers Resources Act of 1981, the Nuclear Waste Policy Act of 1981, the International Environmental Protection Act of 1983, and the Emergency Planning and Community Right to Know Act of 1986.

Federal environmental lawmaking in the late 1970s and early 1980s followed a similar pattern. These new environmental laws adopted a command-and-control style model, ie, they set specific goals, uniform standards, firm deadlines, and narrow methods of compliance, and created heavy-handed federal enforcement provisions. The laws consistently targeted major polluting sources, mandated high standards of protection without regard for the economic costs, and built in substantive mechanisms for public participation in the regulatory and review stages of the rulemaking process. In addition, many of the environmental laws mandated a new form of federal–state cooperation; they created overarching federal rules, regulations, goals, and oversight but delegated significant implementation responsibilities to states.[26] In this way, Congress maintained primary responsibility for enacting legislation, setting high environmental standards, and leading the way forward, but put much of the burden on states to develop detailed implementation strategies within the federal parameters.[27]

Shifting Tides

As with most complex social and scientific problems, bi-partisan support for environmental issues lived a short life. In the early 1970s, environmental problems gained societal recognition and dominated the federal agenda largely because there was a strong public, and thus, political consensus that coordinated legislative response was required. By the late 1970s, however, this consensus began to crumble, in large part due to the looming oil crisis and the ensuing juxtaposition of environmental and economic priorities—a prominent link in the current federal climate change debate.

By the mid 1980s, what was left of the early environmental momentum eroded amidst the regulatory rollbacks of the Reagan era.[28] During President Ronald Reagan's tenure, environmental policy was characterized by budget cuts for administrative agencies, lax enforcement, administrative incompetence, power shifts to state governments, and regulatory relief for big business.[29] President Reagan is often cited as the godfather to President George W. Bush's approach to

---

[26] Eg, Federal Water Pollution Control Act, 33 USC §§ 6921, 6929 (2000) (authorizing federal government to promulgate effluent limitations for pollutants but allowing states to apply to the EPA to administer a permit program authorizing sources to discharge pollutants within the effluent limits); Resource Conservation and Recovery Act, 42 USC §§ 6921, 6929 (2000) (authorizing EPA to promulgate standards identifying and listing hazardous wastes, but authorizing states to administer a permit program to treatment, storage and disposal facilities handling hazardous waste).

[27] One of the best examples of this is the Clean Air Act State Implementation Plan (SIPs) provisions.

[28] Robert F Durant, *The Administrative Presidency Revisited: Public Lands, the BLM, and the Reagan Revolution* (SUNY Press, Albany NY, 1992). [29] Ibid.

environmental policymaking, much of which is evident within the microcosm of the climate change debate.[30]

The attack on environmental laws and policies was not universal. Congress repeatedly butted heads with President Reagan over questions of environmental law. As will be discussed in the section that follows, a similar intra-governmental battle emerged over proposed climate change legislation during President George W. Bush's administration—with Congress once again serving as the proponent for more progressive environmental laws.

Partisan politics defined environmental law during the 1980s. By the time President George H.W. Bush came into office in 1989, the federal government no longer possessed a strong political mandate to deal with issues of environmental law. In fact, the Clean Air Act Amendment of 1990, arguably, represents the last time the US federal government adopted or revised a major, new environmental law.

The 1990s heralded a long-term shift away from command-and-control style laws towards a new environmental agenda focused on regulatory relief for industry, sharp budget reductions for environmental agencies and activities, and a renewed focus on market-based policies. The move towards market-based policies encouraged the development of new environmental decision-making tools including risk assessments, cost-benefit analyses, and market-based incentives—tools that still characterize much environmental regulation today. Environmental law—and thus, climate change law—is now intricately tied to market-based tools and economic analysis.

The past decade and a half has witnessed much of the same. President Bill Clinton advocated a more extensive environmental agenda than his two predecessors, but his leadership was sporadic and inconsistent at best, and his efforts were repeatedly rejected by a Republican Congress that was unsympathetic to progressive, federally-driven environmental laws. In the context of climate change, Vice-President Al Gore attempted to broker US participation in the Kyoto Protocol. His efforts were thwarted by an unreceptive Senate that opposed American participation in an agreement that could hamper economic growth and global competitiveness. The Clinton administration ultimately failed to change Congressional opinion toward the Kyoto Protocol, and may have even deepened political opposition to the regime. Only in the last breath of his term was President Clinton able to realize—largely through Executive Orders—many of the environmental policies he had long promoted as central to his political agenda.[31] As President Clinton stepped down and President George W. Bush took office in 2001, environmentalists prepared themselves for

---

[30] Ibid.
[31] Eg, Patrick Parenteau, 'Anything Industry Wants, Environmental Policy Under Bush II' (2004) 14 *Duke Env L & Poly F* 363, 394 (describing the Roadless Rule that Clinton pushed through in the last days of his presidency).

what they foresaw as their greatest political challenge since the days of President Reagan. As one commentator describes President George W. Bush's approach to environmental law and policy:

[f]rom day one, the Bush [a]dministration has set about the task of systematically and unilaterally dismantling over thirty years of environmental and natural resources law. It started with the 'Card Memo' and the Anything But Clinton ('ABC') rule, which first quarantined and then quietly put to sleep, scores of regulations issued by the previous administration—everything from arsenic in drinking water, to fuel efficiency standards, to snowmobiles in Yellowstone National Park. Since then, the anti-Clinton reaction has grown into a full-fledged ideological crusade to deregulate polluters, privatize public resources, limit public participation, manipulate science, and abdicate federal responsibility for tackling national and global environmental problems.[32]

Throughout his presidency, Bush attempted to 'roll back' environmental laws, regulations, and policies[33] perceived to be economically detrimental or onerous for industry. One of Bush's most egregious 'roll backs' was his withdrawal of support for the Kyoto Protocol.[34] President Bush's approach to environmental policies during the era of critical Kyoto negotiations sets the stage for the analysis of US climate change laws and policies that follows.

Federal–State Relations in US Environmental Law

US environmental policymaking is in a new phase of evolution; on-going changes are highlighted by trends in climate change policymaking. Before examining these changes, it is worth briefly reviewing the intra- and inter-state dynamics that have characterized much of US environmental policymaking to date, as the roles of the federal and state governments are beginning to change noticeably in the context of climate change.

As discussed, prior to the 1970s, states played a central role in defining environmental policies within their territories; federal environmental policies were generally limited to designating protected areas, controlling use of federal lands, and encouraging—rather than mandating—that states take action to address specified types of pollution and resource depletion. In the 1970s, however, Congress began to legislate how, when, and where states should act on the environment.[35]

---

[32] Ibid 363; Jonathan Cannon and Jonathan Riehl, 'Presidential Greenspeak: How Presidents Talk about the Environment and What it Means' (2004) 23 *Stan Envtl LJ* 195, 210–11 (discussing how President Bush came out early attacking environmental laws and was forced to rethink his political approach due to public backlash).
[33] Parenteau (n 31) 364.   [34] Ibid 365.
[35] Elizabeth R DeSombre, *Domestic Sources of International Environmental Policy: Industry, Environmentalists, and U.S. Power* (MIT Press, Cambridge MA, 2000).

The shift was sudden and enduring, and has shaped US environmental law and policymaking for over thirty years.[36]

The federal government has formed the *framework* for environmental law in the US. Within that framework, states have significant responsibility and the competence needed to decide how to meet federal standards. Further, by and large, states also have the right to adopt environmental standards that are *more stringent* than federal standards, so long as the state standards do not conflict with or usurp federal law—in practice, as will be explored in Chapters 3 and 4, it is often difficult for states to adopt environmental standards that exceed federal standards.

State implementation of federal environmental laws has varied over time and across environmental mediums. In the context of air pollution, for example, the EPA relies on states to create and execute State Implementation Plans (SIPs) to implement the Clean Air Act National Ambient Air Quality Standards. The EPA has grappled with states both failing to develop and implement SIPs, and with legal challenges to state SIPs on grounds that the SIP is overly-stringent. Similarly, with the Endangered Species Act, states have challenged decisions both to list and not to list threatened and endangered species. The relationship between the federal government and the states in the area of environmental governance is varied and complex, as the climate change debate reveals.

Despite tight federal control over the environmental agenda, sub-federal governmental entities have often adopted state and local environmental laws that compliment but tighten federal environmental standards. As one commentator notes, '[m]any U.S. federal environmental laws and multilateral international environmental agreements came about partly in reaction to the regulatory measures implemented by lower-level jurisdictions'.[37] There are a number of examples of the federal legislature adopting policies that are modeled on policies at the state level, the best example being the development of the federal social security system.[38] In the environmental realm, the federal legislature embraced state-based models when it enacted the Clean Air Act, the Clean Water Act, the Surface Mining Reclamation Act, the Comprehensive Environmental Response Compensation and Liability Act,[39] and the Toxics Release Inventory.[40]

Overall, however, until very recently, state and local efforts to be nationally influential leaders in environmental lawmaking on issues of an interstate or international nature and efforts to *concertedly*—either independently or in regional blocks—push for more progressive federal environmental lawmaking and

---

[36] Eg, Daniel C Esty, 'Revitalizing Environmental Federalism' (1996) 95 *Mich L Rev* 570, 600–03.   [37] Engel (n 11) 64.
[38] Barry G Rabe, 'Greenhouse & Statehouse: The Evolving State Government Role in Climate Change [2002] *Pew Center on Global Climate Change* 36–39, 152.
[39] Robert B McKinstry Jr, 'Laboratories for Local Solutions for Global Problems: State, Local and Private Leadership in Developing Strategies to Mitigate the Causes and Effects of Climate Change' (2004) 12 *Penn St Envtl L Rev* 15, 16.   [40] Ibid.

participation in international diplomacy has not been the norm.[41] In the past, the occasions where states have had the most impact on influencing environmental law have either been (1) when individual states have generated successful models for addressing an environmental problem, or (2) when states have adopted regulatory regimes that differ from state to state, and have thus prompted the private sector to push for a comprehensive federal regulatory—to avoid being subject to multiple state regulatory regimes. Equally, coordinated state efforts to influence the federal government to adopt a stronger role in international environmental lawmaking and/or improved federal compliance with international environmental law have been modest. Interest group politics—including both industry and environmentalist actors—have often been more influential than state and local politics in influencing the federal environmental agenda.[42] As one commentator notes, the US:

typically pushes to internationalize those domestic environmental policies that would be advantageous on the international level for both economic and environmental reasons. When these two incentives coincide, the chances for action are greatest. Without agreement between the two camps—industry and environmentalist actors—the government is not willing to undertake any but the simplest political action to convince other states to address international environmental problems.[43]

As this comment suggests, economic factors[44] and interest group influence—rather than pressure from state and local governments—have tended to drive federal efforts to engage with environmental problems at the level of international law. The same has been true at the domestic level. State interests are expressed through Congressional representation. There is, however, a crucial difference between state representation within the federal government and the ability of state-level politics to influence federal decision making. While the former maneuvers from the inside, the latter works from the outside; in political terms, the difference between a political insider and a political outsider should not be underestimated.

At present, the relationship between the US federal government and the states in environmental policymaking is in a state of flux.[45] US and global environmental lawmaking is at a crossroads. Local and global environmental problems are no longer distinct; neither national boundaries nor national politics can prevent

---

[41] Ibid.
[42] Kristen H Engel and Scott R Saleska, 'Subglobal Regulation of the Global Commons: The Case of Climate Change' (2005) 32 *Ecology LQ* 183, 224–28 (examining the influence of interest groups on prompting federal environmental regulation).   [43] DeSombre (n 35).
[44] The potential 'economic harm suffered by the regulated domestic industries and the potential for economic gain offered by internationalization', eg, higher production costs, substitute substances, is a key driver in choosing international policies for the US. Ibid 9.
[45] Eg, Ron Scherer and Alexandra Marks, 'New Environmental Cops: State Attorneys General' [2004] *Christian Science Monitor* <http://www.csmonitor.com/2004/0722/p03s01-usju.html?s=rel> accessed 22 October 2005.

the two—local and global—from meeting. Similarly, while the Nation State continues to reign supreme, the distance between local action and international consequences is narrowing, and Nation State laws and policies are increasingly influenced both by sub-national actors pushing for action from the inside, and international actors pushing for action from the outside. Where these internal and external forces converge, national governments experience unprecedented levels of pressure. Nowhere is this better demonstrated than in the context of climate change law and policymaking in the US.

US Federal Climate Change Policy

The climate change debate is still relatively young. The IPCC[46] was formed in 1988; the UNFCCC was adopted in 1992 and came into force in 1994;[47] the Kyoto Protocol was adopted in 1997, but did not come into force until 2005—making it one of the youngest international environmental law agreements. Less than 20 years after its political birth, climate change governance now dominates domestic and international political agendas.

In the US, politicians, academics, interest groups, media, and civil society organizations have rallied around the questions of climate governance—both for and against raising its place on the political agenda. The discussion that follows provides an overview of the US federal climate change policy. Chapters 3 and 4 will then explore the increasingly important role of sub-federal governance, litigation, regulation, and international pressure in federal climate change decision making.

The US is a party to the UNFCCC,[48] but it is not a party to the Kyoto Protocol.[49] Although the UNFCCC is largely aspirational in tone, commentators have argued that many of its provisions can be understood to create legally binding obligations, including commitments to prepare greenhouse gas emission inventories, to cooperate globally to prepare adaptation strategies, to share

---

[46] The IPCC is 'the leading body for the assessment of climate change, established by the United Nations Environment Programme (UNEP) and the World Meteorological Organization (WMO) to provide the world with a clear scientific view on the current state of climate change and its potential environmental and socio-economic consequences.' IPCC, 'Organization' <http://www.ipcc.ch/organization/organization.htm> accessed 16 April 2010.

[47] UNFCCC, 'Status of Ratification' <http://unfccc.int/essential_background/convention/status_of_ratification/items/2631.php> accessed 27 February 2006.

[48] United Nations Framework Convention on Climate Change (adopted 9 May 1992, entered into force 21 March 1994) 31 ILM 849 (UNFCCC).

[49] As previously mentioned, Al Gore signed the Kyoto Protocol on behalf of the US during the Clinton Presidency. However, the US signature and participation was later rescinded by President George W. Bush. Eg, Dana R Fisher, *National Governance and the Global Climate Change Regime* (Rowman & Littlefield Publishers Inc, Lanham, 2004) 127–31; Michael Lisowski, 'Playing the Two-Level Game: US President Bush's Decision to Repudiate the Kyoto Protocol' (2002) 11[4] *Environmental Politics* 101–19.

research and technologies, and to implement national policies to mitigate climate change. Consistent with general principles of international treaty law,[50] as a party to the UNFCCC, the US is bound to uphold, and to avoid contravening the terms of the Convention.[51] The Framework Convention does not bind the US—or any other State Parties—to achieve specific, measurable greenhouse gas emission reductions, but it does require the US to take efforts to control greenhouse gas emissions seriously.

The Kyoto Protocol supplements the UNFCCC by specifying the emission reductions that will be required of each State that is party to both the Framework Convention and the Protocol. Together, the UNFCCC and the Kyoto Protocol establish the overarching goals and specific emission reduction obligations necessary to create an ambitious but focused international climate change framework. On its own however, the Framework Convention cannot be interpreted to require the US to reduce its emissions by a given percentage on a given date. Consequently, the US has the dubious status of being the world's second largest greenhouse gas emitter—surpassed only by China—while abstaining from committing to any legally binding international emission reduction obligations.[52]

The US engagement and participation in international efforts to create an obligatory and enforceable international agreement on climate change has varied over time. This chapter will not provide an exhaustive review of the history of US participation in the international climate change debate,[53] but instead will focus on analyzing modern efforts to develop a US national climate change strategy.

In analyzing US climate change policy, there is a tendency to say that it is non-existent or, at least, legally irrelevant. Although not completely true, this characterization is not far from the truth. While the US federal government does have an official climate change policy, this policy lacks the form, substance, and direction that one would expect from a political, economic, and—formerly—environmental world leader. In particular, the US climate change policy lacks substantive legal content. The US federal government has, arguably, abdicated its role as the national leader in many spheres of environmental law, certainly in climate change law and policy. As two legal commentators characterize the trend:

[r]ecently, the federal government's withdrawal from its role at the forefront of environmental policy has opened up opportunities for states to fill the vacuum. And for no issue is the contrast between the current federal and state stance on environmental issues more stark than on climate change.[54]

The previous federal climate change policy, as enunciated by President George W. Bush, reflected a stark and undeniable shift away from the federal environmental

---

[50] Vienna Convention on the Law of Treaties (adopted 29 May 1969, entered into force 27 February 1980) 8 ILM 689 (VCLT). [51] McKinstry (n 39) 17–19.
[52] Rabe (n 38) 36–39.
[53] For a detailed historic overview of US participation in international climate change negotiations, see Fisher (n 49) 105–41; McKinstry (n 39) 20–26.   [54] Engel and Saleska (n 11) 216.

leadership that characterized much of environmental law and policymaking in the modern era. This is true, not least of all, because the strategy was based merely on a climate change *policy,* lacking the *laws or enforceable obligations/regulations* necessary to substantiate it and give it form.

In the run-up to the 2000 presidential administration, then-candidate George W. Bush ran on a campaign that included a promise to regulate CO_2.[55] Upon election, however, President Bush 'reversed [his] position on the regulation of carbon dioxide as an air pollutant', and shortly thereafter issued a now infamous letter[56] stating that he would not commit to any protocol to the UNFCCC that would exempt developing countries from obligatory emission reduction obligations.[57]

With an expressly anti-Kyoto Protocol Republican administration in the White House and an anti-Kyoto Protocol Republican majority in Congress, the executive and legislative branches merged and 'temporarily unified the U.S. government's position on the issue of climate change—albeit a position different from any other country in the world'.[58] The President brought the executive and legislative branches together in a common political position. From there, however, he stalled on the question of climate change, leaving the US without a formal position for almost a year after he assumed the Presidency.

Despite reversing his campaign promise and failing to offer an alternative to the Kyoto Protocol in a timely manner, President Bush claimed that he had 'consistently acknowledged climate change is occurring and [that] humans are contributing to the problem'.[59] This is the strongest statement one can find on the former President's approach to climate change. In fact, this is one of only three statements that the White House explicitly released to support the claim that, between 2000 and 2006, the President engaged in the climate change debate. Neither this statement nor any other statement the White House released through mid-2007 suggested that the President would support the enactment of a *legally* binding climate change framework, either domestically or internationally. As discussed in more detail below, the US climate change *policy/strategy* under President Bush was largely based on carbon sequestration, voluntary obligations, 'business challenges', non-binding international partnerships, and the *possibility* of developing obligatory emission reduction policies in the future.[60] It contains many essential elements, including research and capacity building, but lacks the critical elements of regulatory infrastructure and legally binding commitments to reduce greenhouse gases.

---

[55] Fisher (n 49) 134–35.
[56] Letter from President George Bush to Senators Hagel, Helms, Craig, and Roberts (13 March 2001). [57] Fisher (n 49) 134–35.
[58] Ibid 135.
[59] White House, 'Open Letter on the President's Position on Climate Change' (7 February 2007) <http://georgewbush-whitehouse.archives.gov/news/releases/2007/02/20070207-5.html> accessed 26 February 2007.
[60] White House, 'Global Climate Change Policy Book' (2002) <http://georgewbush-whitehouse.archives.gov/news/releases/2002/02/climatechange.html> accessed 1 June 2010.

Further, when reviewing the former President's record—outside the statements explicitly released by the White House—it is evident that President Bush repeatedly challenged the veracity of global climate change, eg, alleging that the Kyoto Protocol is based on the 'unproven science' of global warming.[61] It was only towards the end of his second term that he modified his tone to be more conciliatory. Mounting domestic and international pressure ultimately forced President Bush to modify his 'presidential green speak'[62] on the issue of climate change and to develop a national strategy for addressing climate change.

President Bush's National Climate Change Strategy: An Overview

President George W. Bush launched the US Global Climate Change Initiative in February 2002,[63] thus bringing to an end a year of political silence on global climate change. The Global Climate Change Initiative, including the Global Climate Change Policy Book as its centerpiece, articulated the new US strategy for reducing greenhouse gas emissions.[64]

President George W. Bush's climate change plan was based on reducing greenhouse gas 'intensity'[65]—the ratio of greenhouse gas emissions to economic output. The initiative *called for* an 18% reduction in greenhouse gas intensity by 2012. Nowhere did the plan commit to, or even support, absolute reductions in greenhouse gas emissions. The former President's reliance on greenhouse gas intensity rather than *absolute* reductions in greenhouse gas emissions purportedly allowed the US to 'reduc[e] the growth of GHG emissions, while sustaining the economic growth needed to finance investment in new, clean energy technologies'[66] in the longer term. This assumed, of course, that the global community had the time to develop and implement these technologies gradually without short-term absolute emission reductions.

This new climate change framework was based on President Bush's firm belief that 'economic growth and environmental protection go hand in hand'[67] and that 'economic growth is essential to fostering societal preferences for environmental protection as well as providing the means for that protection'.[68] As one commentator characterized President Bush's climate change policy:

Bush's global climate change policy proposed to address what many perceived as the most significant global environmental threat with only further study and a voluntary

---

[61] Parenteau (n 31) 366.   [62] Cannon and Riehl (n 32) 226–27.
[63] George W. Bush, 'Remarks Announcing the Clear Skies and Global Climate Change Initiative in Silver Spring, Maryland' (Speech at the National Oceanic and Atmospheric Administration, 14 February 2002) <http://frwebgate.access.gpo.gov/cgi-bin/getdoc.cgi?dbname=2002_presidential_documents&docid=pd18fe02_txt-18.pdf> accessed 20 October 2009.   [64] Ibid.
[65] Global Climate Change Policy Book (n 63).   [66] Ibid.
[67] Remarks Announcing the Clear Skies and Global Climate Change Initiative (n 63).
[68] Cannon and Riehl (n 32) 227.

incentives program for industry. He invited us to assume with him that protection would follow from affluence.[69]

As a party to the UNFCCC, US climate change policy must be analyzed against the provisions of the Convention. Although the Framework Convention did not establish State-specific emission reduction obligations, it established the 'aim of returning [greenhouse gas emissions] individually or jointly' to 1990 levels.[70] The UNFCCC emission reduction goal is written in terms of absolute emission reductions. Despite committing to the terms of the UNFCCC and enumerating national goals in the Global Climate Change Policy Book, US greenhouse gas emissions are projected to increase by another 14% by 2012, meaning that—in absolute terms—the US will be 28% over the target levels it agreed to meet in the UNFCCC.[71]

Despite these projections, President Bush claimed victory, citing the fact that greenhouse gas 'intensity' was projected to decrease by 18%.[72] Using this intensity metric is deceptive. If the US bases its greenhouse gas emissions on an intensity metric, so long as US economic output increases over the next decade, the intensity metric means that greenhouse gas intensity will automatically decrease, even if the US does nothing to actually reduce absolute emissions.[73] In fact, the Bush administration's own figures showed a projected net increase in greenhouse gas emissions of 14% over the next decade.[74] Despite manifestly acknowledging that the strategy did nothing to reduce absolute emissions and would not meet the US obligations under the UNFCCC, the Bush administration promoted the greenhouse gas intensity approach as a 'serious, but measured mitigation response', and a way to avoid 'harming the economy in the short term'.[75]

President Bush's National Climate Change Strategy: The Detailed Provisions

By early 2010, the US federal government had still not enacted any primary climate change legislation. Unlike the environmental era of the 1970s, when the federal government set the tone for domestic and international responses to fundamental environmental problems by enacting framework environmental legislation, by the middle of the first compliance period for the Kyoto Protocol, the US had failed to develop either piecemeal or comprehensive climate change legislation.

Aside from early efforts to discourage US participation in the Kyoto Protocol, until very recently, Congress has remained unusually silent on the question of climate change. Similarly, until the election of President Obama, federal

---

[69] Ibid.
[70] UNFCCC (n 48) art 4 § 2(b).
[71] Parenteau (n 31) 368.
[72] Global Climate Change Policy Book (n 60) 2.
[73] Further, as one commentator points out: 'the intensity concept is a two-edged sword, since it would require greater levels of emissions control at times of lowest economic growth, when less money may be available for investment in technologies which might increase energy efficiency or control emissions'. McKinstry (n 39) 24–25.
[74] Parenteau (n 31) 368.
[75] Global Climate Change Policy Book (n 60) 6.

administrative agencies, including the EPA, displayed similar reluctance to use existing environmental legislation to regulate greenhouse gas emissions.[76]

During the Bush administration, aside from Congressional decisions to appropriate funding for climate change research, technology development, and the occasional renewable energy tax credits,[77] the core of the US federal climate change strategy was found in the President's global climate change initiative.

The key components of the formal US climate change policy under the Bush administration, as expressed in the Global Climate Change Policy Book, sought to:

- Substantially Improve the Emission Reduction Registry;[78]
- Protect and Provide Transferable Credits for Emission Reduction;
- Review Progress Toward Stated Goals and Take Additional Action if Necessary;
- Increase Funding for America's Commitment to Climate Change;
- Take Action on the Science and Technology Review;
- Implement a Comprehensive Range of New and Expanded Domestic Policies;
- Promote New and Expanded International Policies to Complement Our Domestic Program.[79]

The White House strategy promoted commendable objectives, but it lacked specificity and even the possibility of enforceability. Rather than focusing on or encouraging mandatory commitments, regulatory programs, and legal requirements, the objectives were presented using such non-committal terms as: '*propos[ing]* improvements', '*recommend[ing]* reforms', '*challeng[ing]* American businesses', and '*promot[ing]*... development'.[80] The White House Program even qualified future domestic measures by reference to 'sound science' and, by using non-mandatory, non-specific, and non-binding terms.[81] When, for example, the Policy addressed the question of reviewing progress toward meeting stated policy goals, it stated that, 'if...we find that we are not on track toward meeting our goal, and *sound science justifies* further policy action, the US will respond with additional measures that *may* include a broad, *market-based* program as well as additional *incentives* and *voluntary* measures...'.[82] Even the conditions were conditioned. The Policy created a scenario whereby current and future climate change programs were heavily qualified and

---

[76] *Massachusetts v Environmental Protection Agency*, 549 US 497 (2007) (US Supreme Court); Thomas F Reilly, 'Brief for the Petitioners Massachusetts et al' (Brief submitted to the US Supreme Court) <http://www.abanet.org/publiced/preview/briefs/pdfs/06-07/05-1120petitioners.pdf> accessed 19 December 2006.
[77] Energy Policy Act of 2005, 42 USC §§ 15801–16524 (2005).
[78] The federal greenhouse gas reduction registry was established under § 1605(b) of the Energy Policy Act of 1992, in accordance with the US commitments as a party to the UNFCCC.
[79] Energy Policy Act of 2005 (n 77). [80] Global Climate Change Policy Book (n 60).
[81] Ibid 8. [82] Ibid.

largely dependent on good faith efforts by governmental and industrial actors, with little hope or promise for more concrete and ambitious future actions.

Proposals for New and Expanded Domestic Policies

The lack of structure and enforceability was particularly apparent in the section of the US Global Climate Change Policy Book that addressed the implementation of a comprehensive range of new and expanded domestic policies. This section was where one would expect to find the outline for a robust climate change policy framework—even if it was up to Congress to implement the proposals. The section, however, failed to outline any mandatory programs or regulatory regimes for government, industry, or the public sector.[83] Central objectives for future domestic climate change policies included:

- Tax Incentives for Renewable Energy, Cogeneration and New Technology;
- Business Challenges;
- Transportation Programs;
- Carbon Sequestration.[84]

As with much of the Policy, this section outlined a series of objectives that were central components of any effort to address climate change. First among these proposed policies was the tax incentive scheme, which aimed to commit US$4.6 billion to clean energy tax incentives over the subsequent five years.[85] The goal of these tax credits was to encourage investments in new and existing renewable technologies. Shifting away from a carbon-based economy is essential to combating climate change in the long-term. These incentives encouraged this shift. However, it remains to be seen whether the government will follow through on its commitment to clean energy incentives, whether the level of these incentives will be sufficient—absent a regulatory backdrop, and whether the incentive programs will be effectively designed and implemented.

Significantly, one essential component that the tax incentive scheme omitted was either a voluntary or legally obligatory commitment to increasing the market share of renewable energy by a defined percentage. Commitments to renewable energy—whether voluntary or mandatory—are increasingly common. European States and US states and cities, for example, have put in place regulatory regimes to increase the market share of renewable energy in electricity production. The European Union has adopted a directive on increasing the market share of renewable energy for electricity production and the UK has adopted a Renewable Obligation Commitment (ROC), requiring all electricity producers to provide at least 10% of their electricity from renewable sources.[86] Similarly, California,

---

[83] Ibid.   [84] Ibid.   [85] Ibid.
[86] Utilities Act 2000 c 27 (England). The ROC program will be discussed in detail in a later section, but it demonstrates how obligations can be incorporated along with incentives in crafting climate change regulatory regimes.

Texas, and Massachusetts have adopted standards for increasing the amount of electricity generated from renewable energy sources.[87] The 2005 US Energy Act provides generous production tax credits for various renewable energy options, but it fails to provide for a national renewable portfolio standard or for anything comparable to the UK ROC program.[88] The federal renewable energy strategy is ill-defined; initiatives to increase the market shares of renewable energies, eg, biofuels, wind, solar, and other sources of renewable energy, are poorly coordinated and lack regulatory structure.

The second type of domestic policy the Report established was a 'challenge to business'.[89] Rather than encouraging the creation of a regulatory regime or mandatory commitments for industry, President Bush's plan 'challenge[d] American businesses to make specific commitments to improving the greenhouse gas intensity....'[90] Throughout his two terms in office, President Bush frequently emphasized the economic costs of implementing the Kyoto Protocol and addressing climate change. In the context of climate change, as in other areas of environmental policy, the former President opposed creating new regulatory regimes that burdened the activities of the private sector. Here, President Bush's focus on *challenging* rather than *regulating* business was likely a bid to avoid imposing short-term costs on businesses in a way that would impair American competitiveness or cause political ill-will. Even when *challenging* business, the Report framed the challenge in terms of greenhouse gas intensity rather than absolute reductions in greenhouse gas emissions.[91]

The third category of proposed domestic policy centered on improving domestic transportation programs.[92] The transport sector constitutes one of the fastest growing sources of carbon dioxide emissions in the US and the EU. Reducing absolute emissions from the transport sector is essential to reducing global greenhouse gas emissions. The goals for domestic transport fell short of proposing any significant changes in US transport strategies. The transportation policy goals included support for the 'FreedomCAR' initiative,[93] incentives for fuel cell technology research, tax credits for hybrid and fuel cell vehicles, modifications for the corporate average fuel economy standards (CAFE),[94] a tire pressure monitoring

---

[87] Eg, Rabe (n 38).
[88] Energy Policy Act of 2005 (n 77).
[89] Global Climate Change Policy Book (n 60).
[90] Ibid.
[91] Ibid.
[92] Ibid 3, 8.
[93] 'On January 9, 2002, Energy Secretary Abraham, with the heads of General Motors, Ford Motor Co. and the Chrysler arm of DaimlerChrysler, announced a new partnership, FreedomCAR (Cooperative Automotive Research), to promote the development of hydrogen as the primary fuel for cars and trucks. The "FreedomCAR" program embraces the long-term strategic goal of developing a new breakthrough technology—the hydrogen-powered fuel cell—with a vision of ultimately eliminating our reliance on foreign oil.' Global Climate Change Policy Book (n 60).
[94] 49 USC §§ 32901–32919 (2009) (Regulating automobile fuel economy). The CAFE statute designates a CAFE standard of 27.5 miles per gallon for passenger cars in model years 1984 and beyond, and gives the Department of Transportation the authority to revise the standard to a maximum feasible average fuel economy level for a given model year, subject to a Congressional veto. Ibid § 32902.

system, and new agreements with private industry to develop more efficient automobiles.[95]

The US is profoundly dependent on automobiles; the policy neither acknowledged nor addressed this fact. The US has the technological and financial resources to lead the way in developing alternative transportation technologies if it chose to make this a policy priority. Here, again, the US policy goals failed to create even the bare bones for a comprehensive transport policy. Research initiatives were healthily funded, but tax credits lagged and US CAFE standards were weak.[96] In fact, US CAFE standards, while effective in the early years of development, have been frozen for many years and have failed to keep pace with technological development; until 2009, with the election of Barack Obama, very little had been done to improve this situation.[97] Further, the strategy focused on technological change to improve energy efficiency in the transport sector; it did not propose initiatives to reduce dependency on private automobiles and to increase public transportation. The strategy was single-mindedly focused on technological solutions to greenhouse gas emissions from the transport sector. It steered clear of suggesting that humans change their behavior to reduce dependency on personal automobiles. Much of the debate around transport focuses on biofuels, with both the executive and legislative branches pushing for significant increases in the market share of biofuels.

Carbon sequestration is the fourth area where the plan recommended new domestic policies.[98] Carbon sequestration[99] provides an important but, at best, a limited solution to climate change. It is often viewed as a less costly and less economically disruptive method of reducing greenhouse gas concentrations, as opposed to measures that regulate economic activities.[100] The US was the State that pushed vigorously to include carbon sequestration provisions in the Kyoto Protocol.[101] Despite winning the right to count carbon sequestration towards national greenhouse gas emission obligations under the Kyoto Protocol, the US chose not to participate in the Kyoto Protocol.[102] The US, unlike the European Union, has made carbon sequestration a key component of its domestic climate change strategy from the onset.[103]

---

[95] Ibid 8–20.
[96] 49 USC §§ 32901–32919 (n 94).
[97] Robert R Nordhaus and Kyle W Danish, 'Assessing the Options for Designing a Mandatory U.S. Greenhouse Gas Reduction Program' (2005) 32 *BC Envtl Aff L Rev* 97, 106.
[98] Global Climate Change Policy Book (n 60) 3–9, 14–17.
[99] The US Environmental Protection Agency defines carbon sequestration as '[t]he uptake and storage of carbon. Trees and plants, for example, absorb carbon dioxide, release the oxygen and store carbon. Fossil fuels were at one time biomass and continue to store the carbon until burned.' EPA, 'Glossary of Climate Change Terms' (EPA comments on global warming) <http://www.epa.gov/climatechange/glossary.html> accessed 1 June 2010.
[100] For a discussion of carbon sequestration, see eg, Allen Keiser, 'Carbon Sequestration Options Under the Clean Development Mechanism to Address Land Degradation' (2000) 92 *World Soil Res Rep* 7, 7–11.
[101] Parenteau (n 31) 365–66.
[102] Ibid.
[103] Global Climate Change Policy Book (n 60).

Carbon sequestration will be achieved primarily through land use changes. US carbon sequestration policies target the agricultural sector and wetland protection. The Conservation Reserve Program, for example, is a voluntary program that provides incentives for farm owners and operators to set aside environmentally sensitive land,[104] while the Environmental Quality Incentives Program helps farmers make environmentally sensitive decisions on how to manage their lands.[105] Similarly, the Wetland Reserve Program is a voluntary program that aims to increase the amount of wetland being set aside for protection each year.[106]

All of these programs represent important steps in re-thinking land use decisions. However, all of the programs are voluntary, and the scope of the projects and the resources committed to the projects are modest compared to the magnitude of the problem. Further, the land use changes must be cast against the backdrop of land use policy generally in the US. The US lacks overarching land use laws or policies, leaving land use decisions primarily up to distinct federal agencies, states, and local governments. As a result, land use decisions in the US lack common underlying principles and consistency. As the environmental consequences of unplanned growth and urban sprawl have become more evident, land use management has become a more prominent part of the environmental agenda; however, there is still no comprehensive regulatory regime for land use management at the federal level, and very often, not even at the state level. Land use management continues to be primarily the domain of local governments, with mixed results and with mixed implications for the ability of local governments to make land use a tool for addressing climate change.

Overall, the carbon sequestration programs provide excellent stepping stones. However, to increase the impact of the programs and to make them an integral part of a climate change scheme, they will need to be expanded in scope and need to be better funded. Also, the federal government must promote wider participation by providing incentives and by adding obligations. Finally, unless the federal or state governments develop consistent and cohesive land use planning regulatory regimes or a coherent set of decision-making principles, the US will struggle to make land use planning an integral part of its climate change policy.

For all of these reasons, the domestic policies the federal government promoted and facilitated during the Bush administration represented essential but small steps in formulating an effective domestic climate change strategy. The policies advanced commendable objectives; but, compared to the resources at the disposal of the US and the heavy responsibility the US bears for past and present greenhouse gas emissions, the programs were minimal and legally empty—espe-

---

[104] Farm Security and Rural Investment Act of 2002, 7 USC § 7901 (2002); Conservation Reserve Program, 16 USC § 3843 (2002). [105] Ibid § 3839aa–9.
[106] Ibid § 3837.

cially in their ability to encourage behavioral changes and to produce concrete, measurable, or enforceable reductions in greenhouse gas emissions.

Significantly, US policies during this period completely failed to address a critical component of climate change policy—the domestic sector. These former policies focused on research and changes in the way governmental bodies and businesses thought about and responded to climate change. They did not, however, address the role of the larger, general public. While industry may account for the most significant point sources of greenhouse gas emissions, the public sector—especially transport and domestic homes—also contributes to domestic emissions. To shape climate change laws and policies that are sustainable in the short-term, the public must be given a more central role and more responsibility in addressing climate change.[107] To this end, domestic climate change policies need to provide the public with the information, tools, and incentives necessary to ensure active public participation. Involving the public in efforts to halt climate change is a necessary step towards decreasing the carbon intensity of the economy and one that the US government still has not taken.

Engagement with the International Community

As a final component, the US climate change strategy under the Bush administration proposed that the nation commit to promoting international cooperation. Specifically, the strategy proposed committing to international cooperation that compliments US policies.[108] Rather than working through the UNFCCC mechanisms and promoting Kyoto Protocol policies, the US sought to promote international policies that were complimentary to its existing approach to climate change. Many of the policies supported UNFCCC goals, including investments in climate observation systems, increased funding for 'debt-for-nature' conservation programs, expanded technology transfer, and cooperative research initiatives.[109] However, the US balked at any initiatives that required absolute greenhouse gas emission reductions or that excluded developing countries from any binding obligations to which developed countries committed—two principles at the heart of the UNFCCC and the Kyoto Protocol.

While the Global Climate Change Initiative and subsequent efforts to implement the Initiative promoted international cooperation to address climate change, the US insistence on operating outside the auspices of the UNFCCC and the Kyoto Protocol undermined the legitimacy and efficacy of an authoritative—albeit flawed—international legal framework for climate change.

Since the release of the Global Climate Change Initiative in 2002, the federal government has attempted but struggled to develop a comprehensive global

---

[107] Eg, International Climate Change Taskforce, 'Meeting the Climate Challenge: Recommendation of the International Climate Change Taskforce' (2005) <https://www.tai.org.au/file.php?file=web_papers/WP70.pdf> accessed 3 April 2010.
[108] Global Climate Change Policy Book (n 60).   [109] Ibid.

climate change legal regime. Despite ongoing efforts, by the time of the UN Climate Change Conference in Copenhagen in December 2009, it still had not succeeded in implementing a single robust law that could provide the cornerstone for the regime. Only the judicial branch and the administrative branch had taken concrete steps towards defining the legal responsibilities of the federal government.

Other Federal Activity: The Climate Change Technology Program

Technology is at the heart of US efforts to address climate change. On 20 September 2000, pursuant to President Bush's stipulation that technology should serve as the cornerstone of climate change policies in the US, the US Department of Energy released the US Climate Change Technology Program: Strategic Plan ('Strategic Plan').[110] The Plan 'details measures to accelerate the development and reduce the cost of new and advanced technologies that avoid, reduce, or capture and store greenhouse gas emissions'[111] and provides the groundwork for the Climate Change Technology Plan (CTTP).

The CCTP is a multi-agency research and development coordination activity, organized under the auspices of the Cabinet-level Committee on Climate Change Science and Technology Integration (CCCSTI), established by President George W. Bush on 14 February 2002.[112] The CCTP was established to implement President Bush's National Climate Change Technology Initiative. The initiative was launched on 11 June 2001, to develop new and improved technologies to address climate change.[113] The goal of the CCTP is to coordinate and prioritize the multi-agency federal climate change technology research and programs. Its ultimate aim is to 'provide abundant, clean, secure, and affordable energy and other services needed to encourage and sustain economic growth, while simultaneously achieving substantial reductions in emissions of GHGs'.[114] The initiative seeks neither to stabilize emissions, nor to create policy suggestions for implementation; it is purely a research program.

The CCTP's Strategic Plan is just that—a plan. It is not a policy document, and it does not contain any mandatory or enforceable provisions. The tone, focus, and recommendations, nevertheless, are influential and edifying.

Offering a comprehensive analysis of technologies that can help prevent climate change, the Strategic Plan focuses on assigning priorities for technological

---

[110] US Department of Energy, 'U.S. Climate Change Technology Program: Strategic Plan, DOE/PI-0005' (September 2006) <http://www.climatetechnology.gov/stratplan/final/CCTP-StratPlan-Sep-2006.pdf> accessed 1 June 2010.
[111] US Department of Energy, 'DOE Releases Climate Change Technology Program Strategic Plan' (Press Release 20 September 2006) <http://www.climatetechnology.gov/library/2006/pr20sep2006.htm> accessed 1 June 2010.
[112] US Climate Change Technology Program, 'About the U.S. Climate Change Technology Program' (2006) <http://www.climatetechnology.gov/about/index.htm> accessed 1 June 2010.
[113] Ibid. [114] Strategic Plan (n 108) 10.

research and development across all federal agencies.[115] It makes no attempt to discuss how these technological solutions might be implemented, or to review the policy measures available to encourage the use of *existing* technologies. The Strategic Plan adopts President Bush's carbon dioxide intensity measure and frames climate change as one of many federal policy goals, which also include economic growth, energy security, and affordable energy prices. The Plan does not prioritize policies to address climate change. In fact, the mission and vision of the Strategic Plan reflect political prioritization of economic growth, energy security, and affordable energy prices over aggressive climate change policies.

Although the Strategic Plan sets out a complex and sophisticated array of technology programs, these programs revolve around an ever-present emphasis on the uncertainty of climate change. The document consistently emphasizes scientific uncertainties relating to: (1) what constitutes 'dangerous' levels of atmospheric greenhouse gases, ie, how much reduction must occur, (2) what factors contribute most to human-induced forcing, and (3) what technologies would best mitigate climate change.[116] The Strategic Plan's effort to identify and respond to uncertainty is wise; uncertainty characterizes any complex scientific endeavor.[117] Ignoring scientific uncertainty is unwise and unhelpful; however, the Plan uses the language of uncertainty, at times, in a defensive, if not cynical manner.[118]

Ultimately, the Strategic Plan uses the uncertainties associated with climate change and with technological development as a reason to research everything. That is, the report contends that, because of uncertainties, the US should maintain a diverse portfolio—generally, a wise choice. It therefore advocates further research of every technology it covers. Its only form of prioritization is temporal, ie, which technologies come first, and the *de facto* priorities, for instance, funding, are left to be determined by other sources such as the National Climate Change Technology Initiative.[119] A diverse research agenda is wise; abdicating the responsibility for prioritizing scientific research is not.

---

[115] Ibid 9.   [116] Ibid 208 (discussing 'Long Term Planning Under Uncertainty').
[117] Cinnamon Carlarne, 'Climate Change—The New "Superwhale" in the Room: International Whaling and Climate Change Politics—Too Much in Common?' (2007) 80 *Southern California LR* 101.
[118] For example, under the heading of 'Long Term Planning Under Uncertainty,' where the Plan specifies that:

> CCTP operates within a planning environment characterized by *uncertainty*. First, the complex relationships between population growth; economic development; energy demand, mix, and intensity; resource availability; technology advancement; and many other societal variables make it difficult to estimate with confidence future global GHG emissions over CCTP's long-term planning horizon. This creates *uncertainty* about the scope and scale of the technological challenge. Second, evolving climate science, as well as the *uncertain* nature of the (as yet undetermined) UNFCCC's stabilization objective, adds *uncertainty* about the appropriate pace of technology development. Finally, research and development itself is risky, such that the future readiness, cost, and performance characteristics of the many advanced technologies envisioned to facilitate GHG emissions reductions are unknown. This adds *uncertainty* about deployment and which, if any, technologies will ultimately emerge as successful. Ibid.

[119] Ibid 231.

Despite the fact that the Strategic Plan is a comprehensive, well-organized, and thorough compilation of information, unfortunately, it is one with which most informed people are already familiar. Its failure to provide new insights, progressive ideas, and specific suggestions, and its ultimate inability to sort through the morass of options, deal with uncertainties methodically, or directly lay out a focused and thoughtful technology strategy is a lost opportunity.

President Bush's 2007 State of the Union Address

Following the release of the Global Climate Change Initiative and the CCTP Strategic Plan, the issue of climate change appeared with increasing frequency in federal discourse. The debate has focused on US engagement in international negotiations, the responsibility of the federal government to regulate greenhouse gases, and the role of sub-federal governments in shaping US climate change policies. Despite rapid changes at the sub-federal level, federal discourse and response has evolved slowly, only really taking off following the election of President Obama in 2008.

In early 2007, however, during the waning months of President Bush's Presidency, domestic and international concerns about global climate change escalated following the release of the Stern Review on the Economics of Climate Change[120] and anticipating the release of the IPCC's Fourth Assessment Report.[121] Mounting concerns about climate change and rumblings in Congress over the enactment of a legislative framework for limiting greenhouse gas emission sparked great interest in President Bush's forthcoming State of the Union Address.[122]

Leading up to President Bush's sixth State of the Union on 23 January 2007, expectations were high. Many people expected the President to directly address climate change, reversing his previous refusal to mention, much less engage with the issue of climate change in his previous State of the Union Addresses.[123] Hopes were high among many people—domestically and internationally—that

---

[120] HM Treasury, *Stern Review on the Economics of Climate Change* (Report, 30 October 2006) <http://www.hm-treasury.gov.uk/independent_reviews/stern_review_economics_climate_change/stern_review_report.cfm> accessed 1 June 2010.

[121] IPCC, *The Summary for Policymakers of the IPCC's Fourth Assessment Report* (2 February 2007) <http://www.ipcc.ch/publications_and_data/ar4/syr/en/spm.html> accessed 1 June 2010; IPCC, 'IPCC Adopts Major Assessment of Climate Change Science' (Media Advisory, 2 February 2007) <http://www.ipcc.ch/pdf/press-releases/pr-02feburary2007.pdf> accessed 1 June 2010.

[122] Peter Baker and Jon Cohen, 'Bush To Face Skeptical Congress: Iraq Overshadows Domestic Outreach' *Washington Post* (Washington DC, 23 January 2007) <http://www.washingtonpost.com/wp-dyn/content/article/2007/01/22/AR2007012200236.html> accessed 1 June 2010 (stating that, 'amid much talk about climate change and energy security, the president and his aides have promised unspecified "bold" ideas. Officials have ruled out binding caps on emissions of greenhouse gases, despite support among Democrats and some corporate executives who came to Washington yesterday. But they told allies that Bush will advance ideas for greatly expanding ethanol as an alternative to oil, and some insiders expect changes in fuel efficiency standards for vehicles.').   [123] Ibid.

the President's State of the Union Address might signal a change in the federal government's political approach to climate change.

In his speech, President Bush did, in fact, utter the words climate change— exactly once.[124] He approached the issue of climate change through the vehicle of domestic and energy security. Taking an issue close to the heart of the American public, the President began by opining that '[e]xtending hope and opportunity depends on a stable supply of energy that keeps America's economy running and America's environment clean'.[125] Setting the tone thus, the President continued by lamenting US reliance on foreign oil and the nation's concomitant vulnerability to 'hostile regimes, and to terrorists'.[126] After emphasizing the links between energy and terrorism, he suggested that the US needs to diversify its energy supply and declared that 'the way forward is through technology'.[127] He bypassed any discussion of conservation or efficiency and went directly to challenging the way the nation 'generates electric power'.[128]

After a gentle introduction, the President then made the bold statement that the US needs to reduce gasoline use 'by 20 percent in the next 10 years', couching this potentially divisive goal in security terms by stressing the relationship between reduced gasoline use and reduced oil imports from the Middle East.[129] Expanding on how the US can reduce gasoline use, the President gathered momentum to speak the oft-ignored and highly anticipated language of climate change. Repeating his emphasis on technology as the panacea to our energy problems, President Bush stated:

America is on the verge of technological breakthroughs that will enable us to live our lives less dependent on oil. And these technologies will help us be better stewards of the environment, and they will help us to confront the serious challenge of global climate change.[130]

In this way, the President broke the long trend of avoiding climate change in his State of the Union Addresses. Afterwards, he accepted the applause and promptly moved on to discuss the American justice system.

Following the President's State of the Union Address, the White House issued a variety of reports supporting the President's energy objectives and corroborating the federal government's commitment to 'develop cleaner, cheaper, and more reliable energy sources'.[131] Here, as in the President's speech, climate change is cast as a secondary issue to the over-arching problems of security and clean air. As in the Global Climate Change Initiative, energy goals during the Bush administration were primarily cast in terms of security, with benefits to climate

---

[124] The White House, 'President Bush Delivers State of the Union Address' (News Release, Washington DC, 23 January 2007) <http://georgewbush-whitehouse.archives.gov/news/releases/2007/01/20070123-2.html> accessed 1 June 2010.   [125] Ibid.
[126] Ibid.   [127] Ibid.   [128] Ibid.   [129] Ibid.   [130] Ibid.
[131] The White House, 'Energy Security for the 21st Century: Reliable, Affordable, Environmentally-Sound Energy' <http://georgewbush-whitehouse.archives.gov/infocus/energy/> accessed 1 June 2010.

change being secondary. For example, in the context of the President's call to reduce gasoline usage, the White House stressed that reduced gasoline use will 'help increase our Nation's energy security by reducing our dependence on oil. Achieving this goal will also help address climate change concerns'.[132] Neither the President's speech nor subsequent White House documents made any effort to disguise either the prioritization of security and stable energy supplies or the fact that climate change is neither an administrative priority nor a primary justification for any proposed changes in domestic energy use or production. Similarly, the President's speech and subsequent White House reports maintained the emphasis on technology—rather than life style changes—as the primary avenue for addressing energy issues.

Prioritization and technological dependency aside, President Bush's call for the US to reduce gasoline use 'by 20 percent in the next 10 years' was an important breakthrough. The fact that this relatively constrained and modest call for reform received the attention it did reflected the lethargic state of climate politics in the US.

Recent Political Activity: Congressional Engagement and President Barack Obama

For better or for worse, President Bush and the executive branch dominated the federal climate change agenda between 2001 and 2008. This trend began to change as the end of President Bush's term neared. During the 110th session of Congress, several members proposed multiple climate change bills in both the House and the Senate.

The new Speaker of the House, Nancy Pelosi (D-Cal) quickly took the helm of efforts to redirect US climate politics, stating that by July of 2007, she planned to pass legislation designed to address global climate change. To this end, she co-sponsored the Safe Climate Act bill with Henry Waxman (D-Cal).

Complimenting Pelosi's efforts in the House, four major bills were proposed in the Senate during the 110th Congress—with a fifth circulated as a discussion draft.[133] All four of the bills called for some type of mandatory cap-and-trade program for greenhouse gas emissions. The shared goal of the bills was to create an emissions trading program, and thus, to draw heavily on market-based mechanisms to reduce greenhouse gas emissions. First, Senators Barbara Boxer (D-Cal) and Bernie Sanders (I-VT) re-introduced a bill known as the Global Warming Pollution Reduction Act.[134] This bill sought to gradually reduce emis-

---

[132] Ibid.
[133] Pew Center for Global Climate Change, 'Senate Greenhouse Gas Cap-and-Trade Proposals in the 110th Congress' (2007) <http://www.pewclimate.org/docUploads/Economy-wide%20 bills%20110th%20Senate%20-%20August%202_0.pdf> accessed 1 June 2010.
[134] The bill was originally introduced and rejected in the previous session by Barbara Boxer and Jim Jeffords of Vermont, who had retired by the 110th session.

sions by 80% from 1990 levels by 2050, and thus, stabilize greenhouse gases at 450 ppm by 2050. Commenting on the Global Warming Pollution Reduction Act, Julia Bovey of the Natural Resources Defense Council said, 'We think of the Boxer/Sanders bill as the gold standard because that bill would really set us on a path toward slowing, stopping and reversing global warming'.[135]

Second, Senators Lieberman (I-Conn) and McCain (R-AZ) proposed a re-invigorated version of a bill—the Climate Stewardship Act—that was rejected during previous sessions of Congress. The new version of the Lieberman-McCain bill was more stringent than previous versions, calling for a cap on greenhouse gas emissions that would gradually be tightened until emissions are stabilized at 450 ppm by 2050.

Third, Senators Kerry (D-Mass) and Snowe (R-Maine) introduced a bill that was similar in many ways to the Boxer-Sanders bill, but that proposed different—and slightly more lenient—long-term greenhouse gas emissions goals for the period 2030-2050.

Fourth, Senators Dianne Feinstein (D-Calif) and Tom Carper (D-Del) proposed another cap-and-trade bill. The Feinstein-Carper bill was based on lower emission reduction commitments than the other Senate proposals.

Fifth, Senators Bingaman (D-NM) and Arlen Specter (R-Pa) introduced a discussion draft for a bill that sought to: (1) decrease the rate of emissions before instituting a cap, (2) create a much less stringent cap on emissions than the bills listed above, and (3) build in a 'safety valve' that would free industry from its requirements under certain circumstances.[136]

The first three bills were favored by environmental groups while the latter two bills were criticized as not being strong enough, providing too many loopholes, inadequately promoting clean energy, and failing to protect consumer participation.[137]

Similar bills by the likes of John Kerry and Barbara Boxer had been proposed in Congress in the past. However, the 110th session was characterized by more rigorous proposals and mounting political momentum to act decisively. None of the bills proposed in the 110th session generated consensus, however, and the session closed without any climate change legislation reaching a full vote. The unprecedented flurry of activity, however, laid the groundwork for more focused deliberations in the 111th session.

Congressional action during the last days of the Bush administration extended beyond proposals for climate change legislation. On Tuesday, 30 January 2007,

[135] National Resources Defense Council, 'How Low Should We Go? A Legislative Guide for Reducing Global Warming Pollution' (2006) <http://www.nrdc.org/legislation/factsheets/leg_061218.pdf> accessed 1 June 2010.
[136] Felicity Barringer and Andrew C Revkin, 'Bills on Climate Move to Spotlight in the New Congress' *New York Times* (New York, 18 January 2007) A-24.
[137] Eg, ibid; National Resources Defense Council (n 135); Tara Lohen, 'Climate Change Heats up Washington' (1 June 2010) <http://www.alternet.org/environment/47891> accessed 1 June 2010.

the House Committee on Oversight and Government Reform held a hearing on global warming censorship where it heard the results of a major investigation conducted by the Union of Concerned Scientists (UCS) and the Government Accountability Project.[138] The UCS survey documented an 'epidemic' of political interference in federal global warming science.[139] It found, among other things, that nearly half of the 279 federal scientists it surveyed were pressured to drop references to global warming in their research.[140] Representatives were split in their response to the UCS survey; Representative Waxman emphasized that they '[knew] that the White House possesse[d] documents that contain[ed] evidence of an attempt by senior administration officials to mislead the public by injecting doubt into the science of global warming and minimiz[ing] the potential danger', while Representative Darrell Issa (R-Calif) 'criticized the survey as self-selecting and flawed'.[141] Partisan responses aside, the confluence of legislative proposals and oversight of executive interference in climate change science suggested Congress was coming closer to acting on climate change.

As the 2008 Presidential race loomed, politicians became emboldened to discuss climate change and propose federal legislation to limit US greenhouse gas emissions. Climate change was still nowhere near becoming a bi-partisan issue, as many environmental issues were in the 1970s. It is unlikely that the Senate will ever approve a major piece of environmental legislation unanimously, as it did in 1970 when it passed the Clean Air Act. However, unanimity is not necessary. Climate change will remain politically divisive indefinitely; but as the political tide began to shift in 2007, it became increasingly apparent that the Congressional deadlock of the past ten years was no longer watertight.

With the election of President Barack Hussein Obama to the position of US President in 2008, climate politics experienced an abrupt change of gears. As will be discussed in some detail in Chapters 7 and 8, upon taking office, President Obama initiated a series of changes that created political and cultural space for US domestic and international re-engagement with climate change law and policymaking.

At a very basic level, picking up on the trend started by President Bush towards the end of his Administration, President Obama reframed climate change as part of a larger problem of energy security and national security. In so doing, he removed climate change from its radical roots and tied it to mainstream American politics. At the same time, President Obama sought to realign the US with the

---

[138] Union of Concerned Scientists, 'Atmosphere of Pressure: Political Interference in Federal Climate Change Science' <http://www.ucsusa.org/scientific_integrity/abuses_of_science/atmosphere-of-pressure.html> accessed 1 June 2010. [139] Ibid.
[140] Eric Niiler, 'House Probes Global Warming Censorship' (Marketplace broadcast, 30 January 2007) <http://marketplace.publicradio.org/shows/2007/01/30/PM200701305.html> accessed 1 June 2010.
[141] MSNBC Staff, 'Bush Administration in Hot Seat Over Warming: Lawmakers Get Survey of Scientists, Half of Whom Report Political Pressure' (Report, Washington, 30 January 2007) <http://www.msnbc.msn.com/id/16886008/?GT1=8921> accessed 20 October 2009.

international community and, in particular, with its European counterparts by setting an ambitious goal of reducing US greenhouse gas emissions by 80% by 2050. Here, again, he softened the domestic blow by stressing that he sought to meet this goal by implementing an economy-wide cap-and-trade program and by making the US a leader in clean technologies, thus framing climate change in economic rather than regulatory terms.

President Obama acted early to fulfill campaign rhetoric on climate change. During the first 100 days of his Presidency he took direct and indirect actions to instigate US action on climate change. First, President Obama signaled his commitment to elevating the role of science in political decision-making through his picks for his political Cabinet, including appointing Nobel Prize in Physics winner and climate science expert Steven Chu as the Secretary for the Department of Energy; Oregon State University professor Jane Lubchenco, a marine ecologist and environmental scientist, as the head of the National Oceanic and Atmospheric Administration; and renowned Harvard scientist John Holdren as Assistant to the President for Science and Technology. Similarly, many of President Obama's other political appointees possess rich scientific and 'green' credentials. For example, President Obama created a new Cabinet position, Assistant to the President for Energy and Climate Change, to which he appointed former EPA Administrator Carol Browner. Additionally, his nominee for EPA Administrator, Lisa Jackson, previously served as head of the New Jersey EPA and was a former EPA employee. Signaling his intention to re-engage with international negotiations, President Obama appointed Todd Stern as the US climate change negotiator; Stern had previously held the same position under President Clinton.

Beyond his political appointments, President Obama continued to push the 'New Energy for America Plan' that he and Vice President Joe Biden had unveiled during his Presidential campaign. As the extent of the economic recession became clear, Obama touted the plan not as part of a focused climate policy, but as a strategy for creating millions of new jobs through investments in clean energy technologies, reducing dependency on unstable regions of the world, increasing renewable energy production, and reducing greenhouse gas emissions in an economically sound manner. During the fall 2008 Congressional negotiations for an economic recovery package, President Obama pushed hard for a bill that included core components of his energy plan. The resulting American Recovery and Reinvestment Act (ARRA),[142] enacted by the US Congress and signed into law by President Barack Obama on 17 February 2009, created an economic stimulus package that was based in significant part on President Obama's proposals. Of the total measures estimated to be worth US$787 billion, the President claimed that in excess of US$60 billion would go towards 'clean energy investments that will jump-start our economy and build the clean energy

---

[142] American Recovery and Reinvestment Act of 2009, Pub L No 111-5, 123 Stat 115 (2009).

jobs of tomorrow'.[143] Noticeably, neither ARRA nor the President's assessment of ARRA's 'green' measures characterize any of the provisions as mechanisms to address global climate change. Despite this political sidestepping, many of the green provisions of the bill offer opportunities for reducing greenhouse gas emissions. Social and political focus on economic recovery, however, discouraged the President or Congress from framing the economic stimulus package as 'climate friendly'.

On 26 January 2009, as negotiations for ARRA neared completion, President Obama signed two Presidential memorandums designed to jump-start legislative and administrative action on climate change and energy policy. First, characterizing his effort as 'a down payment on a broader and sustained effort to reduce our dependence on foreign oil',[144] President Obama issued a memorandum directing the Department of Transportation (DOT) to increase the CAFE standards for carmakers' 2011 model year, in line with provisions of the previously enacted Energy Independence and Security Act of 2007. Second, President Obama issued a memorandum requesting that the EPA assess whether its 2008 decision to deny California's petition for a waiver[145] to adopt limitations on greenhouse gas emissions from motor vehicles was appropriate under the Clean Air Act, ultimately enabling California and over a dozen other states to initiate the nation's first tailpipe greenhouse gas emission regulation schemes, and paving the way for future schemes at the federal level. In issuing this second memo, the President sought to distance himself from the actions of the previous administration noting that 'Instead of serving as a partner, Washington [D.C.] stood in [the states'] way. The days of Washington dragging its heels are over.'[146]

Less than ten days later, on 5 February 2009, the President issued a third energy/climate-related memorandum. In this third memo, the President directed the Department of Energy to implement more aggressive efficiency standards for common household appliances, like dishwashers and refrigerators, in compliance with the Energy Policy Act of 2005 and the Energy Independence and Security Act of 2007.

Beyond these specific actions, the Obama Administration has outlined an energy and environment policy program that includes developing renewable energy projects that produce electricity from wind, wave, and ocean currents in the US Outer Continental Shelf; promoting the development of new, green jobs that cannot be outsourced; investing US$150 billion over ten years in energy research and development in order to transition to a clean energy economy;

---

[143] President Barack Obama, 'Energy & Environment' (Statement of Policies on Energy and Environment Issues) <http://www.whitehouse.gov/issues/energy-and-environment> accessed 1 June 2010.

[144] Macon Philips, 'From peril to progress (Update 1: Full Remarks)' (Post on the White House Blog 26 January 2009) <http://www.whitehouse.gov/blog_post/Fromperiltoprogress> accessed 1 June 2010.   [145] This decision will be discussed in more detail in Ch 3.

[146] Philips (n 144).

overcoming dependence on foreign oil; producing more energy domestically; promoting energy efficiency; closing the carbon loophole and cracking down on polluters by stemming carbon pollution through a market-based cap-and-trade program; and protecting domestic manufacturing and global competitiveness by securing comparable actions by the US main trading partners. The contours of this ambitious domestic policy program depend in large part on Congressional support. Absent Congressional enactment of framework climate change legislation that embeds these objectives in legislative form, the President is unlikely to be able to push his progressive energy and climate agenda. Congress remains divided on climate change; yet majority support for climate legislation grows.

During the fall 2009 session of the 111th Congress, US Representatives Henry Waxman (D-Calif) and Ed Markey (D-Mass) released a widely heralded 'discussion draft' of a climate change bill that eventually went on to be approved by the House of Representatives. The 932-page bill, titled the 2009 American Clean Energy and Security Act (ACES), was the first bill to clear the House Committee on Energy and Commerce in the 111th Congressional Session following President Obama's occupation of the White House.[147] The House passed ACES by a vote of 219 to 212 on 26 June 2009.

The Act outlines a detailed two-part plan of action for mitigating and adapting to climate change. In the first part of the plan, ACES creates a comprehensive mitigation strategy, the core of which is a cap-and-trade program that establishes mandatory caps on 87% of US greenhouse gas emissions, including heavy industry and the electric power and oil and gas sectors. The cap-and-trade program is complimented by a series of measures designed to increase energy efficiency and encourage the development and uptake of clean energy technologies. These provisions include new clean energy requirements for utilities; energy efficiency requirements coupled with funding for energy efficiency initiatives and for upgrading of building codes and product efficiency standards; funding for studies and incentives for carbon capture and sequestration technologies; as well as supplemental measures to ensure that the transition to a low carbon economy does not detrimentally impact US industry, jobs, and low-income households. The second part of ACES creates a multi-part adaptation strategy, including a National Climate Change Adaptation Program, an International Adaptation Fund, and separate proposals for addressing questions of public health and natural resource adaptation.

If implemented, ACES would mandate a reduction of emissions from covered sources to 17% below 2005 (baseline) levels by 2020 and 83% below 2005 levels

---

[147] 'Global warming poses a significant threat to the national security, economy, public health and welfare, and environment of the United States, as well as of other nations.' American Clean Energy and Security Act of 2009, HR 2454, 111th Congress Title VII, pt A (2009).

by 2050.[148] These levels exceed President Obama's goal of reducing emissions to 1990 levels by 2020. ACES has been criticized for setting the cap for the emissions trading program above levels recommended by scientists, for failing to address questions of environmental justice and cross-sectoral environmental harms, for marginalizing the role of states in addressing climate change, and for failing to address key questions of how emissions trading revenues will be used. It has also been widely criticized by the energy, oil and gas, and coal lobbies as imposing overwhelming economic burdens on US industry and US consumers.

Despite receiving criticism from all sides, ACES represents a significant turnaround in US climate politics. After 17 years as a party to the UNFCCC, at least one chamber of the US Congress has managed to pass a climate bill. Much of the impetus for ACES can be attributed to President Obama's active campaigning for legislation. Possibly even more important, however, were President Obama's actions to prompt the EPA to regulate greenhouse gas emissions in the absence of Congressional action. President Obama's memorandum requesting the EPA to reassess the decision to deny California a waiver to regulate greenhouse gas emissions under the Clean Air Act, coupled with the agency's progress towards regulating greenhouse gas emissions under the Clean Air Act pursuant to the Supreme Court's decision in *Massachusetts v EPA*,[149] initiated a turf war between Congress and the EPA.

As it became evident that the EPA was laying the groundwork for erecting a greenhouse gas regulatory regime under the Clean Air Act, Congress came under increasing pressure to act so as to exercise authority over US climate change politics and preclude piecemeal regulatory approaches to addressing climate change. House approval of ACES followed closely on the heels of the release of a proposed endangerment finding which, when finalized in December 2009 laid the groundwork for EPA to regulate greenhouse gas emissions from a variety of sources under the Clean Air Act.[150] If passed by the Senate, ACES or any modified version of ACES would constrain EPA's authority to regulate greenhouse gas emissions under the Clean Air Act.

There remains considerable doubt, however, whether the Senate will pass climate legislation. The climate debate in the Senate was long overshadowed by proposals for health care reform and undercut by internal disagreements among Democratic senators over the form and function of the climate bill. The 111th Senate began holding hearings on climate change in early July 2009, with Senator Barbara Boxer—head of the influential Senate Environment and Public Works Committee—expressing her intent to release a bill to the full Senate by the end of September, with the hopes of passing legislation before the commencement of

---

[148] John Larsen, 'A Closer Look at the American Clean Energy and Security Act' (2 July 2009) <http://www.wri.org/stories/2009/07/closer-look-american-clean-energy-and-security-act> accessed 1 June 2010. [149] Discussed in more detail in Ch 4.
[150] Proposed Endangerment and Cause or Contribute Findings for Greenhouse Gases under section 202(a) of the Clean Air Act, 74 Federal Register 18886 (24 April 2009).

international negotiations in Copenhagen in December 2009. While Senators Boxer and Kerry managed to pass the Clean Energy Jobs and American Power Act through the Environment and Public Works Committee on 30 September 2009, the bill had not been debated by the full Senate by the close of 2009. Following intense partisan debate, numerous false starts, and the loss of the Republican sponsor (Senator Lindsey Graham, SC), on 12 May 2010, Senators Kerry and Lieberman rolled out a 987-page draft climate bill, the American Power Act. The American Power Act closely mirrors the emissions reduction goals embodied by ACES and similarly employs a national cap-and-trade program as its primary vehicle for reducing greenhouse gas emissions, while differing significantly in supporting mitigation and adaptation provisions, including provisions to curtail the power of the EPA to regulate greenhouse gas emissions under the CAA. By early summer 2010, the Act remained on the Senate floor for debate with no vote yet taken. In the interim, while Congress continues to debate climate legislation, the EPA remains empowered to continue crafting a regulatory framework for greenhouse gas emissions.

Less than a year after President Obama's election, federal-level US climate politics evolved from a political no man's land to a battlefield for political control. US climate politics under the Obama administration reached an active, high-level sphere. However, despite escalating political activity, leading up to and following the 2009 Copenhagen negotiations, the US continued to lack any type of finalized or implemented legal regime for regulating greenhouse gas emissions. Following eight years of political inactivity under the Bush administration, as well as escalating greenhouse gas emissions, the US Congress was unable to immediately reach a consensus on the issue of how to address climate change, or to formulate legislation that would enable the US to achieve emissions reductions comparable to EU targets.

The total absence of an enforceable climate policy however, poses potentially intractable political obstacles not just for the US, but for the entire international community. The US inability to point to domestic action to mitigate greenhouse gas emissions not only undermined efforts to re-engage with international climate politics but also restricted global efforts to create a more comprehensive post-Kyoto legal agreement because developing countries were unwilling to undertake binding emissions reduction obligations in the absence of US domestic efforts. US re-engagement and leadership efforts were further undercut by amorphous and non-committal rhetoric in international forums preceding the Copenhagen negotiations.

In March 2009 in Bonn, Germany, Todd Stern, the new US chief climate negotiator, made his first appearance at an interim meeting of the UNFCCC Conference of the Parties. He was greeted with a standing ovation. As negotiations proceeded, however, it became clear that the US was neither ready to reveal the contours of a US climate proposal nor truly capable of determining what its domestic constituency would support. Based on the few details Stern shared, it became evident that the US proposal differed significantly from what Europe and the big developing economies of China and India were advocating. First, Stern's comments suggested that the US would focus on long-term emission reduction goals (2050), in contrast to Europe's decision to press for more aggressive short-term emission reduction obligations (2020). Second, Stern's statements confirmed previous reports from the Obama administration that the US proposal would require developing nations, with particular regard to the emerging economies of China and India, to undertake legally binding obligations to reduce greenhouse gas emissions. Thus, while the international community initially greeted Stern with eager applause, by the end of the Bonn meeting, optimism faded as it became apparent that active US participation in international negotiations amplified rather than ameliorated existing differences among developed countries and between developed and developing countries over proposals for a post-Kyoto international legal agreement.

Moving forward with efforts to reform climate politics, the US faces an uphill battle on the domestic and international fronts. The recent burst of activity within the executive and legislative branches suggests that there is increased public support for domestic efforts to address climate change. Yet, it would be a mistake to assume that political change in the executive and legislative branches ensures that the US will support either domestic legislation or a post-Kyoto international legal agreement.[151] Despite pending proposals for legislative action and rhetoric on international re-engagement, federal engagement with climate change continues to remain largely rhetorical and heavily conditioned on economic security. Similarly, it remains far from clear whether the US domestic constituency supports aggressive action at the domestic or international level to address climate change.

The ability of the US to enact a domestic legal framework for addressing climate change controls its ability to influence global climate negotiations. If the US fails to agree on a domestic framework for addressing climate change, it is unlikely either to earn domestic support for participating in an international legal framework or the normative status necessary to influence international negotiations. The success or failure of the US in erecting a domestic legal regime, thus, is of great consequence to efforts to strengthen the international climate regime.

In addition, because the US is not a member to the Kyoto Protocol, it is not bound to report on domestic progress in reducing greenhouse gas emissions. It therefore remains difficult to evaluate the substantive progress the US is making towards reducing greenhouse gas emissions through its various non-regulatory avenues. This information gap is compounded by the fragmented nature of domestic reporting and the failure of the US to fulfil its UNFCCC reporting requirements.

---

[151] Eg, David Driesen, 'Thirty Years of International Environmental Law: A Retrospective and Plea for Reinvigoration' (2003) 30 *Syracuse J Intl L & Com* 353, 362.

Annex I Parties to the UNFCCC, which includes the US, are obligated to submit national communications to the UNFCCC Secretariat every four years. These reports are important sources of information for evaluating national efforts to implement the UNFCCC domestically. Annex I Parties were due to submit their fourth national communication to the UNFCCC secretariat by 1 January 2006. The US submitted its fourth national communication on 27 July 2007, leaving an information gap that hindered transparency and impeded efforts to evaluate US progress in reducing greenhouse gas emissions for over one and a half years. Prior to the submission of the fourth national communication, the most recent US report to the UNFCCC secretariat was the third national communication, submitted in 2002, which was extremely outdated by the time of the submission of the fourth national communication.

Conclusion

The direction of American climate politics began a slow and steady shift towards re-engagement even before the election of President Obama. In the twilight years of his Presidency, President George W. Bush softened his critique of the science and economics of climate change in response to escalating domestic and international pressures. Leading up to the G8 meeting in Germany during the week of 4 June 2007, President Bush surprised the international community by calling for the world's 15 biggest polluters—including both developed and developing countries—to agree on a target for reducing greenhouse gases by the end of 2008. In doing so, he emphasized the role of technology in addressing climate change, saying, '[t]he way to meet this challenge of energy and global climate change is through technology, and the United States is in the lead' and suggesting that '[t]he world is on the verge of great breakthroughs that will help us become better stewards of the environment'. In focusing on the critical role of technology, President Bush consistently avoided any talk of lifestyle changes or limiting greenhouse gas emissions through mandatory regulatory regimes. Further, he proposed collaborative national strategies—whereby individual countries would create 'midterm management targets and programs that reflect their own mix of energy sources and needs'—rather than working through an international legal framework such as Kyoto to set legally binding national commitments.

President Bush spoke in terms of creating a new 'global framework' and a 'long-term global goal' for reducing greenhouse gas emissions but, at all times, he very carefully avoided committing to any form of mandatory greenhouse gas emission reduction. Instead, he spoke of the important roles of 'safe nuclear power', new technologies to reduce vehicle emissions, and of increasing the market share of home-grown bio-fuels because it 'makes sense to have our farmers growing energy, so that we don't have to import it from parts of the world where they may not like us too much'. Throughout his discussion of climate change, President Bush's comments revolved around two core elements: the importance of national sovereignty to choose the best methods for addressing climate change, and the central role of technology in addressing climate change.

Despite the constrained nature of his comments, President Bush's speech surprised many people and provoked a range of responses—both positive and sceptical. Very few people expected President Bush to follow up on his comments with a groundbreaking proposal; yet his initiative in instigating a discussion on the topic and his willingness to enter into negotiations appeared to signal a change in attitude.

During the G8 meeting, President Bush shied away from making any legally binding commitments. He rejected proposals for an international carbon-trading scheme; he rejected specific energy efficiency targets; and he rejected a German proposal for a 'two-degree' strategy, aiming to limit the increase in world temperatures to 2°C this century, which would require a halving of greenhouse gas emissions by 2050. Avoiding making any strong statements or commitments at the G8 meeting, President Bush instead focused on his proposal to initiate a series of meetings over the next year to bring together major polluting countries to negotiate a post-Kyoto agreement.

In the end, the G8 Chair issued a summary statement on climate change, declaring that climate change is a major challenge that threatens both the natural environment and the global economy, and stating that '[w]e are convinced that urgent and concerted action is needed' and that the members of the G8 'accept...responsibility to show leadership in tackling climate change'.[152] The summary statement acknowledged the decisions made by the EU, Canada, and Japan to halve global emissions by 2050, but was unable to declare this as a goal of the combined G8 members. The statement also established the UN climate process as the appropriate forum for future climate change negotiations and called on the major emitters to agree on a new global framework by the end of 2008, based upon the principle of common but differentiated responsibilities and capabilities, and acknowledging the importance of technology, market mechanisms, energy efficiency, and emissions trading as central components of any future agreement.

In addition to the Chair's summary, the G8 President, Angela Merkel, on behalf of the other Heads of State, issued a joint statement that included a statement on climate change.[153] Here, again, the members of the G8 acknowledged the seriousness of the problem, affirmed their commitment to working through

---

[152] The Federal Government of Germany, 'Chair's Summary' (G8 Summit, Heiligendamm, 8 June 2007) <http://www.g-8.de/Content/EN/Artikel/__g8-summit/anlagen/chairs-summary,templateId=raw,property=publicationFile.pdf/chairs-summary.pdf> accessed 1 June 2010.

[153] The Federal Government of Germany, 'Joint Statement by the German G8 Presidency and the Heads of State and/or Government of Brazil, China, India, Mexico and South Africa on the occasion of the G8 Summit in Heiligendamm, Germany, 8 June 2007' <http://www.g-8.de/Content/EN/Artikel/__g8-summit/anlagen/o5-erklaerung-en,templateId=raw,property=publicationFile.pdf/o5-erklaerung-en.pdf> accessed 1 June 2010.

the UNFCCC, agreed 'to contribute our fair share to tackle climate change', and called for a 'flexible, fair and effective global framework and concerted international action'[154] on climate change.

In joining this statement, President Bush *de facto* acknowledged the severity of the problem of climate change and consented to engage in future efforts to develop a post-Kyoto framework. He did not, however, agree to any new commitments that would obligate the US to modify its current climate change strategy.

With the conclusion of the G8 summit, the instigation of a new round of climate change talks and the looming Presidential election, the window opened for the federal government of the US to re-engage with the international community on the issue of climate change. This meeting represented a turning point in US climate politics. President Bush did not encourage the enactment of domestic climate legislation but he did initiate a series of secret, high-level meetings with developing countries—most notably with China—in an attempt to improve political cooperation among the world's biggest greenhouse gas emitters. President Bush's softened stance on climate change created room for a more open political debate and paved the way for climate change to become less of a partisan topic during the 2008 Presidential race.

Since that time, the US has elected a new Democratic President; the US House of Representatives has passed a climate change bill; the US Senate has begun debating a climate bill; the US EPA has initiated a climate change regulatory regime; and climate change has become a regular topic in mainstream social and political debate. The US has seemingly made great strides in aligning its climate politics with those of the EU and much of the international community in a very short time.

Progress thus far, however, remains largely symbolic. The US has not yet committed to any concrete or enforceable measures at the domestic or international levels. The US, along with the entire global community, stands at a political crossroads. The decisions the US government makes in the near term will chart the course of global efforts to address climate change for many years to come.

Yet, it is necessary to look beyond the actions of the federal government to envisage the future pathway of US climate politics. Federal initiatives only constitute a fraction of the legal and political actions underway to address climate change in the US. Climate change is unique among other environmental issues in that the federal government has not led domestic efforts to craft a robust legal framework. Chapters 3 and 4 explore the role sub-federal governments, litigation, regulation, and extra-legal mechanisms play in influencing the direction of climate change law and policymaking in the US.

---

[154] Ibid.

3
Sub-Federal Laws and Policies in the United States

Introduction

States have long been the 'laboratories of democracy'[1] for US lawmaking. In the field of environmental law and beyond, states are often the test grounds for new modes of law and regulation and it has long been recognized that sub-federal lawmaking can catalyze federal action.[2]

In particular, states have frequently led the way for the federal government in experimenting with and promoting new environmental laws and regulations. Federal environmental regulation mimicking state regulation, thus, is not a new phenomenon. Rarely, however, have states—either directly or indirectly—embarked on such a widespread and coordinated campaign to develop effective environmental laws in the absence of federal leadership as in the current case of climate change governance.

Traditional economic theory would argue against the rationality of independent or regional state efforts to regulate the global commons in the absence of national coordination and oversight to prevent problems of competitive disadvantage.[3] Nevertheless, the flurry of activity at the state level is undeniable. States have adopted a variety of legal, regulatory, and policy measures to address climate change, to include: 46 states have greenhouse gas inventories; 33 states have adopted climate action plans; 20 states have public benefit funds; 20 states have adopted greenhouse gas emission targets; 28 states have renewable energy portfolio standards; 16 states are in the process of adopting greenhouse gas emission standards for automobiles; 18 states have mandatory CO_2 reporting programs; 24 states have formed climate change advisory boards; 33 states are participating

---

[1] *New State Ice Co v Liebmann*, 285 US 262, 311 (1932) (Brandeis J, dissenting that '[i]t is one of the happy incidents of the federal system that a single courageous state may, if its citizens choose, serve as a laboratory....').
[2] Kristen H Engel, 'Mitigating Global Climate Change in the United States: A Regional Approach' (2005) 14 *NYU Envtl LJ* 54, 64.
[3] See generally Kristen H Engel and Scott R Saleska, 'Subglobal Regulation of the Global Commons: The Case of Climate Change' (2005) 32 *Ecology LQ* 183.

in one or more of eight existing regional climate change initiatives.[4] Rarely, in the absence of a federal mandate, have states established such an extensive and multifaceted set of environmental policies and regulations.

The impetus for widespread state action is not immediately apparent. The driving force may be genuine concern for a global environmental problem that threatens the health and well-being of state citizenry, the desire to garner the political support and goodwill of concerned constituencies, or the idea that state actions will eventually create models for the federal government to use in creating a comprehensive climate change framework and, thus, 'trigger regulatory action by governments that contribute a larger percentage of global greenhouse gas emissions'.[5] Genuine concern, political maneuvering, and prompting of federal action are not mutually exclusive drivers. States are likely motivated by all of these factors to different degrees. The relative importance of political motivations is not always clear. Unlike federal regulation, which has been extensively researched and written about, state regulation has received far less attention and analysis, particularly in the context of environmental law.[6] Thus, there is very little data suggesting why states adopt particular environmental regulatory strategies. What is clear, however, is that a variety of factors are driving US states to act in the absence of a federal mandate and that, whatever the motivation, federal law is 'rife with examples of federal legislation that has drawn heavily from ideas being developed at the state level'.[7]

In the context of climate change law and policy in the US, state and local entities are creating momentum for progressive climate change policymaking despite a rigid federal superstructure that delimits state lawmaking powers. Unlike EU member states, US states are not sovereign entities. The US federal government could, in theory, enact a command-and-control style climate change statute—in the same vein as the Clean Air Act,[8] the Clean Water Act,[9] or the Endangered Species Act[10]—that requires states to meet federal requirements in a specific, inflexible, legally binding way. This style of environmental lawmaking is unfeasible in the context of the EU. Even in the US, this type of command-and-control style law has become both politically unpopular and increasingly difficult to adopt and implement. Further, the US federal government has demonstrated little interest in enacting any type of comprehensive and binding regulatory

---

[4] Pew Center on Global Climate Change, 'Regional Initiatives' <http://www.pewclimate.org/what_s_being_done/in_the_states/regional_initiatives.cfm> accessed 16 April 2010. Greenhouse gas inventories are currently one of the most popular ways that states seek to engage with the climate change debate. See Dana R Fisher, *National Governance and the Global Climate Change Regime* (Rowman & Littlefield Publishers Inc, Lanham, 2004) 216, citing Randall Freed and others, 'Greenhouse Gas Inventory Tools for States' <http://www.epa.gov/ttn/chief/conference/ei11/poster/freed.pdf> accessed 1 June 2010.   [5] Engel (n 2) 64.
[6] Paul Teske, *Regulation in the States* (The Brookings Institution, Washington DC, 2004) 8.
[7] Engel (n 2) 64, citing Joseph A Ranney, 'The Rise of Labor and Wisconsin's "Little New Deal"' [Oct 1994] *Wisc Law* 22 (describing Wisconsin's contribution to the creation of the federal social security program).   [8] 42 USC §§ 7401–7671(q) (2000).
[9] 33 USC §§ 1251–1376 (2000).   [10] 16 USC §§ 1531–1544 (2000).

regime, much less one that is based on strict federal guidelines, rigid deadlines, and firmly established methods of meeting federal objectives; a regulatory regime based on market mechanisms and economic incentives is much more feasible. At the moment, however, the federal government has chosen to abdicate the right to legislate on climate change. In the absence of federal legislation, states have become the arbiters of legal and political change.

Facing apparent federal reluctance to address climate change, states such as California and New York, and cities such as Portland and Philadelphia, are choosing to follow in the footsteps of the EU to try to create robust climate change laws and policies. The policies and ideologies of these state and local entities increasingly have more in common with one another and with European nations than they do with their own national government. Leaving the US federal government standing still, these cosmopolitan states and cities are moving forward by learning from and mimicking climate change experiences and policies from abroad.

California is a prime example of this phenomenon. As examined in detail in the discussion that follows, California's climate change policies more closely resemble the policies of state parties to the Kyoto Protocol than US federal policies. The following section reviews several case studies that reveal how US state and local policymakers are finding ways to implement climate change laws and policies and examines how these initiatives promote concrete and measurable steps towards developing a comprehensive climate change regime that moves US sub-federal politics closer in form and method to the EU than to the US federal government.

Overview of State Climate Change Initiatives

As compared to other areas of environmental policy, US states are playing a particularly active role in adopting greenhouse gas emission policies. As early as 1997, states began addressing global climate change. Oregon initiated state policymaking efforts.[11] In 1997, the Governor of Oregon, John Kitzhaber, passed the first law in the nation to set carbon dioxide standards for new energy facilities in the state.[12] And, in 2000, New Jersey established state-wide targets for reducing greenhouse gas emissions based on voluntary public and private programs. A year later, in 2001, Massachusetts upped the ante by becoming the first state to establish mandatory carbon dioxide emissions caps for power plants. Massachusetts's regulation mandates that the six largest and dirtiest power plants in the state cut sulfur dioxide and nitrogen oxide emissions by 50–70%, and

---

[11] Progressive Policy Institute, 'State and Local Governments and Climate Change' (2003) <http://www.ppionline.org/ppi_ci.cfm?knlgAreaID=116&subsecID=900039&contentID=251285> accessed 3 April 2010.
[12] Barry G Rabe, 'State Competition as a Source Driving Climate Change Mitigation' (2005) 14 *NYU Envtl LJ* 1, 17.

carbon dioxide emissions by 10%, as well as reducing mercury releases.[13] Thus, by 2001, Massachusetts had managed to do what the federal government had, at the time, taken no steps to do—regulate carbon dioxide emissions.

In setting limits on greenhouse gas emissions, numerous states, including California, Texas, and Massachusetts, have implemented standards for increasing the amount of electricity generated from renewable energy sources.[14] States are also taking the lead in developing methods for identifying and tracking sources of greenhouse gas emissions.

These are just a few examples of how states are adopting measures to address global climate change by curbing greenhouse gas emissions; numerous other states have adopted or plan to adopt greenhouse gas emission regulations, including New York, New Hampshire, Florida, and Illinois, to name a few.[15] States are also taking a leading role in addressing greenhouse gas emissions from motor vehicles, going so far as to adopt legislation that exceeds the minimum regulatory requirements set by the US Clean Air Act.[16]

The emergent role of states in setting climate change policy reflects growing awareness of how climate change will affect state economies and natural environments, including 'the likely effects of climate change on agriculture, forestry, the availability of water, public health, and other areas of traditional state responsibility'.[17] In response to these concerns, states are beginning to take independent and collaborative action to address climate change.

Regional Climate Change Initiatives

Individual states, especially keystone states such as California and New York, create legislative models and regional momentum for addressing environmental problems. In the context of a problem of the global commons, however, independent state action can only go so far to address the problem. Collective action can exponentially increase the effectiveness of single state responses and 'may function as a mechanism by which individual governments can help trigger the implementation of the preferred solution' at the federal or international level.[18]

Recognizing the trans-boundary nature of the problem and the benefits of collective action, US states are joining forces to create regional climate change partnerships. Currently, 32 states are collaborating on three major regional climate change initiatives. On-going regional collaborations include the following: the Midwestern Greenhouse Gas Reduction Accord (MGGRA), the Western

---

[13] Ibid; see also 'State and Local Governments and Climate Change' (n 11) 61. The portions of Massachusetts's regulations that applied to the six largest power plants were eventually phased out and transitioned to the RGGI cap-and-trade system.
[14] Rabe (n 12) 17.
[15] Ibid.
[16] See generally Rachel L Chanin, 'California's Authority to Regulate Mobile Source Greenhouse Gas Emissions' (2003) 58 *NYU Ann Survey Am L* 699.
[17] John C Dernbach, 'Facing Climate Change: Opportunities and Tools for States' (2004) 4 *Widener LJ* 1, 2.
[18] Engel and Saleska (n 3) 184.

Climate Initiative (WCI), and the Regional Greenhouse Gas Initiative (RGGI). Additionally, the Powering the Plains Initiative (PPI) is an on-going regional collaboration between state leaders, and utility industry leaders.[19] Other early regional initiatives include the New England Governors' Conference Climate Change Action Plan 2001 (NEG/ECP), the Western Governors' Association (WGA), Southwest Climate Change Initiative (SCCI), Diversified Energy Initiative (DEI), West Coast Governors' Global Warming Initiative (WCG), Eastern Climate Registry (ECR), Midwest Greenhouse Registry (MGR), as well as Powering the Plains Initiative (PPI), which was an early collaboration between state and industry leaders.[20]

The Climate Change Action Plan

In 2000, the New England Governors[21] passed Resolution 25-9 which sets the goal of addressing global climate change and its impacts on the environment.[22] Resolution 25-9 created the regional Climate Change Action Plan.[23] The Climate Change Action Plan establishes three primary goals: (1) reduce regional greenhouse gas emissions to 1990 levels by 2010; (2) further reduce regional greenhouse gas emissions to 10% below 1990 levels by 2020; (3) finally, reduce regional greenhouse gas emissions to 75–85% below current levels in the long-term.[24]

---

[19] Pew Center on Global Climate Change, 'Regional Initiatives' <http://www.pewclimate.org/what_s_being_done/in_the_states/regional_initiatives.cfm> accessed 16 April 2010.

[20] As a complement to the initiatives, the Climate Change Registry, a non-profit organization, was created to provide meaningful information to reduce greenhouse gas emissions by establishing consistent, transparent standards throughout North America for businesses and governments to calculate, verify, and publicly report their carbon footprints in a single, unified registry. See The Climate Registry, 'About The Climate Registry' <http://www.theclimateregistry.org/about/> accessed 20 December 2009.   [21] The members of the NEGC are also members of the RGGI.

[22] See The New England Governors and The Eastern Canadian Premiers, 'Resolution 25-9' (Resolution concerning climate change 2001) <http://www.negc.org/02En003.html> accessed 3 March 2007. Resolution 25-9 was subsequently complemented by Resolution 32-4, passed on 16 September 2008. Resolution 32-4 reiterated the intent of the members to facilitate climate change mitigation by improving energy efficiency and conservation measures and encouraging greater use of renewable energy resources within the region. In addition, on 15 September 2009, the NEGC passed a Resolution inviting the federal government to work with them to help support the region's low-carbon resources.

[23] The Committee on the Environment and Northeast International Committee on Energy of the Conference of New England Governors and Eastern Canadian Premiers, 'Climate Change Action Plan 2001' (August 2001) <http://www.negc.org/documents/NEG-ECP%20CCAP.PDF> accessed 20 December 2009 (Climate Change Action Plan).The member states of the New England Governors' Climate Change Action Plan are as follows: Connecticut, Maine, Massachusetts, New Hampshire, Rhode Island, and Vermont. Ibid.

[24] Ibid. In addition to these primary goals, the Climate Change Action Plan mandates that each of the member states implement nine specific action items: developing a regional standardized greenhouse gas emission inventory, creating state-specific greenhouse gas reduction strategies, promoting public awareness, encouraging government authorities to lead by authority, reducing emissions from the electricity sector, improving energy conservation measures, preparing climate change adaptation strategies, decreasing the growth in emissions in the transportation sector, and creating a regional emissions registry and exploring the possibility of creating an emissions trading mechanism. Ibid 6–18.

The NEG/ECP was one of the first multi-state initiatives that, by its own terms, recognized the potentially negative consequences of climate change for environmental and economic well-being[25] and the importance of crafting a climate change strategy that 'supports and complements other regional, state and provincial initiatives'.[26] Further, the NEG/ECP addresses the inherently international nature of climate change by seeking to complement the Canadian National Implementation Strategy for Climate Change.[27]

The NEG/ECP Climate Change Action Plan offers an early example of how regional partnerships, in the absence of strong national programs, can devise innovative transboundary mechanisms for addressing climate change. The Climate Change Action Plan further demonstrates how regional compacts, even if they are voluntary rather than mandatory, can prompt states to integrate the commitments of the compact into legislation. For example, following the adoption of the Climate Change Action Plan, the state of Maine implemented the Maine Act to Provide Leadership in Addressing the Threat of Climate Change.[28] The Act translates the Climate Change Action Plan's overarching commitments and certain of the action items into legal obligations. Massachusetts and New Hampshire have also taken concrete steps to integrate their regional commitments into state policy through legislation, regulation, and public–private initiatives.[29]

The NEG/ECP has been heralded by many as a model for encouraging state action and for linking the efforts of US Governors and Canadian Premiers in order to increase the efficacy of individual climate change commitments. It has also been criticized as exceeding state authority under the US Constitution.[30] At least one commentator has argued that agreements, such as the NEG/ECP, between US states and foreign states violate 'Article I, Section 10 of the Constitution, which bars states from entering into a treaty, alliance, confederation, agreement or compact with another state or nation'.[31] This argument

---

[25] Ibid Preamble. [26] Ibid.
[27] Ibid. The New England Governors further refined the Climate Change Action Plan in August 2002 by passing 'Resolution 27-7 Concerning Climate Change'. The resolution expands upon the 2001 Climate Change Action Plan by encouraging active participation of the academic sector, introducing initiatives for LED traffic lights throughout the region, and promoting energy efficient vehicle use in state and provincial fleets. In addition, the plan contains various state and provincial measures to achieve specific greenhouse gas reduction goals. The New England Governors and The Eastern Canadian Premiers, 'Resolution 27-7' (Resolution Concerning Climate Change 2002) <http://www.negc.org/02En003.html> accessed 3 March 2007.
[28] Me. Rev. Stat. Ann. tit. 38, §§ 574–578 (2003).
[29] See Robert B McKinstry Jr, 'Laboratories for Local Solutions for Global Problems: State, Local and Private Leadership in Developing Strategies to Mitigate the Causes and Effects of Climate Change' (2004) 12 *Penn St Envtl L Rev* 1535–45. [30] Engel (n 2) 78–79.
[31] Ibid 79, citing Senate Committees on Foreign Relations and Environmental and Public Works, 'Statement for the Record of the Joint Committee Hearing on US Environmental Treaties' (Statement of Jon Reisman, Associate Professor of Economics and Public Policy at the University of Maine, 2002) <http://epw.senate.gov/107th/Reisman_072402.htm> accessed 20 December 2009.

highlights potential tension between state efforts to move beyond the entrenched position of the federal government on climate change. The validity of the argument, however, is tenuous, as the agreement between the US Governors and the Canadian Premiers represents a joint statement of commitment to meet greenhouse gas emission reduction obligations, but it 'does little more than expres[s] their mutual intent to reduce greenhouse gases; nothing they are doing commits either nation as a whole to reduction targets or to any other requirement of the Kyoto Protocol'.[32] It is unlikely that the federal government will challenge the Constitutionality of the NEG/ECP. Nevertheless, the potential incompatibility of regional climate change agreements with US federalism, and the willingness of commentators to raise federalism concerns as a mechanism for hindering progressive state-based climate change policies reflect on-going tensions between state efforts to fill the political void and federal desire to maintain control over the domain of climate change.[33]

Western Governors' Association and Diversified Energy Initiative

Unlike the New England Governors' Climate Action Plan, the Western Governors' Association is not explicitly focused on climate change. However, the Western Governors' Association[34] has launched the Diversified Energy Initiative[35] to address growing regional concerns about energy security and climate change. The states participating in the Diversified Energy Initiative have 'agreed to examine the feasibility of: (1) [d]eveloping 30,000 Megawatts of clean and diverse energy by 2015; (2) [i]ncreasing energy efficiency 20 percent by 2020; (3) [p]roviding adequate transmission to meet the region's needs through 2030'.[36] The Diversified Energy Initiative is modest in its willingness to directly engage with the cross-sector dimensions of climate change or to create emission reduction commitments. The Initiative, however, represents a starting point for regional coordination on energy security and energy diversification. In this regard, the Diversified Energy Initiative reflects a common dimension of many state and regional initiatives, which is the reality that many state and regional climate change policies 'are more show than substance'.[37] Subsequent to initiating the Diversified Energy Initiative, in 2008, the Western Governors' Association together with the US Department of Energy launched the Western Renewable Energy Zones initiative as a complementary measure. The goal of the Western

---

[32] Engel (n 2) 79.
[33] See Barry G Rabe, 'North American Federalism and Climate Change Policy: American State and Canadian Provincial Policy Development' (2004) 14 *Widener LJ* 121, 144–51.
[34] Member states in the Western Governors' Association include: Alaska, Arizona, California, Colorado, Hawaii, Idaho, Kansas, Montana, Nebraska, Nevada, New Mexico, North Dakota, Oregon, South Dakota, Texas, Oklahoma, Utah, Washington, Wyoming. See Regional Initiatives (n 4).
[35] All of the member states of the Western Governors' Association except for Arizona and Oregon are participating in the Diversified Energy Initiative. Ibid.   [36] Ibid.
[37] Engel and Saleska (n 3) 215.

Renewable Energy Zones is to identify those areas in the West with vast renewable resources in order to better facilitate the development and delivery of renewable energy.[38] Together, these initiatives have prompted a series of advancements in renewable energy development that facilitates region-wide energy diversification efforts.

West Coast Governors' Global Warming Initiative

The West Coast Governors' Global Warming Initiative was launched in September 2003. The regional initiative commits the three member states, California, Oregon, and Washington 'to act individually and regionally to reduce greenhouse gas emissions below current levels through strategies that promote long-term economic growth, protect public health and the environment, consider social equity, and expand public awareness'.[39] On 18 November 2005, the Governors of California, Oregon, and Washington—three states known for leadership on environmental issues—approved a series of recommendations designed to improve regional efforts to address global climate change.[40] The press release for the West Coast Governors' Initiative emphasized the economic, health, and environmental consequences of global climate change as well as the economic *benefits* of adopting a leadership position in the effort to combat global climate change by developing renewable energy and energy-efficient technologies.[41]

The recommendations the Governors approved in 2004 include:

1. Set new targets for improvement in performance in average annual state fleet greenhouse gas emissions.

2. Collaborate on the purchase of hybrid vehicles.

---

[38] The initiative will analyze renewable energy resources in 11 states, two Canadian provinces, and areas in Mexico that are part of the Western Interconnection. See Western Governors' Association, 'Western Renewable Energy Zones Initiative' <http://www.westgov.org/index.php?option=com_content&view=article&id=219:western-renewable-energy-zones&catid=102&Itemid=81> accessed 3 April 2010.

[39] California Environmental Protection Agency, 'West Coast States Strengthen Joint Climate Protection Strategy' (Press Release, 18 November 2005) <http://www.ef.org/westcoastclimate/WC_Climate.pdf> accessed 1 June 2010. For example, to meet the commitment to include aggressive energy efficiency measures in updates of state building codes, in 2008, the California Energy Commission adopted changes to the Building Energy Efficiency Standards. See California Energy Commission, '2008 Building Energy Efficiency Standards' <http://www.energy.ca.gov/title24/2008standards/> accessed 20 December 2009.

[40] California Environmental Protection Agency, 'West Coast States Strengthen Joint Climate Protection Strategy' (Press Release, 18 November 2005) <http://www.ef.org/westcoastclimate/WC_Climate.pdf> accessed 1 June 2010.

[41] See ibid, stating that 'Global warming will have serious adverse consequences on the economy, health and environment of the West Coast states. These impacts will grow significantly in coming years if we do nothing to reduce greenhouse gas pollution. Fortunately, addressing global warming carries substantial economic benefits. The West Coast region is rich in renewable energy resources and advanced energy-efficient technologies. We can capitalize on these strengths and invest in the clean energy resources of our region.'

3. Establish a plan for the deployment of electrification technologies at truck stops in each state on the I-5 corridor, on the outskirts of major urban areas, and on other major interstate routes.
4. Set goals and implement strategies and incentives to increase retail energy sales from renewable resources by 1% or more annually in each state through 2015.
5. Adopt energy efficiency standards for eight to 14 products not regulated by the federal government, establishing a cost-effective efficiency threshold for all products sold on the West Coast.
6. Incorporate aggressive energy efficiency measures into updates of state building energy codes, with a goal of achieving at least 15% cumulative savings by 2015 in each state.
7. Organize a West Coast Governors' conference in 2005 to inform policymakers and the public of climate change research concerning the West Coast states.[42]

The recommendations are soft in character. They are neither legally binding nor based on measurable obligations. However, in addition to adopting the recommendations, the Governors also committed to adopting more concrete obligations at the state and regional levels, including commitments to adopt comprehensive state and regional goals for greenhouse gas emission reductions and standards for reducing greenhouse gas emissions from vehicles, as well as to develop a market-based carbon allowance program and to expand markets for energy efficiency, renewable resources and alternative fuels. Further, the regional commitments made by the West Coast Governors' Global Warming Initiative are supplementary to existing individual state efforts—both binding and non-binding—to address climate change.

Cooperative efforts on the part of California, Washington, and Oregon are not of minor consequence. If these three states were a Nation State, their cumulative greenhouse gas emissions would rank the region as the 7th highest emitter in the world.[43]

Southwest Climate Change Initiative

The West Coast is an expected and long-term leader in environmental protection. The same cannot be said for the Southwest. Thus, the signing of the Southwest Climate Change Initiative by the Governors of Arizona and New Mexico on 28 February 2006 suggested that climate change is permeating new corners of American politics. The Southwest Climate Change Initiative 'establishes a

---

[42] Ibid 2.
[43] Washington, Oregon, and California Governors, 'West Coast Governors' Global Warming Initiative' (Statement on recommendations to reduce global warming pollution, 2004) <http://www.ef.org/westcoastclimate/> accessed 1 June 2010.

framework for the two states to collaborate on strategies to address the impacts of climate change in the Southwest and reduce greenhouse gas emissions in the region'.[44] The Initiative commits the two states to collaborating 'to identify options for reducing greenhouse gas emissions and promoting climate change mitigation, energy efficient technologies, and clean energy sources'.[45]

The Southwest Climate Change Initiative, like the West Coast Initiative, is based on cooperation and gubernatorial commitments to making progress on the issue of climate change. It does not create any legally binding commitments, but it facilitates cooperation between Arizona and New Mexico on a variety of activities, to include, developing consistent emission measuring and reporting mechanisms, creating a greenhouse gas credit trading program, identifying and promoting effective ways to reduce greenhouse gas emissions, encouraging the development and marketing of new technologies for improving energy efficiency and renewable energy sources and, finally, promoting 'regional and national climate policies that reflect the needs and interests of Southwestern states'.[46]

Arizona
The Southwest Initiative, like the West Coast Initiative, resembles a framework international treaty that creates collective goals and commitments but relies on subsequent individual or collaborative efforts to refine and implement measures to meet the objectives of the treaty. In Arizona, former Governor Janet Napolitano took an important step to confirming that the Initiative has substantive value when, on 8 September 2006, she signed an Executive Order adopting recommendations made by the Arizona Climate Change Advisory Group in its Climate Change Action Plan.[47] Governor Napolitano's Executive Order establishes the goal of reducing state-wide emissions to 2000 levels by 2020 and to 50% below the 2000 level by 2040. The Executive Order also creates a Climate Change Executive Committee tasked with developing an implementation plan

---

[44] See Arizona Department of Environmental Quality, 'Welcome to the Arizona Climate Action Initiative' (Statement of Purpose) <http://www.azclimatechange.gov> accessed 27 September 2009.
[45] Pew Center on Global Climate Change, 'Southwest Climate Change Initiative' <http://www.pewclimate.org/node/4651> accessed 16 April 2010.
[46] Renewable Energy Access, 'Arizona, New Mexico Launch Southwest Climate Change Initiative' (20 March 2006) <http://www.renewableenergyworld.com/rea/news/article/2006/03/arizona-new-mexico-launch-southwest-climate-change-initiative-44387> accessed 9 February 2007. See also US Department of Energy: Energy Efficiency and Renewable Energy, *Arizona and New Mexico Launch Southwest Climate Initiative*, 1 March 2006, available at <http://apps1.eere.energy.gov/states/state_news_detail.cfm/news_id=9798/state=NM> accessed 1 June 2010.
[47] Governor J Napolitano, 'Governor Napolitano Issues Executive Order to Promote Energy Efficiency: Order Steps up Efforts to Reduce "Greenhouse Gas" Emissions' (Press Release, 8 September 2006) <http://summits.ncat.org/energy_climate/index.php> accessed 1 June 2010.

for the goals set out in the Climate Change Action Plan, charging the Committee to try and find ways to achieve 2000 levels as early as 2012.[48]

New Mexico

Preceding the creation of the Southwest Climate Change Initiative, on 5 June 2005, Governor Richardson passed Executive Order 05-033 establishing the New Mexico Climate Change Action Council and the New Mexico Climate Change Advisory Group.[49] The Executive Order charges the Council with providing recommendations to the Governor on climate change while the Advisory Group was tasked to provide a report to Governor Richardson by 1 December 2006 including 'proposals for reduction of GHG emissions to reduce New Mexico's total greenhouse gas emissions to 2000 levels by the year 2012, 10% below 2000 levels by 2020, and 75% by 2050; an inventory of existing and planned actions that contribute to GHG emissions reductions', as well as a greenhouse gas inventory, consideration of the costs and benefits of the Group's proposals, and a review of on-going and future efforts to create effective regional and national climate change policies.[50]

Following the release of the Advisory Group's report, on 28 December 2006, Governor Richardson followed in the footsteps of Governor Napolitano by signing an Executive Order outlining a specific greenhouse gas emission reduction strategy for New Mexico.[51] Stating that '[c]limate change is the major environmental issue of our time', and that '[n]othing poses a bigger threat to our water, our livelihood and our quality of life than a warming climate', the Governor committed to 'taking the first step toward implementing as many of these recommendations as are possible, feasible and effective'.[52] Governor Richardson's Executive Order creates a state government implementation team to implement the Governor's proposed policies and, of central importance, commits the state to

---

[48] Ibid. The Climate Change Action Plan recommendations include (but are not limited to): (1) improving energy efficiency for buildings and appliances; (2) reducing energy demand; (3) increasing the use of renewable energy sources; (4) providing incentives for low-emissions vehicles; and (5) using land and forest management practices to increase biological carbon sequestration. Governor Janet Napolitano, 'Executive Order 2006-13: Climate Change Action Plan' (Executive Order, 8 September 2006) <http://www.azclimatechange.gov/download/EO_2006-13_090806.pdf> accessed 3 February 2007.

[49] Governor Bill Richardson 'Climate Change and Greenhouse Gas Reduction Executive Order' (Executive Order, 9 June 2005) <http://www.governor.state.nm.us/orders/2005/EO_2005_033.pdf> accessed 28 December 2009.

[50] New Mexico Climate Change Advisory Group, 'Climate Change Action Plan' <http://www.nmclimatechange.us/> accessed 1 June 2010.

[51] California Office of the Governor, 'Governor Schwarzenegger Commends Governors Napolitano and Richardson for their Efforts to Reduce Greenhouse Gas Emissions' (Press Release, 2 January 2007) <http://gov.ca.gov/index.php?/press-release/5185> accessed 1 June 2010.

[52] New Mexico Office of the Governor, 'Governor Bill Richardson Signs Historic Climate Change Executive Order' (Press Release, 28 December 2006) <http://www.governor.state.nm.us/press/2006/dec/122806_01.pdf> accessed 1 June 2010.

reducing greenhouse gas emissions to 75% below 2000 levels by the year 2050.[53] As with the US climate change strategy, however, Governor Richardson's proposed policies rely largely on incentives and voluntary initiatives to achieve this goal, and the proposed policies omit an overall emissions cap.

New Mexico, however, did not stop with the ambitious but relatively toothless Executive Orders. On 5 March 2007, Governor Richardson signed two new pieces of state legislation promoting the use and development of clean and renewable sources of energy.[54] First, Senate Bill 418 improves upon New Mexico's current Renewable Portfolio Standard[55] by mandating that electric utilities must obtain at least 15% of their power from renewable sources by 2015, and 20% from renewable sources by 2020. Second, House Bill 188 establishes a Renewable Energy Transmission Authority 'to promote clean energy jobs and help New Mexico both develop [its] clean energy resources and market them to other states'.[56]

The Southwest Climate Change Initiative demonstrates the value of regional climate change agreements, even in the absence of legally binding greenhouse gas emission reduction obligations. The creation of regional agreements, such as the Southwest Climate Change Initiative, encourages cooperative action, promotes innovative thinking and prompts state governments to enact increasingly progressive climate change legislation.

Regional Greenhouse Gas Initiative

The Regional Greenhouse Gas Initiative (RRGI) brings together ten Northeast and Mid-Atlantic states in an effort to reduce carbon dioxide emissions.[57] The states participating in the RGGI include Connecticut, Delaware, Maine, Maryland, Massachusetts, New Hampshire, New Jersey, New York, Rhode Island, and Vermont. In addition to the member states, the District of Columbia, Pennsylvania, the Eastern Canadian Provinces, and New Brunswick participate as formal observers to the RGGI.[58]

---

[53] Ibid.
[54] New Mexico Office of Governor Bill Richardson, 'Governor Bill Richardson Enacts Landmark Clean Energy Bills to Create Jobs, Keep Air Clean' (Press Release, 5 March 2007) <http://www.governor.state.nm.us/press/2007/march/030507_01.pdf> accessed 1 June 2010; Pew Center for Global Climate Change, 'Latest News' <http://www.pewclimate.org/what_s_being_done/in_the_states/news.cfm> accessed 1 June 2010.
[55] New Mexico's previous renewable portfolio standard mandated that 10% of the state's energy come from such sources by 2011. Pew Center for Global Climate Change, 'Latest News' <http://www.pewclimate.org/what_s_being_done/in_the_states/news.cfm> accessed 1 June 2010.
[56] New Mexico Office of Governor Bill Richardson, 'Governor Bill Richardson Enacts Landmark Clean Energy Bills to Create Jobs, Keep Air Clean' (Press Release, 5 March 2007) <http://www.governor.state.nm.us/press/2007/march/030507_01.pdf> accessed 1 June 2010; Pew Center for Global Climate Change, 'Latest News' <http://www.pewclimate.org/what_s_being_done/in_the_states/news.cfm> accessed 1 June 2010.
[57] Regional Greenhouse Gas Initiative, 'About RGGI' <http://rggi.org/about> accessed 1 June 2010.
[58] Ibid.

The core of the RGGI is its mandatory, multi-state carbon dioxide cap-and-trade program, which is the first of its kind.[59] The RGGI cap-and-trade scheme became operational in fall 2008. While the program initially focuses on reducing carbon dioxide emissions from power plants, it will eventually be expanded to include other emission sources. The RGGI is designed to function in two distinct phases. In Phase I (2009–2014), the goal of the Initiative is to stabilize carbon dioxide emissions to 2000–2004 levels. In Phase II, the aim is to reduce emissions to 10% below Phase I levels, which would approximate 1990 levels.[60] The first compliance period for each state's linked CO2 Budget Trading Program began on 1 January 2009.

The RGGI was initiated by New York Governor George Pataki in April 2003. Governor Pataki's initiative capitalizes on growing interest in greenhouse gas trading programs and the reality that a healthy and functioning market for carbon dioxide emissions will benefit from bringing together multiple states to cover more power plants and, thus, more emissions. The experience gained through the RGGI played a fundamental role in shaping the climate change bills proposed by the US Congress in 2009 and 2010.

Eastern and Midwest Climate Registry Initiatives

Greenhouse gas inventories and registries are necessary cornerstones for climate change laws and policies. Recognizing the centrality of inventorying and registering emissions in order to effectively regulate them, 46 states have developed greenhouse gas inventories, and four states have operational emission registry systems.[61] Taking advantage of economies of scale, ten Northeast and Mid-Atlantic states[62] banded together to form two regional registry systems. The goal of the Eastern Climate Registry was to create the foundations for multi-state registry and emissions reporting programs.[63] The success of the Eastern Climate Registry prompted the creation of a larger non-profit collaboration, The Climate Registry.[64] The Climate Registry brings together the Northeast and Mid-Atlantic states involved in the Eastern Climate Registry with other US states, Canadian provinces, territories, and Native Sovereign Nations to create a single registry

---

[59] Ibid.
[60] Regional Initiatives (n 4).
[61] Jonathan L Ramseur, 'CRS Report for Congress: Climate Change: Action by States to Address Greenhouse Gas Emissions' 9–10 (18 January 2007) <http://fpc.state.gov/documents/organization/80733.pdf> accessed 1 June 2010. Registry systems are currently being used in New Hampshire, California, Vermont, and Wisconsin while 37 other states are in the process of designing registries. See US Environmental Protection Agency, 'State and Local Climate and Energy Program' (Overview of state and local climate change initiatives) <http://www.epa.gov/statelocalclimate/state/tracking/reporting.html#a01> accessed 16 April 2010.
[62] Prior to merging with The Climate Registry, the Eastern Climate Registry included Connecticut, Delaware, Maine, Massachusetts, New Hampshire, New Jersey, New York, Pennsylvania, Rhode Island, and Vermont.
[63] Ibid.
[64] Ibid.

using 'consistent and transparent standards to calculate, verify and publicly report greenhouse gas emissions'.[65]

State and regional emission reporting and registry programs are largely voluntary in nature.[66] Despite lacking mandatory participation mechanisms, there are tangible and intangible incentives for private parties to participate, including the possibility of obtaining good PR for being 'environmentally friendly' and of gaining future emission credits when/if mandatory emission reduction programs are put in place.

Powering the Plains Initiative

The Northeast and the West Coast states are natural leaders in environmental change and are leading the way in state-based climate change law and policy-making. Increasingly, however, not only the Southwest but also the Midwest and Plains states are joining in the climate change game. The Powering the Plains Initiative (PPI), an initiative of the larger Great Plains Institute, evidences this trend. The PPI began as a public–private coalition designed to bring together North and South Dakota, Iowa, Minnesota, Wisconsin, and Manitoba to create a sustainable energy future for the Plains region. The overarching goal of the PPI was to help the region meet its full energy potential by creating an energy transition roadmap.[67]

The PPI facilitated voluntary, collaborative efforts to create regional economic opportunities for the energy and agricultural sectors while simultaneously addressing the causes and consequences of climate change. The most significant accomplishment of the PPI was the release in June of 2007, of the first-ever 50-year Energy Transition Roadmap for the Upper Midwest region of the United States.[68] This Roadmap offered a comprehensive plan for dealing with energy and agricultural challenges over the next half century. Publication of the Roadmap also stimulated subsequent complementary actions by, for example, prompting nine Midwestern governors and the premier of Manitoba to enter into a series of energy and climate agreements in November 2007 and by spawning complementary and on-going initiatives under the umbrella of the Great Plains Institute.[69]

---

[65] The Climate Registry, 'North America's Leaders Solving Climate Change Together' (Homepage) <http://www.theclimateregistry.org/> accessed 26 December 2009.
[66] Connecticut and Maine, on the other hand, have mandatory reporting requirements.
[67] Great Plains Institute, 'History' <http://www.gpisd.net/index.asp?Type=B_BASIC&SEC={364A2F81-987B-4FB2-8878-0D3987A88CD4}&DE=> accessed 16 April 2010.
[68] Great Plains Institute, 'Powering the Plains Energy Transition Roadmap' <http://www.gpisd.net/index.asp?Type=B_BASIC&SEC=%7B69A636DC-F9CD-4166-B21D-5BDD00DD4032%7D> accessed 18 April 2010.
[69] Great Plains Institute, 'History' <http://www.gpisd.net/index.asp?Type=B_BASIC&SEC={364A2F81-987B-4FB2-8878-0D3987A88CD4}&DE=> accessed 16 April 2010; Midwest Governors Association, '2007 MGA Energy Summit' <http://www.midwesterngovernors.org/energysummit.htm> accessed 18 April 2010.

Although focused on research, collaboration, and strategic planning rather than establishing legal frameworks, the PPI and its successor programs serve an important role by bringing together interested parties across traditional private–public sector lines. In this way, the program facilitated dialogue on the most effective and economical ways to address climate change in a coal and agriculture-heavy region of the country, and it helped create a forum for launching new enterprises and pilot programs. In this regard, the PPI mirrors the US federal approach to climate change—ie, focused on research, voluntary projects, and long-term planning. In another way, however, the PPI proved more vital and effective than on-going federal policies by facilitating the implementation of mandatory legislation and resolution within its member states. North Dakota, South Dakota, Minnesota, and Iowa, for example, have all adopted legislation promoting renewable energy credit tracking and developing hydrogen as an alternative energy source.

The Plains states lag behind the Coastal and Southwest states in developing clear and measurable climate change goals. However, the diffusion of state and regional collaboration to previously reticent areas of the country—due in large part to resource and economic constraints—suggests that states are beginning to recognize the inevitability of mandatory greenhouse gas emission reductions in the future and the importance of creating models for easing the social and economic burdens associated with the anticipated long-term changes.

Western Climate Initiative

Most recently, on 26 February 2007, the Governors of Arizona, New Mexico, Oregon, Washington, and California came together—bridging the pre-existing Southwest Climate Change Initiative and West Coast Global Warming Initiative—to form the Western Regional Climate Action Initiative. The Western Climate Initiative commits the five original states, as well as Montana, Utah, British Columbia, Manitoba, Ontario, and Quebec, who have since joined, to:

- setting an overall regional goal, within six months of the effective date of this initiative, to reduce emissions from our states collectively, consistent with state-by-state goals;
- developing, within 18 months of the effective date of this agreement, a design for a regional market-based multi-sector mechanism, such as a load-based cap and trade program, to achieve the regional GHG reduction goal; and
- participating in a multi-state GHG registry to enable tracking, management, and crediting for entities that reduce GHG emissions, consistent with state GHG reporting mechanisms and requirements.[70]

---

[70] Western Climate Initiative, 'WCI Governors' Agreement' (26 February 2007) <http://www.westernclimateinitiative.org/component/remository/func-download/12/chk,ec5958c233c949b2be738f6d4ea1dcca/no_html,1/> accessed 3 April 2010; see also Oregon Office of the Governor,

The governors further committed to collaborative efforts to promote renewable energy, improve energy efficiency, and develop adaptation strategies.

This Initiative is important less because it creates a new, innovative strategy for addressing climate change—most of its provisions mirror existing regional initiatives—and more because it brings together the efforts of five states that have all demonstrated a tangible commitment to addressing climate change. All five states have 'adopted or committed to adopting state reduction goals, clean tail pipe standards, mandatory emissions reporting, and renewable portfolio standards'.[71] When fully implemented in 2015, this Initiative will cover nearly 90% of the GHG emissions in WCI states and provinces.[72] In addition, the Initiative has taken advantage of its political scope and success to petition Congress to adopt an economy-wide cap-and-trade program.

The Western Regional Climate Action Initiative, like its regional counterparts across the country, takes advantage of individual strengths to create a strong regional block, which creates economic, political, and absolute emission reduction advantages. From an economic perspective, regional cooperation allows states to take advantage of economies of scale in encouraging the development of markets in energy efficiency products, clean and renewable sources of energy—and energy infrastructure. In addition, regional cooperation facilitates a larger, more viable market for cap-and-trade emission reduction programs. Thus, from an economic perspective, the regional approach 'allows the states involved to capitalize upon their shared environmental resources and interconnected economies both to elevate the importance of climate measures and to address climate change in a cost-effective manner'.[73] The regional initiative also creates greater regulatory uniformity across the region, which allows greater emission reductions and reduces opposition from industry.

Politically, when states operate in regional blocks, they have strength in numbers; their actions put more pressure on other states to establish similar programs – or lose competitive advantages and, possibly, the political goodwill of their citizenry. Equally, burgeoning regional efforts put pressure on the federal government to respond. The state and regional programs, again, serve as the 'laboratories' for the federal government; the regional programs create feasible

---

'Five Western Governors Announce Regional Greenhouse Gas Reduction Agreement' (Press Release, 26 February 2007) <http://governor.oregon.gov/Gov/pdf/letters/022607NGA.pdf> accessed 15 March 2007.

[71] See Western Climate Initiative, 'Designing the Program' (Statement about the strategy to reduce greenhouse gas emissions) <http://www.westernclimateinitiative.org/designing-the-program> accessed 1 June 2010.

[72] The WCI is also working together on complementary policies that support the cap-and-trade program; provide additional opportunities to address climate change; and achieve related co-benefits of increased energy efficiency, increased renewable energy generation, improved air quality, and reduced water pollution and job growth, and increased provincial, state, and local revenue. [73] Engel (n 2) 70.

models that the federal legislature can draw from to address climate change. State and regional actions, thus, 'function as a catalyst for regulatory action by higher jurisdictional levels of government'.[74]

The Western Governors' summed up the situation well, saying: 'In the absence of meaningful federal action, it is up to the states to take action to address climate change and reduce greenhouse gas emissions in this country',[75] and to create 'a model and example for the rest of the nation'.[76]

Regional cooperation does much more than create climate initiatives in name only. By maximizing geographic scope, taking advantage of economies of scale, cultivating widespread public and private support, and demonstrating the viability of legal and political measures, regional agreements promote substantive change while simultaneously providing symbolic demonstrations of the political and technological viability of climate change measures.

California—Climate Policy Driver

California has a long history of leading domestic environmental policymaking. California's leadership on climate policy continues this legacy. The state is renowned nationally and internationally as a driving force in state, regional, and national climate change policymaking.

California has taken an early and bold role in tackling climate change at the state level. In 2001, it created a Climate Action Registry, creating a record of greenhouse gas emission baselines for registrants.[77] The Registry began operating in 2002; it currently has in excess of 300 participants and is 'arguably the most comprehensive, as participants register all of their GHG emissions for operations in California; other state (and federal) registries cover only emission reductions'.[78]

California's creation of a voluntary registry initiated a sustained period of climate change law and policymaking. On 22 July 2002, the former governor of California, Gray Davis, approved legislation that would ultimately make California the first US state to regulate greenhouse gas emissions from motor vehicles.[79]

California's Greenhouse Gas Emissions Law

California's legislation is nationally—and internationally—cutting-edge. The legislation requires that 'no later than January 1, 2005, the state board shall

---

[74] Ibid 64.
[75] 'Five Western Governors Announce Regional Greenhouse Gas Reduction Agreement' (n 69), quoting Arizona Governor Janet Napolitano.
[76] Ibid, quoting Oregon Governor Ted Kulongoski.
[77] California Senate Bill 1771 (2001); California Senate Bill 527 (2001).
[78] Ramseur (n 61) 10.
[79] Assem B 1493, 2002 Gen Assem, Reg Sess (Cal 2002); see also Chanin (n 16) 699–754.

develop and adopt regulations that achieve the maximum feasible and cost-effective reduction of greenhouse gas emissions from motor vehicles'.[80] While an earlier Assembly version of the bill regulated only carbon dioxide, the bill was amended prior to adoption in the Senate to include all greenhouse gases—such broad coverage is still rare among greenhouse gas emission reduction legislation worldwide.[81] California's greenhouse gas regulations were to take effect on 1 January 2006 and were only to apply to motor vehicles manufactured in or after the 2009 model year that were to be sold in California.

The California Air Resources Board (CARB) is charged with implementing the greenhouse gas emissions law.[82] Pursuant to its mandate, CARB has devised a set of regulations to implement the legislation. The regulations apply to fleet averages—as opposed to individual autos—and ultimately require automobile manufacturers to reduce carbon dioxide and other greenhouse gases from *new* cars to 30% below the 2009 level by 2016.[83]

Despite geographic and political confines, California's legislation has significant national implications. To begin, the legislation is significantly more progressive than national efforts to combat climate change. In fact, when enacted, the legislation ran contrary to President George W. Bush's climate change and air quality policies, which focused on voluntary agreements, rather than mandatory regulation, and which have retreated from, rather than expanded or enforced, existing federal air pollution laws, eg, the Clean Air Act.[84] Further, California has always played the role of trailblazer in the field of environmental law. This legislation is no different. As in the past, California's efforts have served as a catalyst for other states to adopt similar climate change policies. Eighteen additional states have either already adopted or intend to adopt similar greenhouse gas emissions legislation for automobiles[85]—representing a paramount case of how progressive state action can produce a ripple effect on nationwide policymaking.

California's greenhouse gas legislation is a practical and a political catalyst. It sets measurable, progressive emission reduction goals and demonstrates the 'growing effort by local, state and foreign governments to seize the initiative since [former] President Bush ha[d] been reluctant to act'.[86] Proponents of the legislation candidly expressed their frustration with the federal government's

[80] 2002 Cal Leg Serv, ch 200, 3(a) (2002).
[81] Assem B 1058, 2001–02 Gen Assem, Reg Sess (Cal 2001) (amended 31 May 2001).
[82] California Air Resources Board, 'Rulemaking on the Proposed Regulation to Control Greenhouse Gas Emissions from Motor Vehicles' (Final rulemaking package effective 1 January 2006) <http://www.arb.ca.gov/regact/grnhsgas/grnhsgas.htm> accessed 1 June 2010.
[83] Engel and Saleska (n 3) 221.
[84] See generally, The White House, Global Climate Change Policy Book (2002) <http://www.usgcrp.gov/usgcrp/Library/gcinitiative2002/gccstorybook.htm> accessed 3 April 2010.
[85] States adopting similar regulations include Oregon, Washington, and eight states in the Northeast. Ramseur (n 61) 14.
[86] See Chanin (n 16) 703, citing Gary Polakovic, 'Assembly Bill Targets Global Warming Trend' *LA Times* (Los Angeles, 26 January 2002) B1.

failure to address climate change and suggested that California is seeking to fill the void and lead the way towards more progressive policies nationwide.[87] California's greenhouse gas legislation demonstrates the increasing importance of state actions in the area of environmental policymaking and states' growing willingness to adopt policies that surpass, or even conflict with national policies.

On its own, California can exert political pressure and make emission reduction gains. As was demonstrated in the previous discussion of regional collaborations, however, California is not trying to go it on its own. Instead, the current Governor, Arnold Schwarzenegger, is coupling California's state-wide efforts with multiple regional partnerships that seek to expand the geographic, economic, and political scope of sub-national climate change policies. And it is these growing, intentionally politically concerted efforts that begin to distinguish state-led climate change policymaking from the previous model of states as laboratories for federal policymaking. In the context of climate change, states are not merely taking individual action to develop unique tools: they are banding together to improve the physical and political efficacy of their efforts. In the words of one prominent commentator, there is a 'stealth-like' effort by state-level politicians to create enough political momentum to overcome federal inertia.

California's efforts have not gone untested. The state has faced ongoing resistance from multiple sectors, including the federal government, industry groups, and its citizenry.

The federal government handed California its greatest challenge to date in its efforts to reduce state-wide greenhouse gas emissions. Under the Federal Clean Air Act (CAA), the federal government retains near exclusive authority over automobile regulation. California, however, is in the unique position of having provisional authority under the CAA to enact automobile standards that are more stringent than existing federal standards.[88] In order to do so, California must apply for and receive a waiver from the EPA under the CAA. If the EPA grants California a waiver, not only can California adopt automobile emissions standards that exceed federal measures, but the remaining 49 states may also opt to follow California's more stringent standards.[89] Absent a grant of permission from the EPA, however, neither California nor any other state is permitted to adopt measures that diverge from existing federal standards.

California applied for a CAA waiver to regulate greenhouse gas emissions from automobile tailpipes in December 2005. Despite numerous personal requests for action from California's Governor Schwarzenegger, an active environmental campaign, and even a lawsuit, the EPA failed to respond to California's petition for nearly three years. In December 2007,[90] the EPA finally responded by denying

---

[87] Ibid.   [88] Clean Air Act of 1970, 42 USC § 7543 (2006).   [89] Ibid § 7507.
[90] The denial was formally issued in the US Federal Register on 6 March 2008.

California's request for a waiver. The denial was met with great frustration and instigated an on-going debate over the proper locale for regulating greenhouse gas emissions. In denying the waiver, then EPA Administrator Stephen Johnson indicated that California did not meet the criteria for receiving a waiver.[91] In particular, Johnson suggested that California 'does not need a waiver in order to meet compelling and extraordinary conditions'.[92] Johnson cited the Bush Administration's efforts to craft a national solution to reducing greenhouse gas emissions from motor vehicles, as well as the global nature of climate change, as evidence that the state of California did not need its own set of standards. Johnson's statement revealed a strong preference on the part of the EPA for a uniform, nationwide emissions standard for motor vehicles in order to avoid 'a confusing patchwork of state rules'. Immediately following EPA's decision, on 21 January 2009, the CARB submitted a request asking the Agency to reconsider its decision.

EPA's decision to deny the waiver undercut one of the most substantive greenhouse gas emission reduction schemes proposed nationwide. In California alone, state officials estimated that the proposed standard would generate emissions reductions of approximately 30 million metric tons by 2020 and over 50 million metric tons by 2030, equating to an 18% reduction in climate change emissions from passenger cars in 2020 and a 27% reduction in 2030.[93] Potential reductions nationwide could exceed 100 million metric tons if California's standards were adopted in the 18 states that have either adopted or intend to adopt California's standards. The scale of the program ensured that EPA's decision to deny the waiver was met both with great relief by many, including the auto industry and automobile producing states; and with outrage by many, including environmental groups and a large group of state governments.

The waiver decision quickly became a contentious and partisan political issue. As discussed in Chapter 2, on 26 January 2009, one of newly-elected President Obama's first formal actions was to direct the EPA to revisit its decision to deny California a waiver to regulate GHG emissions. Subsequently, on 6 February 2009, the EPA issued a notice for public hearing and comment in the Federal Register on its denial of California's application for a pre-emption waiver under the CAA. On 19 May 2009, while the EPA still had the issue under review, President Obama announced a new national standard for automobiles to be established through a joint rulemaking process between the EPA and the DOT.

The new federal standard would achieve, by 2016, improvements in automobile fuel economy equivalent to the proposed California standards as well as

[91] Ibid.   [92] Ibid.
[93] California Environmental Protection Agency, Air Resources Board, 'Climate Change Emissions Standards for Vehicles, actions to reduce greenhouse gases from cars and trucks', Frequently Asked Questions, 30 May 2007 <http://www.arb.ca.gov/cc/factsheets/ccfaq.pdf> accessed 29 December 2009.

including a per-vehicle greenhouse gas emission limit. The new federal standard does not directly duplicate or preempt California's proposed standards. Rather, in order to comply with the new federal standard, California would be required to amend its own proposal to conform to the federal standard from 2012 through 2016 while enforcing its own standards for model years 2009 through 2011, and post-2016. As part of President Obama's new standard, automakers agreed to drop ongoing lawsuits against California's proposed vehicle standard, as discussed further below.

Following President Obama's announcement and a series of intra-agency consultations, on 30 June 2009, the EPA granted California a waiver to regulate greenhouse gas emissions from vehicles within the state.[94] With the waiver granted, California and the other states electing to follow California's standards are free to regulate vehicle emissions from new motor vehicles beginning with model year 2009 until the time the new federal standard takes effect in 2012, and then again after 2016. Regarding the decision to grant the waiver, EPA Administrator Lisa Jackson stated that 'the March 6, 2008 Denial was based on an inappropriate interpretation of the waiver provision', focusing in particular on the findings that 'determined that Congress intended to allow California to promulgate only those state standards that address pollution problems that are local or regional', rather than problems of a global nature. In rejecting the Agency's prior determination that California could not show that its need for the separate standard was 'compelling and extraordinary', Administrator Jackson noted that:

Congress established that there would be only two programs for control of emissions from new motor vehicles—EPA emission standards adopted under the Clean Air Act and California emission standards adopted under its state law. Congress accomplished this by preempting all state and local governments from adopting or enforcing emission standards from new motor vehicles [under Section 209(a) of the Act], while at the same time providing that California could receive a waiver of preemption.[95]

This policy, Administrator Jackson observed, allows California to maintain its long-standing role as a national leader in environmental policymaking and as a laboratory for new motor vehicle emissions, while at the same time protecting automobile manufacturers from being subject to a series of confusing and conflicting state standards that deviate from federal standards.

The resolution of the waiver dispute in 2009 coincided with a burst of activity within the Executive and Legislative branches to address climate change. EPA's decision to grant California the pre-emption waiver alongside federal efforts to structure a national climate strategy revealed both the challenges and

---

[94] US Environmental Protection Agency, 'California State Motor Vehicle Pollution Control Standards; Notice of Decision Granting a Waiver of Clean Air Act Preemption for California's 2009 and Subsequent Model Year Greenhouse Gas Emission Standards for New Motor Vehicles' 74 Fed Reg 32744 (July 8, 2009).   [95] Ibid.

opportunities afforded by the US federalist system. The complexities inherent in federal–state relations delayed California's receipt of the pre-emption waiver. The eventual grant of the waiver, however, set in motion not only a progressive series of intra-state environmental measures, but also a new nationwide climate change regulatory program. The combination of the new federal automobile standards with widespread adoption of California's new auto emissions standards ensures that regardless of whether Congress is able to reach a consensus on climate change legislation, the federal government is committed to addressing greenhouse gas emissions from the transport sector, the fastest growing sector for greenhouse gas emissions in the US.

Resolution of the California waiver dispute charted the course for state and federal regulation of automobile emissions over the near-term. It also intensified ongoing debate over the wisdom of crafting a national climate change strategy through tools available in existing environmental laws, as opposed to through enactment of a new piece of primary legislation. President Obama and EPA's decision to use existing environmental and energy legislation as vehicles for limiting greenhouse gas emissions from automobiles sparked legislative concern over the possibility that US administrative agencies would continue to use provisions of existing legislation to piece together an intricate greenhouse gas regulatory program.

Beyond addressing federal–state tensions, settlement of the waiver dispute accomplished the secondary feat of resolving a long-standing dispute between the state of California—as well as other states employing California's auto emissions standards—and US automakers. Alongside resistance from the federal government, California has faced ongoing legal challenges to its auto emission regulations from automakers. Even before former Governor Davis signed the legislation into law, major automakers expressed entrenched opposition to the bill, fearing that it would cost them billions of dollars to manufacture vehicles that meet California emission standards.[96] The automakers also opposed the creation of a patchwork regulatory regime, where they would face different regulatory requirements from state to state. Amidst growing industrial opposition, in December 2004, 13 California car dealerships and the Alliance of Automobile Manufacturers filed a lawsuit in the US District Court in Fresno alleging that California's legislation violated federal fuel economy laws and seeking to block the legislation from going into effect in 2006.[97] Similar lawsuits were filed in Vermont, Maine, Rhode Island, and New Mexico.[98]

This series of lawsuits involves disputes between state governments and private parties, but also highlights the more fundamental divide between states and

---

[96] See Deborah Keeth, 'The California Climate Law: A State's Cutting-Edge Efforts to Achieve Clean Air' (2003) 30 *Ecology LQ* 715, 719.
[97] *Central Valley Chrysler Jeep v Goldstene*, 563 F Supp 2d 1158 (ED Cal 2008).
[98] See *Green Mountain Chrysler-Plymouth-Dodge v Crombie*, 508 F Supp 2d 295 (D Vt 2007); *Lincoln Dodge Inc v Sullivan*, 588 F Supp 2d 224 (D RI 2008).

the federal government. In October 2002, even before the automakers' lawsuit was filed in California, the Department of Justice expressed support for the decision by the automobile industry to challenge California's program to promote low-emission vehicles.[99] The Department of Justice went so far as to offer a legal brief in support of the automakers' challenge, alleging that California's regulation encroached on federal regulatory authority.[100]

The automakers' lawsuits met with different interim fates from state to state between 2006 and early 2009. In the first instance, in the California case, on 16 January 2007, the federal district judge postponed the trial until the Supreme Court issued its decision in *Massachusetts v EPA*, concerning the EPA's ability and responsibility to regulate greenhouse gas emissions under the CAA. The Supreme Court issued the decision in this case, discussed *infra* in Chapter 4, in April 2007. In finding that greenhouse gases are pollutants, that the EPA has the authority to regulate them under the CAA, and that greenhouse gas emission standards under the CAA are not preempted by the federal fuel economy law, the Court paved the way for the EPA to regulate greenhouse gases under the CAA and for the EPA to grant California's request for a preemption waiver. In December 2007, following the decision in *Massachusetts v EPA*, the federal judge dismissed the automakers' case in California. The plaintiffs filed an appeal in the Ninth Circuit on 30 October 2008. Similarly, on 12 September 2007, the federal district judge in the Vermont case ruled that the state may impose its own greenhouse gas emission standards on automakers. The plaintiffs appealed this decision to the Second Circuit Court of Appeals. The court dismissed the automakers' case in Rhode Island on 25 November 2008, due to their ongoing litigation in California, but the automakers appealed the decision to the First Circuit Court of Appeals and their co-plaintiff auto dealers maintained their suit in federal district court. The New Mexico case, which involves a suit brought by automobile *dealers*, is pending in federal district court. Finally, in Maine, the State Superior court denied the Alliance of Automobile Manufacturers' request to stay the implementation of Maine's standards and refused to send the issue back to the Maine Board of Environmental Protection for reassessment.

The fate of the ongoing cases changed *en masse* in early 2009. In announcing the new federal fuel economy standards, President Obama revealed that the automakers had agreed to drop ongoing state and federal lawsuits challenging the 'California standards'. As part of the negotiated agreement, the individual automakers signed a series of 'Commitment Letters'[101] declaring their intent to stay all litigation pending in state and federal courts that challenged California's

---

[99] Rabe (n 33) 148, citing Tom Doggett, 'White House, EPA Clash on Lower Vehicle Emissions' (Reuters News Service, 17 October 2002) <http://www.planetark.com/dailynewsstory.cfm/newsid/18200/story.htm> accessed 1 June 2010. [100] Rabe (n 33) 148.
[101] The letters are available online, at <http://www.epa.gov/otaq/climate/regulations.htm> accessed 3 April 2010.

regulation of greenhouse gas emissions from automobiles.[102] In June 2009, the California and Vermont cases were stayed upon joint motion by the parties, with status reports due in October 2009. In contrast, however, as of July 2009, automakers had not yet taken steps to stay their lawsuit against Rhode Island. The auto *dealers* lawsuit against the state of New Mexico is similarly ongoing; they did not commit to stay pending litigation, thus, there is no suggestion that their cases in Rhode Island or New Mexico will be stayed.

With most of the automakers' litigation stayed, the new federal standards pending and the CAA waiver granted, the path was cleared for California and like-minded states to implement the new measures, marking a turning point in sub-federal and federal climate policy.

While California was battling federal and private opposition to its automobile standards, it was also engaged in a proactive legal campaign to address climate change. On 20 September 2006, defying the ongoing legal and political opposition from the automobile industry, then-California Attorney General, Bill Lockyer, filed a nuisance lawsuit against key US and Japanese auto manufacturers. In the suit, Lockyer alleged that the defendants' 'vehicles' emissions have contributed significantly to global warming, harmed the resources, infrastructure and environmental health of California, and cost the state millions of dollars to address current and future efforts'.[103] The complaint alleged that, under both state and federal law, automakers have created a public nuisance by manufacturing automobiles that emit massive quantities of carbon dioxide that contributes to global warming, which has 'already injured California, its environment, its economy, and the health and well-being of its citizens'.[104] In June 2009, the lawsuit was withdrawn on the basis that the newly-announced federal automobile standards would achieve the desired reductions in automotive greenhouse gas emissions, thereby rendering the lawsuit moot.

This lawsuit was only one of several that California's Attorney Generals filed against the EPA, the Bush Administration, and the transport industry to address the 'Bush Administration's inaction on global warming'.[105] The growing willingness of not only California's Attorney General, but also state attorney generals nationwide to actively pursue climate litigation across state lines has played an important role in prompting regulatory responses at the state and federal levels.

---

[102] Eg, Alliance of Automobile Manufacturers, 'Letter to Ray LaHood, Secretary of Transportation & Lisa Jackson, Administrator at the Environmental Protection Agency' (18 May 2009) <http://www.epa.gov/otaq/climate/regulations/alliance-of-automobile.pdf> accessed 28 December 2009.

[103] California Office of the Attorney General, 'Attorney General Lockyer Files Lawsuit Against "Big Six" Automakers for Global Warming Damages in California: Impacts of Uncurbed Vehicle Emissions Costing Taxpayers, Economy and Environment' (Press Release, 20 September 2006) <http://ag.ca.gov/newsalerts/release.php?id=1338> accessed 1 June 2010 (Lockyer Press Release).   [104] Ibid, quoting California Attorney General Lockyer.

[105] Ibid, quoting California Attorney General Lockyer.

California's automobile emissions regulations and its judicial activism are neither merely symbolic[106] nor inconsequential in terms of potential emission reductions.[107] The regulations are concrete, mandatory, enforceable, and far-reaching. Further, the automobile regulations constitute only one element of California's burgeoning portfolio of climate change policies. This initial effort has led to the enactment of increasingly stringent state climate change laws and policies, with a ripple effect on federal climate policy.

Recent Climate Change Initiatives in California

Despite confronting resistance from automakers and the federal government until summer 2009, California continued to push forward with its climate change agenda. In 2006 and early 2007, California took steps confirming its status as a national leader on climate change policy.

In March 2006, the California's Climate Action Team issued a report commissioned by Governor Schwarzenegger in June 2005.[108] The report made specific suggestions, including recommendations for: (1) a multi-sector, market-based emissions trading program, (2) mandatory emissions reporting, (3) a public education campaign, (4) detailed cost-effective analysis of possible strategies, (5) analysis of alternative fuels (specifically biofuels), (6) development of low-carbon energy sources, and (7) expansion of the Renewable Portfolio Standard to public utilities (and other recommendations). These recommendations discernibly informed Governor Schwarzenegger's subsequent actions over the course of the year.

Following the release of the Climate Action Team's report, in April 2006, Governor Schwarzenegger signed an executive order establishing targets for bioenergy use and production as part of California's Renewable Portfolio Standards.[109] The bioenergy executive order set minimum targets for biofuel production, use, and renewable generation standards—focusing both on producing more biofuel within the state and making biofuel a more central component of energy consumption.[110] Governor Schwarzenegger followed up the executive order in July 2006 with a 'Bioenergy Action Plan', 'outlin[ing] ways for California to bring alternative energy into the mainstream and reduce dependency on foreign fossil fuels'[111] and committing 'state agencies to take

---

[106] Barry G Rabe, *Statehouse and Greenhouse: The Emerging Politics of American Climate Change Policy* (Brookings Institute, Washington DC, 2004) xii.   [107] Engel (n 2) 56.
[108] California Environmental Protection Agency, 'Climate Action Team Report to Governor Schwarzenegger and the Legislature' (March 2006) <http://www.climatechange.ca.gov/climate_action_team/reports/index.html> accessed 1 June 2010.
[109] California Office of the Governor, 'Executive Order S-06-06' (25 April 2006) <http://gov.ca.gov/index.php/executive-order/183/> accessed 1 June 2010; see also California Office of the Governor, 'Gov. Schwarzenegger Announces Bioenergy Action Plan for California' (Press Release, 13 July 2006) <http://gov.ca.gov/index.php?/press-release/1685> accessed 2 February 2007.
[110] 'Executive Order S-06-06' (n 108).
[111] 'Governor Schwarzenegger Announces Bioenergy Action Plan for California' (n 109).

detailed actions within a specific timeframe to advance the use of bioenergy in California'.[112]

Then, in August 2006, the Governor expanded California's range of renewable energy initiatives by signing Senate Bill 1, which implements final portions of his 'Million Solar Roofs' project, first introduced in 2005. The California Public Utilities Commission is mandated with implementing the Bill,[113] which will result in the addition of one million new solar roofs for California by 2018. Although the California Public Utilities Commission initially lacked full authority to implement the Million Solar Roofs plan, Senate Bill 1 addresses this gap by implementing the section of the plan that would have otherwise lain dormant.

Governor Schwarzenegger's and the California legislature's actions to promote biofuels and solar roofs signaled a perceptible move towards creating a comprehensive climate change framework. These sector specific policy initiatives set the scene for the implementation of a critical piece of legislation.

On 27 September 2006, Governor Schwarzenegger signed into law Assembly Bill 32 (AB 32), which sets an enforceable cap on greenhouse gas emissions.[114] It requires California to reduce its greenhouse gas emissions to 1990 levels by the year 2020 (a 25% reduction), and to reduce emissions to 80% below 1990 levels by 2050.[115]

California's AB 32 mirrors the most rigorous climate change bills proposed in the 110th Congress, and embodies many of the emissions reductions goals long advocated by environmental groups such as the Natural Resources Defense Council. The bill relies on both regulatory and market mechanisms to achieve its goals. The CARB is tasked with developing regulations and market mechanisms to meet its stated goals. Mandatory caps will begin in 2012 and will be incrementally ratcheted down to meet the 2020 goals.[116]

AB 32 puts California at the forefront of national and international climate change policymaking.

In early 2007, California continued to construct its climate change policy framework. On 18 January 2007, Governor Schwarzenegger signed an executive order establishing a Low Carbon Fuel Standard for all transport fuels sold in California.[117] The standard requires that the carbon intensity of transporta-

[112] Ibid.
[113] California Office of the Governor, 'Schwarzenegger Signs Legislation to Complete Million Solar Roofs Plan' (Press Release, 21 August 2006) <http://gov.ca.gov/index.php?/press-release/3588> accessed 1 June 2010.
[114] Assem B 32, 2006 Gen Assem, Reg Sess (Cal 2006) (Global Warming Solutions Act of 2006).
[115] Ibid; see also California Office of the Governor, 'Gov Schwarzenegger Signs Landmark Legislation to Reduce Greenhouse Gas Emissions' (Press Release, 27 September 2006) <http://gov.ca.gov/index.php?/press-release/4111/> accessed 1 June 2010.
[116] Ibid.
[117] California Office of the Governor, 'Executive Order S-01-07' (Press Release, 18 January 2007) <http://gov.ca.gov/index.php?/press-release/5174/> accessed 1 June 2010; see also California Office of the Governor, 'Gov Schwarzenegger Signs Executive Order Establishing World's First

tion fuels for passenger vehicles be reduced by at least 10% by 2020. This is a 'first-of-its kind standard'[118] and it represents the first major action under AB 32. The Low Carbon Fuel Standard 'is expected to replace 20 percent of our on-road gasoline consumption with lower-carbon fuels, more than triple the size of the state's renewable fuels market, and place more than 7 million alternative fuel or hybrid vehicles on California's roads (20 times more than on our roads today)'.[119]

One week later, on 25 January 2007, the California Public Utilities Commission stepped into the unwelcoming territory of electricity producers and providers by banning California's largest utilities from buying electricity from high-polluting sources, including most coal-burning power plants.[120] This action was taken to comply with AB 32, which requires the commission to adopt emissions standards for utilities. Although there are few coal-burning plants in California, about 20% of the state's power comes from coal plants in other states, making the decision potentially politically divisive.[121]

In the fall of 2008, California was hit hard by the global economic recession. As a result of economic decline and a burgeoning state deficit, the state experienced massive budget cuts in early 2009 that weakened air pollution rules for off road vehicles and gutted spending on mass transit. Despite these set-backs, California's climate policy infrastructure remained largely intact.

California has led state efforts to address climate change by implementing laws and policies that mirror the most progressive policies being implemented worldwide. Neither the political nor pragmatic value of the state's climate change policy framework should be underestimated. In the ongoing efforts to address political stagnation in federal environmental policy and to address the physical manifestations of climate change, California is leading the way.

Low Carbon Standard for Transportation Fuels' (Press Release, 18 January 2007) <http://gov.ca.gov/index.php?/press-release/5174/> accessed 1 June 2010. CARB has since approved a regulation implementing Governor Schwarzenegger's Low Carbon Fuel Standard.

[118] Ibid.
[119] Ibid. 'The LCFS will use market-based mechanisms that allow providers to choose how they reduce emissions while responding to consumer demand. For example, providers may purchase and blend more low-carbon ethanol into gasoline products, purchase credits from electric utilities supplying low carbon electrons to electric passenger vehicles, diversify into low carbon hydrogen as a product and more, including new strategies yet to be developed.' Ibid.
[120] Associated Press, 'California Regulators Vote to Ban Utilities from Buying "Dirty" Power' *International Herald Tribune* (25 January 2007) <http://www.climatestrategies.us/news.cfm?Page=1&NewsID=35275> accessed 1 June 2010.
[121] California is not the only state that is regulating electricity purchasing/production. Oregon, New Hampshire, Massachusetts, and Washington State regulate—or are in the process of regulating—carbon dioxide emissions from electric power generators. For example, in 2002, Washington passed a law that requires new power plants to 'offset twenty percent of their carbon dioxide emissions by planting trees, buying natural gas-powered buses or taking other steps to cure such emissions.' Engel and Saleska (n 3) 221, citing David Ammons, 'New Washington Law Boosts Emissions Standards' *Associated Press* (31 March 2004). While the Washington law only applies to new builds, both New Hampshire and Massachusetts have put in place laws that regulate carbon dioxide emissions from existing power plants. See Engel and Saleska (n 3) 221.

State Climate Change Laws and Policies: A Review

California is by no means the only state to adopt an increasingly comprehensive portfolio of climate change policies. Many of the New England, West Coast, and Southwestern states are incrementally creating comprehensive state-level climate change programs. All in all, almost two dozen states have some type of renewable energy obligations, and over a dozen states have enacted, or are in the process of enacting, legislation to control greenhouse gas emissions.

Despite prodigious efforts on the part of California and the Northeast, and increasingly bold state and regional partnerships across the US, not all states are following suit. Many states continue to lag, prioritizing economic and social well-being over climate change concerns.[122] Some traditionally fossil fuel heavy states, such as Texas, have enacted limited climate change policies, but are delving into renewable energy industries as a new sources of economic opportunity.[123] Other states are outwardly hostile towards climate change policymaking and specifically avoid or negate efforts to adopt binding limits on greenhouse gas emissions or mandatory renewable energy programs.[124] Those states stalling or actively opposing progressive climate change policies are exceptions to the norm. On the whole, the discernible trend is for states—for political, economic, environmental, or strategic reasons—to adopt a range of climate change policies that promote renewable energy and to inventory, register, and increasingly limit greenhouse gas emissions.

Sub-national activity is particularly remarkable for three reasons: (1) the breadth and diversity of state-level action, (2) the growing number and variety of regional partnerships, and (3) the persistent increase of state and regional activity in the face of federal inactivity and even, at times, resistance.

At the individual state level, California's leadership in limiting emissions from automobiles offers real possibilities for widespread emissions reductions. If successfully implemented and mimicked by other states, this single piece of legislation could eventually control emissions for up to one-fourth of the US light-duty vehicle market.[125] This is no small feat since transport is the fastest growing sector for greenhouse gas emission growth. In a more dispersed manner, however, the efforts of numerous small and medium-sized states to encourage and/or mandate renewable energy obligations also offers potential for discernible emissions reductions.[126]

While it is imperative that individual states continue to develop, implement, and enforce state-level policies to modify energy use and production, and to limit greenhouse gas emissions, regional partnerships offer important advantages of scale. Regional cooperation 'allows the states involved to capitalize upon their shared environmental resources and interconnected economies[,] both to elevate the importance of climate measures and to address climate change in a

---

[122] See Rabe, *Statehouse and Greenhouse* (n 106) xii–xiii, 47–49.   [123] See ibid 49–62.
[124] See ibid 40–47.   [125] Engel and Saleska (n 3) 212.   [126] Ibid.

cost-effective manner'.[127] By working together, states pool economic, geographic, and political resources to improve the efficacy of their programs and to magnify the political significance of their actions.

While state and regional programs have developed in a 'stealth-like' manner which has been advantageous in many ways, they are now reaching a critical mass. As climate change laws and policies multiply, so does the political pressure exerted at the national level for comparable, if not superior action.

Local Climate Change Laws and Policies

In the recent flurry of attention directed at state and regional climate change policies, it is easy to overlook local and municipal activities. To do so would be a mistake. Municipalities rival states in the diversity—if not the scale—of their efforts to address climate change. For example, in the US, 1,026 mayors from all 50 states, the District of Columbia, and Puerto Rico have united to combat global climate change, with the objective of meeting the Kyoto Protocol target for the US by reducing greenhouse gas emissions in US cities to 7% below 1990 levels by 2012.[128]

Many local efforts function on the basis of cooperative, public-private partnerships. Officials at the local level have the advantage of being able to work directly with their constituents, including environmental, industrial, and civic representatives, to structure carefully tailored, pragmatic climate change programs. These cooperative and collaborative local efforts between interested parties offer three main benefits. First, they facilitate policymaking that is effective in addressing the root causes and specific manifestations of climate change at the local level. Second, local policies provide paradigms for structuring policies at the state and national level. Third, local policies promote public awareness and participation that, in turn, contributes to the grassroots political movement within the US whereby environmental policies and, especially, climate change policies are being driven by political pressure from the bottom up.

Cities for Climate Protection Campaign

Local governments are working individually and collaboratively to develop strategies for addressing and adapting to global climate change. The Cities for Climate Protection Campaign (CCP) is one of the more prominent collaborative efforts designed to assist local governmental efforts to reduce greenhouse gas

---

[127] Engel (n 2) 70.
[128] The United State Conference of Mayors Climate Protection Center, 'Mayors Leading the Way on Climate Protection' <http://www.usmayors.org/climateprotection/revised/> accessed 20 May 2010.

emissions.[129] The CCP program is sponsored and coordinated by the International Council for Local Environmental Initiatives (ICLEI). The ICLEI is an umbrella organization helping local governments to frame, adopt, and implement policies designed to achieve measurable reductions in greenhouse gas emissions, as well as to enhance air quality and community sustainability.[130] Currently, over 560 local governments are participating in the CCP in the US. Each participating local government commits to including climate change strategies in local decision making processes.[131]

The CCP is based on five milestones that local governments commit to when they join the initiative. The five milestones include the following: (1) conducting a baseline emissions inventory and forecast; (2) adopting an emissions reduction target for the forecast year; (3) developing a local action plan; (4) implementing policies and measures, eg, energy efficiency and transport policies; and (5) monitoring and verifying progress on implementing greenhouse gas emission programs.[132] The milestones provide a comprehensive but flexible framework for cities to use to develop their climate change programs. Even in the absence of strong federal leadership, the CCP exemplifies the ability and willingness of municipalities to work jointly to accomplish environmental objectives.[133]

The CCP is not exclusive to the US. The ICLEI manages CCP campaigns at various levels in Australia, Canada, Europe, Japan, Latin America, Mexico, New Zealand, South Africa, South Asia, and Southeast Asia. The ICLEI and the CCP campaign represent a new type of globalized collaboration, ie, one that is founded and coordinated by a non-governmental organization to help local governments across geographical boundaries work from the bottom up to influence national and, ultimately, international policymaking.

US Conference of Mayors

Unlike the CCP, which is coordinated by an external organization, the US Conference of Mayors (USCM) operates as an independently coordinated group of political representatives and 'is the official nonpartisan organization of cities with populations of 30,000 or more. There are 1,204 such cities in the country

---

[129] McKinstry (n 29) 12.
[130] See International Council for Local Environmental Initiatives, 'ICLEI Climate Program' <http://www.iclei.org/index.php?id=800> accessed 20 May 2010. [131] Ibid.
[132] Ibid.
[133] For example, together, CCP participants in the US annually reduce greenhouse gas emissions by 23 million tons, eliminating more than 43,000 tons of local air pollutants, and savings in excess of $535 million in energy and fuel costs. These achievements are the equivalent of annual savings of over 74 million gallons of gasoline, 4,000 gigawatt hours of electricity, 6 million therms of natural gas. See Partnerships for Sustainable Development 'Cities for Climate Protection Campaign' <http://webapps01.un.org/dsd/partnerships/public/partnerships/1670.html> accessed 28 December 2009.

today. Each city is represented in the Conference by its chief elected official, the mayor.'[134]

The USCM has taken a number of actions regarding climate change: it has partnered with ICLEI 'to reduce greenhouse gas emissions in cities through outreach, education, and technical assistance';[135] it has formed the Mayors' Council on Climate Protection;[136] and it has endorsed the US Mayors Climate Protection Agreement. The Climate Protection Agreement commits signatories to meeting or exceeding Kyoto Protocol obligations for the US within their city limits and to supporting the Chicago Climate Exchange.[137]

At the 74th Annual Conference in June 2006, the UCSM approved resolutions designed to reduce the use and consumption of fossil fuels. These measures included resolutions to reduce fossil fuel consumption in building operations and to promote carbon neutral buildings, to encourage the use of landfill gas-to-recovery technologies, to use plug-in hybrid vehicles, to use renewable fuels, to promote clean, renewable energy sources, and to establish a new municipal energy agenda to help address the nation's energy and environmental challenges and to improve local communities.[138]

In November 2006, 31 mayors representing 5 million US citizens met at the second annual *Sundance Summit: A Mayors' Gathering on Climate Protection* to discuss tools and techniques for addressing global warming. The goal of the Summit was to encourage and 'to amplify the leadership mayors have been showing, in the absence of significant federal initiatives pinpointing this issue'.[139] In addition to serving as a meeting place and a mechanism for supporting local action, participating mayors made 'specific commitments and calls to action in their cities'.[140]

---

[134] US Conference of Mayors, 'About the United States Conference of Mayors' <http://www.usmayors.org/about/overview.asp> accessed 20 May 2010.

[135] US Conference of Mayors, Press Release, 5 June 2006, 'The U.S. Conference of Mayors Partners With ICLEI to Combat Global Warming' <http://www.usmayors.org/74thAnnualMeeting/iclei_060506.pdf> accessed 16 April 2010.

[136] Ibid. The goal of the Mayors Council on Climate Protection is 'to provide mayors the tools they need to carry out their mission of GHG emissions reduction and sustainable community design.' Ibid.

[137] The Chicago Climate Exchange is 'the world's first and North America's only voluntary, legally binding rules-based greenhouse gas emission reduction and trading system.' Chicago Climate Exchange, 'About CCX' <http://www.chicagoclimatex.com/about/pdf/CCX_Overview_Brochure.pdf> accessed 16 April 2010.

[138] 'The US Conference of Mayors Partners with ICLEI to Combat Global Warming' (n 135).

[139] ICLEI Global, 'Sundance Summit has US Mayors Reaching Kyoto Goals' (News Release, 23 November 2006) <http://www.iclei.org/index.php?id=1487&tx_ttnews[backPid]=983&tx_ttnews[tt_news]=579&cHash=f18db68bb7> accessed 1 June 2010.

[140] Ibid. The USCM also helped initiate the Energy Efficiency and Conservation Block Grant (EECBG) Program, which is the first program allowing cities, counties and states to receive grants to fund energy-efficiency projects. This program was a top priority of the Mayors' 10-Point Plan and the Mayors' Main Street Recovery Program. In 2009, the Obama Administration distributed $2.8 billion for EECBG, included monies apportioned in the Congressional

At the 77th annual mayor's conference in June 2009, the USCM continued to make climate change a priority, passing a series of resolutions calling on the President and Congress to reform and 'green' the tax code by eliminating or significantly reducing tax preferences to entities emitting large quantities of greenhouse gases; calling on Congress to pass the American Clean Energy and Security Act of 2009 in advance of the United Nations Climate Change Conference 15th Conference of Parties meeting in Copenhagen in December 2009; and, urging the President and Congress to establish a national emissions target of 80% by 2050.[141]

The actions of the USCM and the Mayors' Council on Climate Protection are complemented by a similar endeavor at the international level. The World Mayors Council on Climate Change was initiated in 2005 'to politically promote climate protection policies at the local level, to foster international cooperation of municipal leaders, and to help, through advocacy, to make the multilateral mechanisms for global climate protection effective'.[142]

The coordination and synergy of local mayors between and across state boundaries is indicative of the trend among sub-national governments to utilize new networks of communication to magnify and improve localized efforts to address climate change.

The City of Philadelphia, Pennsylvania

The City of Philadelphia is one of many US cities undertaking local climate change initiatives. In 1999, Philadelphia joined the CCP.[143] As a member of the CCP, the City is obligated to establish a local strategy to create a greenhouse gas emissions inventory. The City received a grant from ICLEI, as well as staff support from the Commonwealth of Pennsylvania, that it used to jump-start its 'Climate Wise' program. Although now defunct, the Climate Wise Program facilitated partnerships between the City and local businesses to encourage voluntary emissions reductions and spawned a number of new climate-based initiatives.[144] Philadelphia is also participating in the Greater Philadelphia Clean Cities Program, which is a US Department of Energy program designed to encourage alternate fuel vehicle programs.[145]

Recovery Package (ARRA). See US Conference of Mayors, 'Mayors Leading the Way on Climate Protection' <http://www.usmayors.org/climateprotection/revised/> accessed 1 June 2010.

[141] See US Conference of Mayors, '"Greening" the Tax Code and Promoting Coordinated Federal Investment in Cities' (Resolution, June 2009) <http://www.usmayors.org/resolutions/77th_conference/environment02.asp> accessed 1 June 2010.

[142] ICLEI Europe, 'CCP Campaign' (2010) <http://www.iclei-europe.org/ccp/ccp-campaign/> accessed 1 June 2010.

[143] Judith Samans-Dunn, 'The City of Philadelphia—The Government and Community Work Together to Reduce Greenhouse Gas Emissions' (2004) 12 *Penn St Envtl L Rev* 207, 208.

[144] Ibid 209.   [145] Ibid 209.

The City of Philadelphia has also committed to reaching the UNFCCC global target of ensuring a 5% reduction in greenhouse gas emissions below 1990 levels by the first Kyoto Protocol commitment period, 2008–2012.[146] The City took this pledge one step further, stating that it will attempt to achieve a 10% reduction in greenhouse gases by that time period.[147] As an added measure, in 2000, Mayor John Street introduced a Green Communities/Sustainable Lifestyles Campaign.[148]

Other climate change initiatives underway in Philadelphia include: inventorying the city's greenhouse gas emissions, creating a Sustainable Energy Management Plan, promoting urban renewal to discourage automobile use, encouraging green building, supporting a 'Million Solar Partnership' that creates incentives for investment in renewable energy, and promoting energy efficiency in the public sector.[149] Philadelphia has also signed onto President Clinton's 'Clinton Climate Initiative',[150] which was launched in August 2006 as a joint venture between the Clinton Foundation and the Large Cities Climate Leadership Group.[151] The aim of the Clinton Climate Initiative is to help cities 'reduce[e] greenhouse gas emissions and increase[e] energy efficiency' using business-oriented, ie, market-based, approaches.[152]

Finally, as part of its continuing campaign to address climate change, Philadelphia launched 'Greenworks Philadelphia'. This initiative creates an ambitious, comprehensive framework designed to make Philadelphia the 'greenest' city in the US by 2015. In order to achieve this status, the city has committed to

---

[146] Ibid 208   [147] Ibid.
[148] Ibid. The program was subsequently renamed the Livable Neighborhood Program.
[149] Seattle Office of Sustainability and Environment, *Current and Planned City of Philadelphia Initiatives* (Report) <http://www.seattle.gov/environment/usm/PhilyInitiatives.pdf> accessed 1 June 2010.
[150] William J. Clinton Foundation, 'Combating Climate Change: Clinton Climate Initiative' <http://www.clintonfoundation.org/what-we-do/clinton-climate-initiative/> accessed 16 April 2010. Other participating cities include: Berlin, Buenos Aires, Cairo, Caracas, Chicago, Delhi, Dhaka, Istanbul, Johannesburg, London, Los Angeles, Madrid, Melbourne, Mexico City, New York, Paris, Rome, Sao Paulo, Seoul, Toronto, and Warsaw.
[151] William J Clinton Foundation, 'Clinton Climate Initiative' <http://www.clintonfoundation.org/what-we-do/clinton-climate-initiative/> accessed 1 June 2010.
[152] Ibid. The Clinton Climate Initiative will promote local climate change action and 'enable its partner cities to reduce energy use and green house gas emissions' by:
1. Creat[ing] a purchasing consortium that will pool the purchasing power of the cities to lower the prices of energy saving products and accelerate the development and deployment of new energy saving and greenhouse gas reducing technologies and products.
2. Mobiliz[ing] the best experts in the world to provide technical assistance to cities to develop and implement plans that will result in greater energy efficiency and lower greenhouse gas emissions.
3. Creat[ing] and deploy[ing] common measurement tools and internet based communications systems that will allow cities to establish a baseline on their greenhouse gas emissions, measure the effectiveness of the program in reducing these emissions and to share what works and does not work with each other. Ibid.

reducing GHG emissions by 20%; improving air quality in order to attain US CAA standards; and diverting 70% of solid waste from landfills.[153]

In these various ways, the City of Philadelphia is adopting measures that are designed with the dual goals of helping to improve the local environment, as well as offsetting the negative consequences of global climate change. Philadelphia's climate change strategy is designed around tangible objectives, making addressing global climate change more manageable for the City and its citizens alike.

Portland, Oregon

Oregon was the first US state to pass legislation setting carbon dioxide standards for new energy facilities to be built within the state.[154] Similarly, the City of Portland, Oregon, was an early leader in the climate change campaign. In 1993, only shortly after the adoption of the UNFCCC, Portland became the very first local government in the US to implement a city-wide strategy for reducing carbon dioxide emissions.[155] In 2001, Portland continued the trend of adopting aggressive climate change policies by setting a goal of reducing greenhouse gas emissions to 10% below 1990 levels by 2010 for its encompassing Multnomah County.[156]

As one method for reducing greenhouse gas emissions, the City has adopted a green building standard for public facilities and buildings financed through the Portland Development Commission.[157] The City is also developing partnerships with the private sector to promote green building standards in the private sector and to encourage voluntary greenhouse gas emission reductions.[158] Using these and other measures, the City's ultimate goal is to reduce greenhouse gas emissions while also creating secondary benefits, including 'reduced energy demand, reduced traffic and congestion, reduced air pollution caused by automobiles, and reduced fiscal demands upon the municipality'.[159]

Portland, like many US cities, is still in the early phases of developing its climate change plan. However, in June 2005, Portland issued *A Progress Report on the City of Portland and Multnomah County Local Action Plan on Global Warming*.[160] The report showed that 'local greenhouse gas emissions in 2004 were only slightly

---

[153] See City of Philadelphia, 'Mayor Nutter Unveils Plan for Making Philadelphia America's Number One Green City' (News Release, 29 April 2009) <https://ework.phila.gov/philagov/news/prelease.asp?id=544> accessed 9 October 2009.

[154] See State and Local Governments and Climate Change (n 11).

[155] Portland Office of Sustainable Development, 'Portland/Multnomah County Local Action Plan on Global Warming' <http://www.portlandonline.com/bps/index.cfm?c=41917&a=112115> accessed 1 June 2010.

[156] Portland Office of Sustainable Development, 'Global Warming Update' (2002) <http://www.portlandonline.com/bps/index.cfm?c=41917&a=112116> accessed 1 June 2010 (Global Warming Update).   [157] Ibid.

[158] Ibid.   [159] McKinstry (n 29) 57; Global Warming Update, ibid.

[160] Maria Rojo de Steffey and others, *A Progress Report on the City of Portland and Multnomah County Local Action Plan on Global Warming* (June 2005) <http://www.portlandonline.com/shared/cfm/image.cfm?id=112118> accessed 1 June 2010.

above 1990 levels, the benchmark year established in the Kyoto Protocol'[161] and that 'on a per capita basis, emissions have fallen by 12.5%, an achievement likely unequalled in any other major U.S. city'.[162] Successful emission reductions are attributed to programs to add light rail lines and streetcars, public purchasing of renewable energy for 10% of city needs, high recycling rates, construction of green buildings, major tree planting campaigns, the 'weatherization' of housing, and the creation of an Energy Trust to fund energy efficiency and renewable energy programs.[163] Further, by 2007, Portland's emissions had fallen below 1990 levels, despite rapid population and economic growth.[164] Following up on these early successes, in 2007, the Portland City Council and the Multnomah County Board of Commissioners passed resolutions directing staff to design a strategy to reduce local carbon emissions by 80% by 2050.

By creating a diversified portfolio of climate change policies and focusing on a mix of regulatory and voluntary programs, Portland has laid the groundwork for an effective climate change policy regime.

Other large US cities, such as New York, are beginning to follow suit, calling for local and regional caps on greenhouse gas emission, implementing renewable portfolio standards, and funding energy efficiency and renewable energy programs.[165]

The climate change strategies of the CCP, Philadelphia, Portland, and New York City reflect the complementary efforts that are taking place all over the US. The buzz of activity at the local and state levels suggests an unprecedented level of impatience among sub-national governments with the federal government's failure to legislate in the context of climate change. This frustration is not confined to climate change policymaking; local governments are similarly keen to encourage sustainability policies, given the absence of national policies directly addressing this issue. The increase in state and sub-state governments adopting climate change, as well as sustainability, policies that exceed federal policies reflects a larger shift in the dynamics of environmental policymaking within the US. As one commentator aptly put it, 'a new federalism is stirring'.[166]

State and Local Climate Change Policies in Review

The flood of independent and collaborative climate change law and policymaking activities taking place at the state and local level is unparalleled in modern

---

[161] Ibid.   [162] Ibid.   [163] Ibid.

[164] On a per capita basis, local emissions had fallen by 17% since 1990. See Portland Bureau of Planning and Sustainability, *What Is Local Government Doing?* (Report on Carbon Reduction 2009) <http://www.portlandonline.com/bps/index.cfm?c=49993> accessed 28 December 2009.

[165] See Christine Van Lenten, *New York Tackles Climate Change: Promoting Renewable Energy and Capping Greenhouse Gas Emissions* (Report by New York Academy of Sciences, 20 October 2005) <http://sallan.org/nyas/NYAS-Climate-Change-preBriefing_Sallan.pdf> accessed 1 June 2010.   [166] Ibid 1.

environmental history. Unlike the EU, where a majority of member states have independent international greenhouse gas commitments as well as obligations under EU law, state and local governments in the US are currently acting in an open legal field. They are neither bound by international greenhouse gas emission reduction commitments nor constrained by federal climate change legislation. The open field creates a situation similar to that present in the 1960s when, prior to the federal government's subsuming of the field, states had primary decision making authority and responsibility for environmental protection. At that time, by and large, states opted for minimal environmental standards in favor of promoting economic activity, creating a race to the bottom scenario. At present, in many cases, the opposite is true. The faith that the public once vested in the federal government is now vested with state and local governments as the potential harbingers of change. Sub-federal governments are responding by taking varied steps—both substantive and symbolic—to fill the regulatory vacuum.

This is not to suggest that state and local governments have developed comprehensive, robust regulatory structures for climate change. On the whole, they have not—with notable exceptions, such as the efforts underway in California and the Northeast. Sub-federal governments are merely in the process of laying the foundations for local and regional climate change strategies. The rate of law and policy evolution at the sub-federal level, however, greatly outpaces the actions of the federal government. With California poised to start regulating greenhouse gases economy-wide in 2011, and many of the Northeast states already regulating carbon dioxide emissions from power plants, there is little doubt that state and local governments are incrementally building the foundations for more comprehensive federal responses to climate change.

Regardless of the progress that state and local governments achieve, or the pace at which they achieve it, ultimately, it will be impossible for sub-federal governments to be the dominant figures in US climate change policy. In the long-term, a comprehensive climate change policy regime requires initiatives that do the following: regulate greenhouse gas emissions; promote/mandate changes in energy production and consumption; facilitate changes in the transport sector; and engage all sectors of public and private society in efforts to address the multitude of social, political, and economic challenges posed by climate change. The US federalist system of governance, coupled with the historic development of federal authority in environmental law, suggests that comprehensive regulatory regimes for greenhouse gas emissions, energy production and consumption, transportation, and coordinated public and private attempts to address trans-boundary environmental problems are not only the domain of the federal government, but would finally prove impossible in a country as diverse as the US without federal coordination.

Further, as a federalist state, only the federal government has the legal authority to enter international negotiations on behalf of the 50 states. Given the global nature of climate change, the influence of the US federal government in international negotiations is as important as the domestic legal and political approach that the US adopts. No level of coordinated sub-federal activity will be able to change the fact that the federal government is the final voice of authority in most domestic regulation and all international negotiations. Sub-federal efforts might, however, influence the attitude the federal government adopts towards climate change, which in turn, will shape federal choices at the domestic and international level.

This is the most important role sub-federal governments play—the role of agenda-setters.

Sub-federal climate change initiatives play a significant role in shaping the federal climate change agenda; however, they only form part of the picture. Chapter 4 moves beyond state and local government activities to explore how public and private parties are using litigation, regulation and international mechanisms to influence US federal climate change policy. Chapter 4 suggests that, in the majority of cases, parties are turning to litigation, regulation, and international mechanisms to compel the federal government to take its 'rightful', or at least traditional, place in environmental policymaking. After examining this trend, Chapter 4 analyses the sum total result of state and local initiatives, litigation, regulation, and international law on determining the course of US climate change law and policymaking.

4
Litigation, Regulation, and International Law as Law and Policy Drivers in the United States

In a country as geographically, politically, and socially diverse as the United States, federal policy is continuously influenced by on-going legal initiatives nationwide. These initiatives can take the form of state laws, local laws, civil and criminal litigation, regulation, and even recourse to international law. In the absence of overarching federal climate change legislation, public and private entities are attempting to influence federal policy using a wide variety of techniques. In the discussion that follows, this chapter examines how public and private actors are using litigation, regulation, and international law to influence federal climate change policy in the US.

Litigation and alternative forms of law and policymaking exist in both the US and the EU. In contrast to the central role such initiatives have played in influencing US climate policy, recourse to litigation and alternative forms of policymaking have played a less central role in shaping the policy agenda in Europe. Instead, in Europe, such measures have primarily served to drive improved implementation, enforcement, and expansion of existing policy regimes or have been more technical challenges to the validity of specific measures.[1] Here, special attention is paid to the role of litigation, regulatory developments and international law in the US in order to demonstrate how variations in the overarching climate change frameworks of the US and the EU have prompted an important series of secondary developments in US law and policymaking.

Climate Change Litigation in the United States

In the US, state and local climate change policies are creating momentum for political change. These political efforts, however, are dispersed, varied, and subject

---

[1] Eg, Navrah Singh Ghaleigh, 'Emissions Trading Before the European Court of Justice: Market Making in Luxembourg', 29, in *Legal Aspects of Carbon Trading: Kyoto, Copenhagen and Beyond,* D Freestone and C Streck, (eds) (Oxford University Press, Oxford, 2009).

to federal pre-emption, making it difficult to bring about rapid federal responses. One traditional avenue for directly engaging the US federal government in areas of environmental protection where legislative and executive attention lags is through the judiciary. Judicial actions can be directly addressed to, or appealed to the federal courts, providing a mechanism for fast-tracking federal consideration of environmental issues.[2]

In the context of climate change, there are three primary ways for petitioners, whether private citizens or state and local officials, to use the judiciary to compel the federal government to address climate change. First, petitioners who believe climate change harms them directly can sue the federal government or its agents in US federal court—challenging statutes, regulations, legal gaps, and failures to act. Second, petitioners can bring actions against private individuals and companies, alleging that the third parties' acts or omissions are contributing to climate change, thereby resulting in specific harm to the parties. Third, petitioners can address climate change complaints against the US government to international tribunals, eg, the International Court of Justice or the Inter-American Commission on Human Rights.[3]

Increasingly, litigation is seen as an avenue for change in the context of climate change policy. More than two dozen lawsuits related to global climate change 'currently sit on the dockets of our federal and state courts'.[4] In the sections that follow, this chapter reviews how increasing numbers and types of petitioners are using the judiciary in new and creative ways to require public and private individuals to respond to climate change.

Private Suit Against the Federal Government: *Friends of the Earth, Inc et al v Spinelli et al*

In 2002, two non-profit environmental groups and four cities sued the US government in federal court, alleging that financial investments made by federal agencies have harmed the US by escalating the intensity of global warming. In *Friends of the Earth v Watson/Mosbacher*[5]—later renamed *Friends of the Earth, Inc et al v Spinelli et al*—the plaintiffs contend that, over the past ten years, the Export–Import Bank (Ex–Im) and the Overseas Private Investment Corporation (OPIC) have granted over US$32 million in financial assistance, in direct financing and insurance, to oil and other fossil fuel ventures without

---

[2] See Kristen H Engel and Scott R Saleska, 'Subglobal Regulation of the Global Commons: The Case of Climate Change' (2005) 32 *Ecology LQ* 183, 219.
[3] See Michael G Faure and André Nollkaemper, 'International liability as an instrument to prevent and compensate for climate change' (2007) 23A *Stan Envtl LJ* 123.
[4] Justin R Pidot, 'Global Warming in the Courts: An Overview of Current Litigation and Common Legal Issues' (2006) Georgetown Environmental Law & Policy Institute <http://www.loe.org/images/070406/GlobalWarmingLit_CourtsReport.pdf> accessed 1 June 2010.
[5] No C 02-4106 JSW (ND Cal 23 August 2005).

first evaluating how the projects will contribute to global warming, or otherwise impact the environment in the US as required by the National Environmental Policy Act (NEPA).[6]

NEPA requires federal agencies to undertake environmental impact assessments for any major federal action likely to significantly impact the environment.[7] The plaintiffs contend that NEPA should apply equally to overseas projects that are financed by US government agencies. Specifically, the petitioners claim that between 1990 and 2003, Ex–Im and OPIC loans sponsored projects that produced cumulative emissions that were equivalent to approximately 8% of the world's annual CO2 emissions, or nearly 1/3 of annual US emissions in 2003, an amount equivalent to in excess of 260 million tons of carbon dioxide emissions annually.[8] These emissions, the petitioners allege, contribute to climate change, and, thus, impair the US environment.[9] Furthermore, the plaintiffs contend that by harming the environment, Ex–Im and OPIC have injured them.

This lawsuit is the first attempt by public or private plaintiffs to hold the US directly accountable for its contribution to global warming. Not surprisingly, the US government has adamantly opposed the lawsuit. In fact, in 2004, Peter Watson, President and CEO of OPIC, and Phillip Merrill, the former Vice Chairman and First Vice President of Ex–Im, sought a summary judgment, alleging that the case lacked the basis for a claim.[10] On 23 August 2005, however, Judge Jeffrey White, for the US District Court for the Northern District of California, denied the motion for summary judgment, allowing the case to proceed to trial.[11] The Court found that based on the lesser standards for standing applicable to a procedural claim, the plaintiffs had adequately demonstrated standing, causation, and redressability.[12]

Judge White's decision marked the first time that a federal court has granted legal standing in a case where the alleged injury is solely based on the impacts of global warming and where the challenge is based on the US government's failure to assess the impact of its actions on the global atmosphere and the US citizenry.[13]

---

[6] See ibid.
[7] The National Environmental Policy Act of 1969, 42 USC §§ 4321–4347 (2002).
[8] Pidot (n 4) 13.   [9] See ibid.
[10] Defendants' Motion for Summary Judgment and Memorandum in Support, *Friends of the Earth Inc v Watson*, No C 02-4106 JSW, (ND Cal 11 February 2005).
[11] See Organic Consumers Association, 'Global Warming Lawsuit Against U.S. Agencies Passes Court Test' <http://www.organicconsumers.org/Politics/globalwarm082505.cfm> accessed 15 April 2006; ClimateLawsuit.org, 'In Landmark Decision Against Bush Administration, Federal Court Recognizes Harm Caused by Global Warming: Lawsuit by Environmental Groups and Cities Goes Forward' (24 August 2005) <http://www.greenpeace.org/usa/press-center/releases2/Ex-ImOPICdecision> accessed 1 June 2010.
[12] 'Global Warming Lawsuit Against US Agencies Passes Court Test' (n 11); see also Stephen L Kass and Jean M McCarroll, 'Litigating Climate Change Via State Regulations, Federal Courts' [28 April 2006] 235 *New York LJ* 3.   [13] See ibid.

In a judgment filed on 30 March 2007, Judge White held in relevant part that OPIC's activities are generally subject to NEPA, and that:

based on the statements in [OPIC and Ex–Im's] climate change reports regarding the effects of GHGs on climate change, it would be difficult for the Court to conclude that Defendants have created a genuine dispute that GHGs do not contribute to global warming.[14]

He concluded that the court could not determine whether OPIC and Ex–Im projects qualify as major federal actions for purposes of NEPA without a 'fact-intensive inquiry' and ordered the parties to prepare for a trial on the merits of the case.[15]

Judge White's decisions on the petitions for summary judgment are noteworthy because of the broad interpretation of standing requirements, the applicability of NEPA to OPIC and Ex–Im activities, and the relevance of NEPA in the context of climate change. Following Judge White's decision, both sides began preparing for a fact-intensive battle, the outcome of which would have ripple effects for a wide range of federal decision making processes previously considered to be outside the scope of NEPA. The precedential role of the case was deferred, however, when on 6 February 2009, after six years of litigation, the parties settled the case out of court. As part of the settlement, the Ex–Im Bank agreed to begin taking CO2 emissions into account when evaluating fossil fuel projects and to create an organization-wide carbon policy, while OPIC agreed to establish a goal of reducing GHG emissions associated with its projects by 20% over the next ten years. In addition, both agencies committed to increasing financing for renewable energy projects.

The settlement agreement represented a significant victory in efforts to promote governmental accountability for its direct and indirect contributions to climate change. It also demonstrated the viability of using the courts to find political pressure points and provided a roadmap for similar climate change litigation to follow. The substantive effect of the settlement hinges on the agencies' compliance with the terms of the agreement and the ability of the settlement to shape agency decision making government-wide.

State Suit against Private Industry: *State of Connecticut et al v American Electric Power Co et al*

While it is imperative that the US federal government play a leading role in addressing climate change, it is equally important that private industry in the US actively participates in these efforts. Private companies are primary sources of greenhouse gas emissions in the US. Ensuring their participation, either through regulatory

---

[14] Order Denying Plaintiffs' Motion for Summary Judgment and Granting in Part and Denying in Part Defendants' Motions for Summary Judgment, *Friends of the Earth Inc v Mosbacher Jr*, Case No 3:02-cv-04106-JSW (ND Cal 30 March 2007). [15] Ibid.

regimes or financial incentives, is essential to a comprehensive climate change policy. Currently, however, there is no federal regulatory system ensuring that private industry substantially reduces its greenhouse gas emissions. Recognizing the importance of private sector participation, on 21 July 2004, eight US states representing over 77 million Americans, including California, Connecticut, Iowa, New Jersey, New York, Rhode Island, Vermont, and Wisconsin—and New York City—filed a public nuisance lawsuit in the US federal court in Manhattan against five of the largest power companies in the US: Cinergy Corp, Southern Company, Xcel Energy, American Electric Power (AEP), and the Tennessee Valley Authority (TVA).[16] Together, these companies and the TVA 'represent 25% of the [total] CO2 emissions from the power sector in the U.S. and 10% of world wide [sic] emissions'.[17]

The plaintiffs allege that the defendants' practices constitute a public nuisance, citing existing pollution-as-public-nuisance cases as precedent for their claim.[18] The lawsuit is unique, however, in that the plaintiffs are not seeking any monetary damages. Instead, the plaintiffs are seeking injunctive relief, demanding that the companies be forced to reduce CO2 emissions at 174 plants by 3% per year over the next ten years.[19] The plaintiffs claim that the companies already have access to technology that will enable them to produce the same amount of electricity while simultaneously ensuring significant CO2 emission reductions.[20]

From the moment the lawsuit was filed, there was little doubt that the plaintiffs would face staunch resistance from the defendant companies, and possibly, even from the federal government. This, in fact, proved to be the case. On 15 September 2005, the US District Court for the Southern District of New York dismissed the plaintiffs' complaints, holding that the action involved a 'non-judiciable [sic] political question' and because 'resolution of the issues presented...requires identification and balancing of economic, environmental, foreign policy, and national security interests, "an initial policy determination of a kind clearly for non-judicial discretion" is required'. The court then concluded that 'questions presented here "uniquely demand [a] single-voiced statement of the Government's views"'.[21]

The decision by the District Court did not bring this case to an end. On 15 December 2005, the plaintiffs and three NGOs filed briefs in the Second Circuit Court of Appeals appealing the dismissal of the case. The arguments for

[16] *Connecticut v American Electric Power*, 406 F Supp 2d 265 (SDNY 2005).
[17] Edna Sussman, 'Climate Change Litigation: Past, Present and Future: ABA Renewable Energy Resources Committee Program' (21 June 2006) <http://www.abanet.org/environ/committees/renewableenergy/teleconarchives/062106/6-21-06_SussmanClimateChange.pdf> accessed 29 December 2009. [18] *Connecticut v American Electric Power* (n 16).
[19] See ibid. [20] See ibid.
[21] Ibid 271 (citing *Baker v Carr*, 369 US 186, 211–12 (1962)); see also ibid 271–72 (discussing what kind of questions are not justiciable and citing the fact that '[a]t oral argument, counsel for AEP and Cinergy argued that by "asking this Court to resolve an environmental policy question with sweeping implications for the nation's economy, its foreign relations, and even potentially its national security", "Plaintiffs have put the cart before the horse"'); ibid 272, citing *Vieth v Jubelirer*, 541 US 267, 278 (2004).

the case were heard in the Second Circuit in June 2006, and the court issued its decision on 21 September 2009 vacating and remanding the District Court's decision. In relevant part, the Second Circuit held that the Appellants' claims did not present nonjusticiable political questions and that the Appellants had standing to bring their federal nuisance law claims.[22]

The outcome of the appeal was highly anticipated as it offered a critical indicator for other common law based climate litigation. While the case has not yet been decided on the merits, the Second Circuit's decision created opportunities for other individuals, non-governmental organizations and sub-national governments to use federal common law to hold members of the private sector accountable for their contributions to global warming and, ultimately, to compel behavioral changes.[23]

Private Suit to Prevent State-led Climate Change Litigation: Alliance of Automobile Manufacturers (et al) v California

In Chapter 3, this book examined efforts underway in California to address global climate change. Central among these efforts was the enactment of California's AB 1493, a regulation implementing new automobile greenhouse gas emissions standards.[24] AB 1493 directed the California Air Resources Board to adopt rules requiring automakers to reduce carbon dioxide emissions from new cars and light trucks beginning with model year 2009.[25]

On 7 December 2004, the Alliance of Automobile Manufacturers and 13 California car dealerships filed suit in federal court in Fresno opposing California's automobile emission regulations.[26] The lawsuit was based on the allegation that the regulation is 'inconsistent with federal law, as well as fundamental principles for sound regulation of motor vehicles'.[27]

The first oral arguments took place on 15 September 2006. On that date, California and six intervening environmental groups asked the judge to dismiss the case. On 25 September 2006, the federal judge ruled that the automaker lawsuit would not be dismissed, and the trial would begin on 30 January 2007. Subsequently, on 11 December 2007, US District Judge Anthony Ishii rejected the claims of the automakers and dismissed the lawsuit. In 2008, the car companies appealed the ruling. As discussed *infra* in Chapters 3 and 7, in 2009, the automakers

---

[22] *Connecticut v American Electric Power*, 582 F3d 309 (2nd Cir 2009).
[23] Following on from the Appellant's victory in the Second Circuit, the Fifth Circuit released a similar decision in the case of *Comer v Murphy Oil*, overturning a district court decision finding that the plaintiff-appellant lacked standing to bring tort-based litigation against insurance, oil, coal, and chemical companies for their contributions to climate change and that the case presented non-justiciable questions. The case was remanded to the district court to consider the merits of the case. *Comer v Murphy Oil*, 585 F3d 855 (5th Cir 2009).
[24] Assem B 1493, 2002 Gen Assem, Reg Sess (Cal 2002). [25] Ibid.
[26] *Central Valley Chrysler-Jeep v Witherspoon*, 456 F Supp 2d 1160 (ED Cal 2006).
[27] Auto Alliance, 'Automakers and Dealers Cite Federal Law, Marketplace Principles in Challenging Carbon Dioxide Law' (Press Release, 7 December 2004) <http://www.robinwood.de/german/verkehr/ausbremsen/autoalliance7_12_4.pdf> accessed 3 April 2010.

agreed to drop all legal action against California and the states as part of a deal that President Obama brokered to create a new nationwide regulatory regime for fuel economy and greenhouse gas emission standards for new automobiles.

As discussed *infra*, the California suit was not the only one of its kind. Just as other states followed California's lead in adopting stringent emissions standards for automobiles, so did automakers and dealers follow the lead taken by industry in the California lawsuit in filing legal challenges seeking to block the efforts of these other states to adopt California's auto emissions standards. Given that 15 other states and the District of Columbia have adopted legislation similar to California's and that automakers have filed suit in at least four of these states, the outcome of this litigation will affect climate policy nationwide. And, while much of this litigation was stayed in 2009, at least two of the automobile dealers' cases are on-going and the stayed cases are due to be revisited, leaving the line of litigation open-ended in the near term.

As previously mentioned, California has not only fought challenges to its emissions legislation brought by automakers and dealers, it has also filed a series of lawsuits of its own, including the now withdrawn nuisance case against auto manufacturers. These dueling lawsuits set the stage for years of intensive wrangling between individual states and the auto industry over the future of state and federal greenhouse gas emission reduction legislation with the dual effect of provoking federal response. With the announcement of new federal automobile emissions standards and the grant of the CAA waiver to California, the automakers eased, but did not put into abeyance, their opposition campaign. Transport will undoubtedly remain one of the most hotly contested regulatory areas in future governmental efforts to address climate change.

Climate Change Litigation Reaches the Supreme Court: *Massachusetts et al v EPA*

In one of the most important environmental cases of the 21st century, on 29 November 2006, the US Supreme Court heard oral arguments in the case of *Massachusetts et al v EPA*.[28]

Massachusetts et al v EPA involved a challenge to the EPA's denial of a petition to regulate carbon dioxide and other greenhouse gases from new automobiles under § 202(a)(1) of the Clean Air Act (CAA), and an EPA general counsel memorandum claiming that the EPA lacks authority to regulate greenhouse gases under the CAA.[29]

---

[28] *Massachusetts v EPA*, 549 US 497 (2007).
[29] See EPA, 'Control of Emissions From New Highway Vehicles and Engines' 68 *Fed Reg* 52, 922 (8 September 2003) (denying petition to regulate CO2 because EPA lacks authority under the CAA to regulate it); Robert E Fabricant, 'EPA's Authority to Impose Mandatory Controls to Address Global Climate Change Under the Clean Air Act' (Memorandum from EPA General

The case began in 1999 when environmental groups petitioned the EPA to regulate greenhouse gases from new motor vehicles. The environmental groups cited § 202 of the CAA to support their claim that the EPA had the mandate and responsibility for regulating greenhouse gases from new vehicles.

In August 2003, the EPA denied the petition, concluding that the CAA does not authorize the Agency to regulate greenhouse gas emissions and, even if it did, the EPA would not exercise such authority.

Upon the EPA's denial of the petition, 12 states, three major cities, and American Samoa joined the environmental groups to challenge the EPA's decision in the federal courts. Similarly, ten states and multiple trade associations joined the EPA as interveners in the case.

The case was first heard in the federal courts by the Court of Appeals for the District of Columbia (DC Circuit). On 15 July 2005, in a 2–1 split vote, the DC Circuit upheld the EPA's position and dismissed and denied the appellants' petitions.[30]

The three written opinions released in this case revealed critical differences in the reasoning underlying each judge's respective decision. Judge Randolph cited scientific uncertainty and policy considerations as justification for the EPA's decision; Judge Sentelle upheld the EPA's decision based on his finding that the petitioners lacked standing; while Judge Tatel dissented on the grounds that the CAA 'plainly authorizes' the EPA to regulate greenhouse gas emissions from autos.[31] The inconsistencies in the DC Circuit Court decision provided fertile ground for an appeal to the US Supreme Court.

On 2 March 2006, the petitioners—*Massachusetts et al*—filed a writ of certiorari with the Supreme Court.[32] The Supreme Court granted certiorari on 26 June 2006, and heard oral arguments on 29 November 2006.[33]

The key issues in this case revolved around the interpretation of § 202(a)(1) of the Clean Air Act, which requires the administrator of the EPA to set emission standards for 'any air pollutant' from motor vehicles or motor vehicle engines 'which in his judgment cause[s], or contribute[s] to, air pollution which may reasonably be anticipated to endanger public health or welfare'.[34] The two central questions before the Supreme Court were: (1) whether the EPA Administrator has

---

Counsel to EPA Administrator Marianne L Horinko, 28 August 2003) <http://www.connellfoley.com/hselaw/pdf/Fabricantmemo.pdf> accessed 1 June 2010.

[30] The judges on the panel issued three separate opinions. In the majority opinion, Judge Randolph dismissed and denied the appellants petitions without directly addressing the question of whether the EPA has authority to regulate greenhouse gases. Joining Judge Randolph in the majority decision, but issuing a separate opinion, Judge Sentelle held that the petitioners failed to establish standing because they had not demonstrated a particularized injury. In an extensive dissenting opinion, Judge Tatel held that the petitioners had established standing, that the EPA had the proper authority to regulate greenhouse gas emissions and denied EPA's argument that it could choose to deny a petition to regulate greenhouse gases based on public policy considerations. See Briefing for Petitioners at 6–8, *Massachusetts v EPA*, 549 US 497 (2007) (No 05-1120, 31 August 2006). [31] *Massachusetts v EPA*, 415 F3d 50, 57–67 (DC Cir 2005).
[32] Petition for Writ of Certiorari, *Massachusetts v EPA*, 549 US 497 (2007) (No 05-1120).
[33] *Massachusetts v EPA* (n 28). [34] 42 USC § 7521(a)(1).

authority to regulate carbon dioxide and other air pollutants associated with climate change under § 202(a)(1) of CAA; and (2) whether the EPA Administrator may decline to issue emission standards for motor vehicles based on policy considerations not enumerated in § 202(a)(1) of CAA.[35]

The case was particularly contentious because it involved a potential conflict over the mandate and authority of the EPA between two consecutive presidential administrations. During President Clinton's administration, 'two different EPA general counsel and EPA Administrator Carol Browner suggested that the agency had some authority to regulate CO2 or other GHGs under the CAA even if the US did not ratify the Kyoto Protocol'.[36] In contrast, under George W. Bush's administration, in 2003, EPA counsel issued a new opinion, finding that the EPA lacked the authority to regulate carbon dioxide and other greenhouse gases under the CAA.[37] The case raised critical questions of statutory interpretation in an undeniably heated and divided political context—creating rifts through presidential administrations, and among federal, sub-federal and public–private relations.

The political rifts were evident in the political debate surrounding the case, with the Massachusetts Attorney General bemoaning that '[i]t's unfortunate that the Commonwealth of Massachusetts and the other states that petitioned the US Supreme Court had to be here today to force the EPA and the administration to do their job', while opponents argued that the petitioner's case 'aim[ed] at nothing less than litigating America into compliance with a non-ratified treaty and/or a non-enacted bill.[38] This is judicial activism in overdrive, perhaps the most audacious attempt ever to legislate from the bench.'[39]

The politics and frustration entangled with the climate change debate did not stop at the door of the Court. The oral arguments before the Supreme Court revealed considerable tensions among the Justices over both the politics and the science of climate change. While Justice John Paul Stevens commented on the fact that the Bush administration had selectively edited quotes to support its argument that there was not enough evidence on the causes of global warming, an argument contested by the scientists upon whose report the EPA relied, Justice Scalia insisted that the petitioners needed to show the imminent harm associated

---

[35] 'Briefing for Petitioners' (n 30) pt I.
[36] Bradford C Mank, 'Standing and Global Warming: Is Injury to All Injury to None?' (2005) 35 *Envtl L* 1, 8–9, citing Arnold W Reitze Jr, 'Global Warming' (2001) 31 *Envtl L Rep* 10253, 10257–59 (discussing memorandum and congressional testimony by two Clinton-era EPA general counsel, Jonathan Z Cannon and his successor Gary S Guzy, indicating that EPA has the authority to regulate CO2 under the CAA); see generally Christopher T Giovinazzo, 'Defending Overstatement: The Symbolic Clean Air Act and Carbon Dioxide' (2006) 30 *Harv Envtl L Rev* 99.
[37] Mank (n 36) 9.
[38] Nina Totenberg, 'High Court Hears its First Global Warming Case' (Report on National Public Radio, 29 November 2006) <http://www.npr.org/templates/story/story.php?storyId=6556413> accessed 29 December 2009.
[39] Marlo Lewis, 'Judicial Activism in Overdrive: *Massachusetts, et al. v. EPA*' (8 September 2006) <http://www.renewamerica.us/columns/mlewis/060908> accessed 1 June 2010.

with climate change by demanding to know '[w]hen...the predicted cataclysm [would occur]'.[40] And, in a moment that reflected the apathy and defiance so characteristic of the policies of the Bush administration, upon being corrected on a point of science, Justice Scalia retorted: 'Whatever. I'm not a scientist. That's why I don't want to have to deal with global warming, to tell you the truth.'[41]

After four months of deliberation, on 2 April 2007, the Supreme Court issued one of the most important decisions in modern US environmental law. In a 5–4 split decision, the Court overturned the DC Circuit and held that *Massachusetts et al* had standing to challenge the EPA's denial of their rulemaking petition, that the Court has the authority to review the EPA's decisions, that the EPA has the authority to regulate greenhouse gases from new motor vehicles under the Clean Air Act § 202(a)(1), and that neither policy grounds nor scientific uncertainty provided permissible considerations for the EPA's decision to avoid regulating greenhouse gases under § 202(a)(1).[42] The Court had 'little trouble concluding' that the CAA 'authorizes EPA to regulate greenhouse gas emissions from new motor vehicles' and then concluded that EPA's decision not to regulate greenhouse gases based on scientific or policy grounds was 'arbitrary, capricious, or otherwise not in accordance with law'.[43]

The Court remanded the case, directing the EPA that in re-evaluating its decision whether to regulate greenhouse gases under the CAA, it 'must ground its reasons for action or inaction in the statute'.

In dissent, Justices Roberts, Scalia, Thomas, and Alito argued that the petitioners' challenges were non-justiciable.[44] In the first dissent, authored by Justice Roberts, the four dissenting Justices claimed that the petitioners lacked standing and declared that the 'grievances of the sort at issue here,' ie, global climate change, should be left to the discretion of the legislative and executive branches.[45] In a second dissent, authored by Justice Scalia and again joined by Chief Justice Roberts and Justices Thomas and Alito, the dissenting Justices reaffirmed their opinion that the Court lacked jurisdiction, but proceeded to address the merits of the case. Here, the Justices upheld the right of the EPA Administrator to withhold a judgment when a petition for rulemaking is filed, defended the EPA's reliance on uncertainty as grounds for denying a rulemaking petition, challenged the finding in the majority opinion that the CAA authorizes the EPA to regulate greenhouse gases, and supported the EPA's statutory interpretation of the CAA that excluded greenhouse gases as agents of air pollution.[46] In concluding, Justice Scalia accused the Court of substituting its 'own desired outcome for the reasoned judgment' of the EPA.[47]

---

[40] Totenberg (n 38). [41] Ibid.
[42] *Massachusetts v EPA* (n 28). In finding that the Commonwealth of Massachusetts satisfied Constitutional standing requirements, the Court found that 'EPA's steadfast refusal to regulate greenhouse gas emissions presents a risk of harm to Massachusetts that is both "actual" and "imminent"', and that '[t]here is, moreover, a substantial likelihood that the judicial relief requested will prompt EPA to take steps to reduce that risk'. Ibid 498–99.
[43] Ibid 528 and 534. [44] Ibid 535–49. [45] Ibid.
[46] Ibid 549–60. [47] Ibid.

The substantive differences and the impassioned tone in the majority opinion and the two dissenting opinions reflect the political schisms that characterize current American climate change policy. The dissenting Justices base their dissent on administrative and procedural grounds. However, the lines between the Justices in the majority—supporting an interpretation of the CAA and the EPA's duties under the CAA that demands more rigorous attention to global climate change—and the Justices in the dissent—supporting an interpretation of the CAA and the EPA's duties under the CAA that precludes judicial involvement in the issue of global climate change and permits the agency more discretion in choosing whether to regulate greenhouse gas emissions—reflect the conflicting political tides that have paralyzed climate change policymaking at the federal level. While numerous sub-federal entities and increasing numbers of federal representatives push for aggressive interpretations of existing environmental laws and implementation of new environmental laws to address global climate change, an equal or greater number of public and private entities use scientific uncertainty and procedural grounds to delay specific and concrete political responses to climate change. In *Massachusetts v EPA*, the proponents of change narrowly won their first major federal victory.

Regulatory Response to Massachusetts v EPA: EPA's Endangerment Finding

The Supreme Court's decision in *Massachusetts v EPA* required the EPA Administrator to determine whether, under CAA § 202(a), emissions of greenhouse gases from new motor vehicles cause or contribute to air pollution that may reasonably be anticipated to endanger public health or welfare, or whether the science remains too uncertain to allow a reasoned decision. Pursuant to this duty, on 17 April 2009, EPA Administrator Lisa P. Jackson issued the long-awaited proposed endangerment finding.[48] In relevant part, the Administrator proposed finding 'that greenhouse gases in the atmosphere endanger the public health and welfare of current and future generations.... These high atmospheric levels are the unambiguous result of human emissions, and are very likely the cause of the observed increase in average temperatures and other climatic changes.'[49]

In the proposed finding, the Agency emphasized that while carbon dioxide is the most prevalent greenhouse gas, methane, nitrous oxide, and hydrofluorocarbons are also emitted by new motor vehicles (the source category under 202(a)) and contribute to the mix of greenhouse gases in the atmosphere. The Agency then discussed the role transport plays in contributing to climate change, noting that the US transportation sector accounted for 24% of total US greenhouse gas emissions in 2006 and 4% of global GHG greenhouse gas emissions in 2005, making it a significant contributor to both domestic and international greenhouse gas emissions.

---

[48] EPA, 'Proposed Endangerment and Cause or Contribute Findings for Greenhouse Gases under Section 202(a) of the Clean Air Act' 74 Fed Reg 18886 (24 April 2009). [49] Ibid.

Next, the Agency analyzed § 202(a), which directs the Administrator to exercise her discretion when determining 'first, whether air pollution may reasonably be anticipated to endanger public health or welfare, and second, whether emissions from new motor vehicles or engines cause or contribute to this air pollution' and concluded that this language allows the Administrator to take precautionary regulatory action where necessary despite the absence of full scientific certainty.

Alongside the proposed finding, the EPA issued a technical support document (TSD) detailing the approach the Agency took to gather, synthesize, and analyze the data on the links between greenhouse gas emissions and human health and well being. In issuing the TSD and the proposed finding, the Agency relied heavily on IPCC and US Climate Change Science Program reports to conclude that there is compelling evidence that climate changes can be attributed to the anthropogenic rise in atmospheric greenhouse gases and that the detrimental effects of climate change in the US alone warrant the endangerment finding. The proposed rule surmises that:

[t]he scientific evidence described here is the product of decades of research by thousands of scientists from the U.S. and around the world. The evidence points ineluctably to the conclusion that climate change is upon us as a result of greenhouse gas emissions, that climatic changes are already occurring that harm our health and welfare, and that the effects will only worsen over time in the absence of regulatory action.[50]

Publication of the proposed finding in the US Code of Federal Regulations did not trigger regulatory action under the Clean Air Act or otherwise create any new regulatory burdens. Rather, publication commenced a 60-day period during which time the Agency accepted public comments on the rule from interested parties. The public comment period for the proposed endangerment finding ended 23 June 2009.

The EPA sent its final endangerment finding to the White House for review on 9 November 2009. Just over one month later, on 7 December 2009, Administrator Jackson signed two distinct findings, which together require the EPA to regulate greenhouse gas emissions from new automobiles under the § 202(a) of the US CAA. In relevant part, the Administrator determined that 'six greenhouse gases taken in combination endanger both the public health and the public welfare of current and future generations',[51] and that 'the combined emissions of these well-mixed greenhouse gases from new motor vehicles and new motor vehicle engines contribute to the greenhouse gas pollution which threatens public health and welfare under CAA section 202(a)'.[52] While the issuance of the final endangerment finding did not automatically generate any new regulatory requirements, it marked the first substantive steps towards the creation of a nationwide greenhouse gas regulatory regime.[53]

---

[50] Ibid 18904. [51] 74 Fed. Reg. 239, 66496 (15 December 2009). [52] Ibid.
[53] The timing of the release of the Endangerment Finding was important, because on 15 September 2009, the EPA and the NHTSA had proposed a new program for regulating

The EPA's endangerment finding lays the groundwork for sweeping changes in US climate policy. Beyond automobiles, the endangerment finding triggers further obligations under the CAA for the Agency to impose permitting and technology requirements on stationary sources of greenhouse gas emissions. For example, anticipating the expansion of its regulatory authority—and mandate—under the CAA, on 30 September 2009, the EPA proposed new greenhouse gas permitting requirements for large stationary sources under two of the CAA's major permitting programs.[54] Upon implementation, the proposed regulatory program would expand greenhouse gas permitting requirements to cover nearly 70% of the US's largest stationary sources for greenhouse gas emissions, including but not limited to power plants, refineries, and cement production facilities. At the same time, by limiting the new regulatory regime to large sources, the EPA would protect small businesses and farms from being subject to excessive new permitting requirements in the early days of the scheme, thus improving the political palatability of the program.

Thus, prior even to the release of the final endangerment finding, the EPA laid the groundwork for a program that would extend the Agency's greenhouse gas emissions regulatory regime beyond mobile sources to include tens of thousands of stationary facilities nationwide. The scale of this regulatory program could potentially dwarf not only existing EU schemes, but also pending US legislative proposals. It would also place enormous administrative burdens on an already overtaxed Agency and create a potentially complex and sprawling new regulatory framework. Mounting pressure and momentum for regulatory action via existing federal environmental laws thus has the compounding effect of exerting pressure on the legislature to formulate a more comprehensive, joined-up response.

Fuel Economy Regulatory Challenge: *Center for Biological Diversity v National Highway Traffic Safety Administration*

The high-profile nature of *Massachusetts v EPA* and the subsequent regulatory proceedings have overshadowed other on-going legal challenges to the federal government's inactivity with regard to climate change. One such challenge is the case of *Center for Biological Diversity v National Highway Traffic Safety Administration*.[55] Here, the Center for Biological Diversity, 11 states, New York

---

greenhouse gas emissions and improving fuel economy for light duty vehicles sold in the United States. The joint program would be the first of its kind. The Administrator's Endangerment finding was a necessary prerequisite to finalizing and implementing the proposed joint regulatory program. 74 Fed. Reg. 186, 49454 (28 September 2009).

[54] 74 Fed. Reg. 212, 57126 (4 November 2009), The rule proposes new thresholds for greenhouse gas emissions (GHG) that define when facilities would have to obtain permits under the CAA to construct new facilities or to undertake major modifications to existing industrial facilities.

[55] *Center for Biological Diversity v NHTSA*, 538 F3d 1172 (9th Cir 2008); see also California Office of the Attorney General, 'Attorney General Lockyer Challenges Federal Fuel Standards for

City, the District of Columbia, and three other environmental groups challenged a National Highway Traffic Safety Administration (NHTSA) final rule, 'Average Fuel Economy Standards for Light Trucks, Model Years 2008–2011', which sets the CAFE standards for light trucks as mandated by the Energy Policy and Conservation Act of 1975 (EPCA).[56] The petitioners challenged the rule under the EPCA and NEPA, alleging in relevant part that NHTSA's new rule fails to evaluate the impact of CAFE standards on the environment and, in particular, on global climate change as required by the NEPA. In devising the new rules, NHTSA found that the new standards 'would have no significant environmental impact and did not warrant an Environmental Impact Statement EIS'.[57] The petitioners challenged this finding on two primary grounds. First, the petitioners contended that the Administration failed to rigorously analyze the environmental benefits of fuel economy regulations as required by law before issuing the new rules. Second, the petitioners argued that NHTSA failed to analyze the impact of gasoline consumption on climate change in the rulemaking process.

The petition was filed in the Ninth Circuit Court of Appeals in May 2006 and the court issued its opinion on 15 November 2007. In a seminal decision, the court found for the petitioners, ruling that:

> the Final Rule is arbitrary and capricious, contrary to the EPCA in its failure to monetize the value of carbon emissions, failure to set a backstop, failure to close the SUV loophole, and failure to set fuel economy standards for all vehicles in the 8,500 to 10,000 gross vehicle weight rating (GVWR) class. We also hold that the Environmental Assessment was inadequate and that Petitioners have raised a substantial question as to whether the Final Rule may have a significant impact on the environment. Therefore, we remand to NHTSA to promulgate new standards as expeditiously as possible and to prepare a full Environmental Impact Statement.[58]

The Ninth Circuit's ruling dealt a sharp blow to the federal government's unwritten policy of ignoring climate change considerations in legislative and administrative decision making. The ruling compels the government both to reassess decision making processes generally, and to re-examine its specific failures to consider climate change in the regulation of automobile emissions. Following the release of the Ninth Circuit's decision in November 2007, NHTSA requested that the court revisit its ruling. On 18 August 2008, the court denied NHTSA's request for a rehearing, but vacated and withdrew its original ordering, replacing it with a revised ruling that affirmed the original decision in relevant part.

The outcomes in both *Center for Biological Diversity* and *Friends of the Earth v Mosbacher* offer positive indications that NEPA provides ample scope for

---

Failing to Increase Fuel Efficiency, Curb Global Warming Emissions' (Press Release, 2 May 2006) <http://ag.ca.gov/newsalerts/release.php?id=1299&year=2006&month=5> accessed 24 April 2007.

[56] 49 USC §§ 32901–32919 (2007).   [57] Pidot (n 4) 12.
[58] *Center for Biological Diversity v NHTSA*, No 06-71891, 14841 (9th Cir 2007).

mandating the inclusion of climate change considerations in federal decision-making.[59] Substantiating these decisions, on 18 February 2010, the White House Council on Environmental Quality (CEQ)—the Agency tasked with coordinating federal environmental efforts—released draft guidance on the consideration of greenhouse gases under NEPA.[60] In key part, the draft guidance confirmed that NEPA requires considerations of greenhouse gas emissions and climate change impacts in decision-making and proposed to advise Federal Agencies to 'consider opportunities to reduce GHG emissions caused by proposed Federal actions and adapt their actions to climate change impacts throughout the NEPA process'.[61] The draft guidance confirmed the findings in the *Center for Biological Diversity v National Highway Traffic Safety Administration* case regarding the applicability of NEPA to climate change considerations and laid the groundwork for making climate change considerations a fundamental component of decision-making across the federal government.

Other On-going Climate Change Litigation

The cases discussed here represent only a handful of the climate change litigation on-going in federal and state courts nationwide.[62] Other on-going cases challenge the jurisdiction of states to adopt greenhouse gases, the construction of private manufacturing facilities that would emit greenhouse and ozone depleting substances without obtaining the requisite CAA permits, and the applicability of common law causes of action to climate change.[63]

Turning to common law causes of action, in *Korinsky v US EPA et al*,[64] one New York plaintiff sued federal and state government entities, claiming that their activities contributed to global climate change and that they failed to act to curb the culprit emissions. The plaintiff claimed that as a result of these failures, the agencies were responsible for the mental health effects

---

[59] In another NEPA case, *Mayo Foundation v Surface Transportation Board*, the Sierra Club and Mid-States Coalition for Progress are challenging a decision by the Surface Transportation Board to approve a new rail line that will transport coal form mines to power plants. The case involves the degree to which the Agency must evaluate the impact of increased coal use on the environment and the outcome of the case will determine whether 'agencies may be able to satisfy NEPA in the global warming context by paying only lip service to their environmental review and disclosure obligations'. Pidot (n 4) 14.
[60] Nancy H. Sutley, Chair, Council on Environmental Quality, Memorandum for Heads of Federal Departments and Agencies, 'Draft NEPA Guidance on Consideration of the Effects of Climate Change and Greenhouse Gas Emissions', (18 February 2010) <http://www.whitehouse.gov/sites/default/files/microsites/ceq/20100218-nepa-consideration-effects-ghg-draft-guidance.pdf> accessed 4 March 2010. [61] Ibid at 1.
[62] See generally Michael B Gerrard and J Cullen Howe, 'Climate Change Litigation in the U.S.' (Chart) <http://www.climatecasechart.com/> accessed 29 December 2009.
[63] See *Central Valley Chrysler-Jeep v Witherspoon*, 456 F Supp 2d 1160 (ED Cal 2006); *Northwest Environmental Defense Center v Owens Corning Corp*, 434 F Supp 2d 957 (D Oregon 2006).
[64] *Korinsky v US Environmental Protection Agency*, Case No 05 Civ 859 (SDNY 2005) (Southern District of New York).

that he suffered due to his fear of global warming. Although the court dismissed Korinsky's complaint due to his inability to show adequate injury or redressability sufficient to establish standing, the *Korinsky* case is a prime example of the creative ways plaintiffs are trying to use litigation as a means of addressing climate change.

Plaintiffs are also turning to nuisance laws as a potentially more successful way to use common law to address injuries associated with climate change. Not only has the state of California lodged a nuisance suit against motor vehicle manufacturers, as previously discussed, US states and individual citizens have used nuisance law as a mechanism for challenging the actions of power, oil and coal companies.[65] Nuisance law has proved difficult to use as a vehicle for forcing change in climate change law and policy; nevertheless, several nuisance suits are still pending and provide plaintiffs with appealing bases for judicial challenges.

In the most high-profile tort-based case, *Comer v Murphy Oil USA Inc*,[66] the owners of property damaged by Hurricane Katrina sued a number of energy companies, alleging that the companies' greenhouse gas emissions contributed to the intensity of the storm by causing climate change. The plaintiffs based their claims on a number of tort theories, including: unjust enrichment, civil conspiracy and aiding and abetting, public and private nuisance, trespass, negligence, and fraudulent misrepresentation and concealment. The plaintiffs' claims rested on a causation chain that made success unlikely. In order to substantiate their claims, the plaintiffs would have had to prove, by a preponderance of the evidence, the degree to which climate change is attributable to anthropogenic greenhouse gas emissions; the degree to which the actions of defendant companies, individually or collectively, contributed to climate change; and the extent to which greenhouse gas emissions, as contributing factors to climate change, intensified or otherwise affected Hurricane Katrina.

Given the complexities involved in climate science, the plaintiffs' ability to establish these facts was unlikely from the start and the defendants quickly filed a series of motions asking the court to dismiss the case based on the court's lack of subject-matter jurisdiction (because it was a nonjusticiable political question), the plaintiffs' lack of standing, pre-emption, and failure to state a claim for relief as a matter of law. On 30 August 2007, the court granted the defendants' motion to dismiss, finding that the plaintiffs lacked standing due to their inability to trace the harm to individual defendants and that the claims alleged were nonjusticiable pursuant to the political question doctrine.[67] The holding of nonjusticiability turned on the Court's finding that the executive and legislative branches of the US government needed to speak to the question of climate change in order for the

---

[65] See *California v GM*, No C 06-05755 MJJ (ND Cal Filed 20 September 2006); *Comer v Murphy Oil USA* (n 23); *Connecticut v American Electric Power Company* (n 16).
[66] *Comer v Murphy Oil USA* (n 23).
[67] *Comer v Murphy Oil USA*, Civ Action No 1:05-CV-436-LG-RHW (SD Mississippi, 30 August 2007).

Court to determine a proper standard of culpability. The plaintiffs filed an appeal with the Fifth Circuit Court of Appeals on 28 September 2007.

Adding further resonance to the Second Circuit's decision in *Connecticut v AEP*, on 16 October 2009, the Fifth Circuit reached a similar outcome in *Comer v Murphy Oil*. On appeal, the Fifth Circuit overturned the district court's decision, finding that the plaintiff–appellant lacked standing to bring tort-based litigation against insurance, oil, coal, and chemical companies and that the case presented nonjusticiable questions. In finding that that the plaintiff–appellant raised justiciable questions and possessed legal standing, the Fifth Circuit confirmed that Comer had alleged an injury that was sufficiently traceable to defendants' alleged conduct. The Court's decision supports the validity of climate-based tort claims but it also foreshadows more difficult questions of causation and redressability that the plaintiffs will face in a trial on the merits.[68]

Nuisance claims and remedies remain unproven. Even if plaintiffs can win on the merits, the remedies they offer are necessarily limited in scope and effectiveness, as with other judicial remedies. Nuisance cases, CAA cases, NEPA cases, and preemption cases all provide potential routes for shaping climate change policy choices at the state and federal level. The routes, however, are expensive, slow, and often limited in impact. As with early 20th century efforts to use common law causes of action to address pervasive environmental pollution, much of the on-going US climate litigation only offers piecemeal solutions to addressing climate change. Thus, while litigation has become a favorite tool of civil society, cities, and states in their attempts to influence federal climate change policy, litigation alone cannot produce the comprehensive policy changes desired by many of the plaintiffs. Nevertheless, litigation is increasingly a favored and—as in the case of *Massachusetts v EPA*—effective choice in a growing arsenal of political tools.

Regulatory Actions for Climate Change

The Endangered Species Act

As a complement to the more antagonistic route of litigation, climate change campaigners also look to existing regulatory procedures as a mechanism for policy change.

One of the most high-profile examples of this is the 2005 petition to list the polar bear as a threatened species under the Endangered Species Act (ESA). On 16 February 2005, the Center for Biological Diversity, an NGO dedicated to protecting threatened and endangered species of flora and fauna and their habitat, filed a petition with the Secretary of the Interior and the US Fish and Wildlife Service (FWS) requesting that the Agency list the polar bear as a threatened

---

[68] *Comer v Murphy Oil* (n 67).

species, and thus accord it all of the concomitant protections accorded to threatened species under the ESA.[69] The basis of the Center for Biological Diversity's petition rested on the argument that the polar bear 'faces likely global extinction in the wild by the end of this century as a result of global warming. The species' sea-ice habitat is literally melting away.'[70]

Petitions to list species of flora and fauna as threatened or endangered under the ESA are common. What sets the polar bear petition apart is the link drawn between the animal's continued existence and the threats posed to the polar bear and its habitat—Arctic sea-ice—by climate change. The petition emphasizes, 'that global warming as a result of anthropogenic greenhouse gas emissions (primarily carbon dioxide, methane, and nitrous oxides) is occurring and accelerating is no longer subject to credible scientific dispute;' the best available scientific evidence establishes 'that Arctic sea ice is melting, and that absent significant reductions in human-generated greenhouse gases, such continued warming and consequent reduction of sea ice will occur that the polar bear will face severe endangerment and likely extinction in the wild by the end of the century.'[71]

The petition to list the polar bear followed from a decision by the IUCN Polar Bear Specialist Group to upgrade the polar bear from a species of 'Least Concern' to a 'Vulnerable' species, one of the three categories of 'threatened' species under the IUCN listing system. The Specialist Group made their decision to upgrade the threat to polar bears 'based on the likelihood of an overall decline in the size of the total population of more than 30% within the next 35 to 50 years. The principal cause of this decline is climatic warming and its consequent negative effects on the sea ice habitat of polar bears. In some areas, contaminants may have an additive negative influence.'[72]

The ESA requires the Secretary and the FWS to respond to a listing petition within 90 days with a determination of whether the petition 'presents substantial scientific or commercial information indicating that the petitioned action may be warranted'.[73] The Agency's decision must be based solely on the 'best

---

[69] The Natural Resources Defense Council (NRDC) and Greenpeace joined the petition on 5 July 2005. Kassie Siegel, Andrew Wetzler, and John Passacantanado, 'Re: Addition of Parties to Petition to List the Polar Bear (Ursus maritimus) as a Threatened Species under the Endangered Species Act' (Letter to Gale Norton and Rowan Gold, 5 July 2005) <http://www.biologicaldiversity.org/species/mammals/polar_bear/> accessed 28 April 2007.

[70] Center for Biological Diversity, 'Petition to List the Polar Bear (Ursus maritimus) as a Threatened Species Under the Endangered Species Act' (Executive Summary, 16 February 2005) ii.

[71] Ibid iii. The petition states that: 'Even short of complete disappearance of sea ice, projected impacts to polar bears from global warming will affect virtually every aspect of the species' existence, in most cases leading to reduced body condition and consequently reduced reproduction or survival'. Ibid v.

[72] IUCN, 'Proceedings of the 14th Meeting of the IUCN/SSC Polar Bear Specialist Group 20–24 June 2005, Seattle Washington USA' (2005) <http://pbsg.npolar.no/export/sites/pbsg/en/docs/PBSG14proc.pdf> accessed 1 June 2010.

[73] 16 USC § 1533(b)(3)(A). Following a positive '90-day' finding, the Secretary and FWS must within one year of receipt of the Petition complete a review of the status of the species and publish

available scientific evidence'.[74] Accordingly, on 9 February 2006, the FWS issued a finding 'that the petition presents substantial scientific or commercial information indicating that the petitioned action of listing the polar bear may be warranted'.[75] The FWS decision triggered a 60-day comment period and initiated a status review of polar bears to determine whether threats to the polar bear warranted a decision to list the species as threatened.[76] The status review culminated in a preliminary decision to list the polar bear as threatened.[77] The decision was published on 9 January 2007 and was open for public comments until 22 October 2007; the Agency was required to make a final listing decision by 9 January 2008.

Using the ESA to protect polar bears and, consequently, their habitat from the forces of global climate change creates new points of pressure and sources of liability for the federal government. Protecting the polar bear under the ESA requires the FWS to determine the links between addressing climate change and protecting the polar bear's habitat—Arctic sea-ice—which creates potential obligations for the government to regulate greenhouse gases. Further, the ESA requires federal agencies to carryout programs to conserve threatened species and prevents the federal government from taking any actions that would 'jeopardize' listed species.[78] In this way, a decision to list the polar bear as 'threatened' under the ESA creates new procedural and substantive requirements for the federal government, many of which involve questions of climate change mitigation or adaptation.

In this way, a simple decision to list a species as threatened under the ESA might force the federal government to make climate change a central part

---

either a proposed listing rule or a determination that such listing is not warranted. 16 USC § 1533(b)(3)(B). The Secretary and FWS then have an additional year to finalize the proposed rule. 16 USC § 1533(b)(6)(A).

[74] 16 USC § 1533(b)(1)(A).
[75] Petition to List the Polar Bear as Threatened, 71 Fed Reg 6745 (9 February 2006).
[76] Scott Schliebe and others , 'Range-Wide Status Review of the Polar Bear' (21 December 2006) US Fish & Wildlife Service.
[77] US Department of Interior, 'Interior Secretary Kempthorne Announces Proposal to List Polar Bears as Threatened Under Endangered Species Act' (Press Release, 27 December 2006) <http://www.fws.gov/home/feature/2006/12-27-06polarbearnews.pdf> accessed 1 June 2010.
[78] 16 USC § 1531(2)(b) (2006) (defining the purposes of the Act as 'to provide a means whereby the ecosystems upon which endangered species and threatened species depend may be conserved, to provide a program for the conservation of such endangered species and threatened species, and to take such steps as may be appropriate to achieve the purposes of the treaties and conventions set forth in subsection (a) of this section'); see also ibid § 1531(c)(1) (providing that 'It is further declared to be the policy of Congress that all Federal departments and agencies shall seek to conserve endangered species and threatened species and shall utilize their authorities in furtherance of the purposes of this Act'). Section 7 applies to all federal actions, both activities undertaken directly by federal agencies and nonfederal actions involving federal authorization or assistance. The relevant federal agent must demonstrate that the proposed action is 'not likely to jeopardize the continued existence of any endangered species or threatened species or result in the destruction or adverse modification' of such species' critical habitat. Ibid § 1536.

of governmental decision-making processes. Andrew Wetzler of the NRDC explored this phenomenon:

Let's say the federal government was going to issue permits for coal-fired power plants in the Midwest, which are major sources of carbon and global warming gases being emitted to the atmosphere. Because those power plants require federal permits and because those emissions are a direct cause of the polar bear's decline, that power plant permit is now subject to the endangered species act in a way that it was not before.[79]

Very few people believed that a decision to list the polar bear would create sweeping changes in federal climate change policy. The ESA was not designed as a tool to circuitously implement comprehensive policy changes. The authority of the ESA, however, should not be taken lightly.

The underlying goal of the ESA is to protect species of flora and fauna and the habitat upon which they depend. This basic goal, however, is grounded in deep moral concerns that we have an obligation to protect and conserve biodiversity. The intersection between moral and environmental concerns creates the foundations for a strong and substantive environmental law.

In one of its earliest outings, the ESA proved its mettle. In the renowned case of *TVA v Hill*,[80] the Supreme Court affirmed far-reaching protections for endangered species under the ESA. When deciding the case, the Court considered whether the completion of the Tellico Dam would violate the ESA by 'jeopardizing' the continued existence of the listed snail darter. The 1978 case revolved around the question of whether the presence of a single species of fish could prevent the completion of a dam that had been under construction since 1967 and which, by the time of the listing of the snail darter as an endangered species in 1975, was 70% completed and had received over US$50 million in federal funding.[81] In upholding the decision by the Court of Appeals to halt the project, Chief Justice Warren Burger said:

One would be hard pressed to find a statutory provision whose terms were any plainer than those in Section 7 of the Endangered Species Act. Its very words affirmatively command all federal agencies 'to insure that actions authorized, funded, or carried out by them do not jeopardize the continued existence' of an endangered species or 'result in the destruction or modification of habitat of such species'. This *language admits of no exceptions.*[82]

Justice Burger continued by stating that:

It may seem curious to some that the survival of a relatively small number of three-inch fish among all the countless millions of [existing] species would require the permanent

---

[79] NPR, Elizabeth Shogren, 'Warming May Put Polar Bear on Threatened List', 27 December 2006, <http://www.npr.org/templates/story/story.php?storyId=90447621> accessed 30 December 2009.   [80] 437 US 153 (1978).
[81] See ibid.   [82] Ibid 173 (emphasis added).

halting of a virtually completed dam for which Congress has expended more than $100 million. The paradox is not minimized by the fact that Congress continued to appropriate large sums of public money for the project, even after... [it knew about the dam's]...impact upon the survival of the snail darter. We conclude, however, that the explicit provisions of the Endangered Species Act require precisely that result.[83]

With the decision in *TVA v Hill*, Justice Burger signaled to public and private entities alike that the Court would take seriously the substance and intent of the ESA. In the years since *TVA v Hill*, the ESA has experienced judicial 'victories' and 'losses' but has maintained its status as an environmental law with teeth. In light of the historic judicial deference accorded to the ESA, a decision to list the polar bear as threatened offered unknown but potentially significant possibilities for provoking widespread federal response to climate change.

Responding to growing expectations over the role of the ESA in addressing climate change, former Interior Secretary Dirk Kempthorne sought to distance the pending polar bear listing decision from federal efforts to regulate greenhouse gases, repeatedly stating that reducing greenhouse gas emissions is beyond the scope of the ESA.[84] Despite repeated reservations and persistent delay, following three years of legal wrangling, on 14 May 2008, one day prior to a court ordered deadline, Secretary Kempthorne announced the Department of Interior's (DOI) decision to list the polar bear as a threatened species pursuant to the ESA.[85] As discussed, pursuant to an affirmative listing decision, the Agency would normally be tasked to ensure the viability of the species by protecting its critical habitat, Arctic sea-ice. In issuing his decision, however, Secretary Kempthorne noted that:

While the legal standards under the ESA compel me to list the polar bear as threatened, I want to make clear that this listing will not stop global climate change or prevent any sea ice from melting. Any real solution requires action by all major economies for it to be effective. That is why I am taking administrative and regulatory action to make certain the ESA isn't abused to make global warming policies.[86]

In his statements, the Secretary acknowledged that anthropogenic greenhouse gas emissions contribute to the degradation of the polar bear's critical habitat, but stressed that the polar bear's threatened status could not be used to convert the ESA into a tool for regulating greenhouse gas emissions. Following up on these qualifications, on the same day that the DOI published its Final Rule to list the polar bear, it also published an Interim Special Rule for the conservation of the polar bear. The Special Rule was finalized on 11 December 2008.[87] In addition to clarifying the relationship between the ESA and other laws

[83] Ibid 172–73.   [84] Shogren (n 79).
[85] US Department of the Interior, 'Secretary Kempthorne Announces Decision to Protect Polar Bears under Endangered Species Act' (News Release, 14 May 2008).   [86] Ibid.
[87] US Fish & Wildlife Service, 'New Rule Unifies Domestic and International Conservation Laws to Manage Polar Bear' (News Release, 10 December 2008) <http://www.fws.gov/news/

applicable to the polar bear, the Special Rule limits the government's ability to invoke the ESA to restrict emissions of greenhouse gases. Noting that activities and federal actions outside of Alaska cannot currently be shown to have a causal connection on Arctic sea ice, the Special Rule determines that federal agencies are not required to analyze any such activities and actions under the ESA for potential effects on the polar bear. This provision, the Agency said, 'ensures that the ESA is not used inappropriately to regulate greenhouse gas emissions'.[88] In the words of Professor Lisa Heinzerling, climate advisor to EPA Administrator Jackson, this decision means that 'the listing will not require anybody but trophy hunters to change their behavior to protect the bear'.[89] The Special Rule sparked controversy with some commentators contending that the rule unduly limits the scope of the ESA while others commended the Agency for preventing the misuse of existing legislation to address climate change.

With the release of the Special Rule occurring during the final days of President George W. Bush's second term, there was widespread speculation over whether the Obama administration would retain the limiting rule. On 8 May 2009, the newly appointed Interior Secretary, Ken Salazar, announced that he would leave the Special Rule in place. Secretary Salazar conceded that climate change indeed poses the greatest threat to the polar bear's habitat, but insisted that the ESA is not the proper tool for addressing climate change. Instead, he suggested that the US government needed to proceed with efforts to enact a comprehensive climate change and energy regime.

Salazar's decision once again highlighted pervasive tensions over the proper pathway towards regulating greenhouse gas emissions and the scope of existing environmental laws to offer regulatory tools in the absence of comprehensive legislative action. In contrast to actions under NEPA and the CAA, however, the DOI succeeded in limiting the applicability of ESA to greenhouse gas emissions for the near term. These limitations remain open to reassessment as multiple pending ESA listing decisions hinge on the relationship between climate change and species survival.

The Clean Water Act

In a similar vein to the petition to list the polar bear as an ESA threatened species, on 27 February 2007, the Center for Biological Diversity petitioned the State of California to add its ocean waters to lists of impaired waters under

NewsReleases/showNews.cfm?newsId=27A58FDE-922A-2B50-ED394D030EE543BD> accessed 29 December 2009.
    [88] Ibid.
    [89] Georgetown Law Faculty Blog, Lisa Heinzerling, 'Climate Contrast: Of Polar Bears and Power Plants', 15 May 2009, <http://gulcfac.typepad.com/georgetown_university_law/2008/05/index.html> accessed 30 December 2009.

the federal Clean Water Act (CWA).[90] Specifically, the Center for Biological Diversity asked that the North Coast Water Quality Control Board include '[a]ll ocean waters under Region 1's jurisdiction... in the state List of Impaired Waters ('303(d) List') under Section 303(d) of the Clean Water Act as impaired for pH due to absorption of anthropogenic carbon dioxide pollution'.[91] The Center for Biological Diversity has also filed similar petitions with all of the California Regional Water Quality Control Boards that have jurisdiction over California's ocean waters.

The Center for Biological Diversity maintains that California's ocean waters are impaired for pH due to ocean acidification, which is a result of 'past, ongoing, and projected absorption of anthropogenic carbon dioxide pollution'.[92]

Ocean acidification results when ocean waters absorb carbon dioxide emissions, resulting in increased levels of acidity that deprive the oceans of key compounds necessary for marine species to build shells and skeletons.[93] As a result, '[m]any sea organisms from phytoplankton to snails and crabs are being harmed as acidic waters dissolve protective structures or inhibit growth', while other marine species 'experience impaired metabolism'.[94] In these ways, ocean acidification due to carbon dioxide emissions negatively affects the functioning of the marine ecosystem.

For these reasons the Center for Biological Diversity suggests that:

> The goals of the Clean Water Act and California's Ocean Plan can only be met by taking steps to slow ocean acidification. The changing pH of the ocean and associated impacts on marine resources are unlike any that have been experienced on this earth for millions of years. California must take actions now to abate carbon dioxide pollution by listing California's ocean segments as impaired on the 303(d) List and establishing a TMDL [total maximum daily load] for carbon dioxide.[95]

Through its petitions to California's Regional Water Quality Control Boards, the Center for Biological Diversity effectively suggests that the State is required to regulate carbon dioxide emissions under existing federal regulatory mechanisms, ie, the CWA.

The State of California has not yet responded to the Center for Biological Diversity petition. However, on 14 April 2009, the EPA published a notice in

---

[90] Center for Biological Diversity, 'Re: Request to Add California Ocean Waters to List of Impaired Waters due to Carbon Dioxide Pollution Resulting in Ocean Acidification; Response to "Notice of Public Solicitation of Water Quality Data and Information for 2008 Integrated Report—List of Impaired Waters and Surface Water Quality Assessment [303(d)/305(b)]"' (Letter to Bruce Gwynne of North Coast Water Quality Control Board, 27 February 2007) (Letter to Bruce Gwynne).
[91] Ibid 1.   [92] Ibid.
[93] Ibid 3; see also Peter M Haugan, Carol Turley, and Hans-Otto Poertner, *Effects on the Marine Environment of Ocean Acidification Resulting from Elevated Levels of CO2 in the Atmosphere* (Report prepared for the OSPAR Commission for the Protection of the Marine Environment of the North-East Atlantic, 2006) <http://en.scientificcommons.org/22142634> accessed 2 April 2010.
[94] Letter to Bruce Gwynne (n 90) 3.   [95] Ibid.

Litigation, Regulation, and International Law as Law and Policy Drivers 121

the US Federal Register announcing that it would begin taking steps to review the threats ocean acidification poses to US waters and the regulatory remedies available to address such threats. The EPA's notice of data availability, 'Ocean Acidification and Marine pH Water Quality Criteria', solicits data on ocean acidification that the agency will use to evaluate water-quality criteria under the CWA.[96] EPA issued the notice in response to a formal petition and pending litigation by the Center for Biological Diversity that sought to compel the agency to impose stricter pH criteria for ocean water quality, and to publish guidance to help states prevent ocean acidification.[97]

With the notice of data availability pending, the EPA took another significant step towards increasing the Agency's regulatory capacity to address domestic greenhouse gas emissions. On 11 March 2010, EPA settled a related lawsuit the Center for Biological Diversity had filed to force EPA to require the state of Washington to list its marine waters as impaired due to increasing acidity.[98] In settling the law suit, EPA agreed to use the CWA as a tool for addressing ocean acidification; as a first step EPA agreed to take public comment on ocean acidity, ways states can determine if coastal waters are affected, and how states might regulate 'total maximum daily loads' of pollutants and to develop guidance for states on these matters. EPA's decision is significant because it lays the groundwork for the EPA to use the CWA to protect acidifying marine waters nationwide, which could ultimately require new restrictions on greenhouse gas emissions that are linked to ocean acidification. The settlement itself does not obligate EPA or the states to begin listing waters as impaired due to acidification or to begin regulating greenhouse gas emissions, but it is nonetheless significant. The CWA has not been previously employed to address ocean acidification or other climate-related threats to water quality. A decision by the EPA to use the CWA as a vehicle for controlling greenhouse gas would further expand EPA's increasingly expansive regulatory control over greenhouse gas emissions under the CAA.

The Supreme Court's decision in *Massachusetts v EPA*, EPA's endangerment finding under the CAA, and the DOI's decision to list the polar bear as a threatened species under the ESA, coupled with EPA's initial steps towards addressing ocean acidification via the CWA, reflect both an increasing interest in using existing legal mechanisms to force the federal government to regulate greenhouse

---

[96] Ocean Acidification and Marine pH Water Quality Criteria, 74 Fed Reg 17484 (15 April 2009).

[97] Center for Biological Diversity, 'Petition for Revised ph Water Quality Criteria Under Section 304 of the Clean Water Act 33 U.S.C. § 1314, to Address Ocean Acidification' (18 December 2007) <http://www.biologicaldiversity.org/campaigns/ocean_acidification/pdfs/2009-08638_PI.pdf> accessed 1 February 2008. The Center for Biological Diversity was also challenging EPA's failure to require Washington state to list its marine waters as impaired by rising acidity.

[98] Press Release, Center for Biological Diversity, 'Legal Settlement Will Require EPA to Evaluate How to Regulate Ocean Acidification Under Clean Water Act' (11 March 2010) <http://www.biologicaldiversity.org/news/press_releases/2010/ocean-acidification-03-11-2010.html> accessed 12 March 2010.

gas emissions and the scope of existing environmental laws to accommodate federal policies to regulate greenhouse gases and otherwise address climate change.

As Congressional proposals, state programs, litigation, and regulatory petitions to address climate change proliferate, the federal government must decide how to respond to individual challenges and whether and how it will modify and adapt the US comprehensive climate change policy regime to respond to growing public and private concerns.

Concerns over climate change policy choices are not, of course, one-dimensional.

The Clean Air Act

In addition to the high-profile clean air challenges raised by *Massachusetts v EPA* and California's CAA waiver petition, on 13 November 2008, the EPA Environmental Appeals Board (EAB) issued a landmark decision in a case questioning the need for the EPA to regulate carbon dioxide emissions from a coal-fired power plant in Utah. The case, *In re Deseret Power Electric Cooperative*,[99] involved a challenge under the CAA Prevention of Significant Deterioration (PSD) rules. The PSD rules dictate that new and modified projects undertaken in areas with acceptable air quality must install best available control technologies (BACT) in order to obtain a PSD permit if the project would lead to an increase in the emission of 'pollutants'. In the case at hand, the Sierra Club challenged EPA's grant of a PSD permit to Deseret Power Electric Cooperative to build a new coal-fired unit at an existing power plant. Sierra Club alleged that EPA erred in granting the permit by failing to require BACT emission limits for carbon dioxide emissions and for adopting positions on this matter that conflicted with agency decisions in other on-going coal-fired power plant proceedings.[100] In response, EPA contended that it was not required to regulate carbon dioxide because, while the gas constituted a 'pollutant' under the CAA, it was not currently subject to regulatory standards.

In its November 2008 decision, the EAB ruled that EPA erroneously exempted the new coal-fired unit from carbon dioxide emissions and remanded the permit decision to EPA Region 8 to reopen the record and reconsider its refusal to impose limits on carbon dioxide emissions under the PSD permit. The EAB rejected EPA's argument that it was not required to regulate carbon dioxide. On remand, EPA must reassess its PSD permitting decision. This does not force EPA to require regulated entities to install BACT for carbon dioxide emissions, rather it mandates that EPA reassess its agency-wide policy regarding the regulation of carbon dioxide emission in PSD areas.

---

[99] PSD Appeal No. 07-03 (USEPA Environmental Appeals Board, 13 November 2008) <http://yosemite.epa.gov/OA/EAB_WEB_Docket.nsf/ce9f7f898b59eae28525707a00631c97/c38150f6c4bba3608525736900687af4!OpenDocument> accessed 29 December 2009.

[100] Petition for Review and Request for Oral Argument, Re Deseret Power Electric Cooperative, PSD Permit Number OU-0002-04 (1 October 2007).

In the ruling, the EAB acknowledges the far-reaching implications of the decision for national climate policy and cautions the Agency against using piecemeal regulatory decisions to cobble together a climate change regulatory scheme, advising that the interested parties 'would be better served by the Agency addressing the interpretation [of CO2 regulation] in the context of an action of nationwide scope, rather than through this specific permitting proceeding'.[101]

The EAB's decision in *In re Deseret Power Electric Cooperative* suggested that, under the CAA, the EPA might have the regulatory capacity to limit carbon dioxide emissions from thousands of sources.

In the wake of the *In re Deseret Power Electric Cooperative* decision and the release of EPA's proposed endangerment finding, as previously discussed, on 30 September 2009, the EPA released a proposed rule establishing new thresholds for greenhouse gas emissions under the CAA's New Source Review and Title V permit program that would require facilities emitting over 25,000 tons of greenhouse gases a year to obtain permits demonstrating that they are using the best practices and technologies to minimize GHG emissions.[102] The EPA suggests that the proposed regulatory regime would target nearly 70% of the nation's largest stationary source greenhouse gas emitters, including power plants, refineries, and cement production facilities, while protecting smaller facilities from excessive regulatory burdens. The proposed rule is still pending.

EPA's proposal to extend the reach of the CAA's new source review permitting program to encompass greenhouse gas emissions is the latest in a long stream of decisions that reveal the scope of the CAA and other existing federal environmental laws to control greenhouse gas emissions. Simultaneously, the proposed rule highlights the enormous economic costs and administrative complexity involved in using these laws to control greenhouse gas emissions through piecemeal regulatory regimes in the absence of a clear and coherent national climate policy.

Reverse Momentum

An overview of state and civil society efforts to influence climate change policy suggests that a majority of the current initiatives focus on strengthening climate change policies. This emphasis is not universal. In addition to the obvious and vocal efforts on the part of certain energy, automobile, and business representatives that challenge both the certainty of climate change and the legality of the toughening up of greenhouse gas regulations, certain members of Congress have lodged proposals to modify environmental laws in ways that would limit their scope to regulate greenhouse gas emissions.

---

[101] Order Denying Review in Part and Remanding in Part, Re Deseret Power Electric Cooperative, PSD Appeal No 07-03 (13 November 2008) 4–5.
[102] See Prevention of Significant Deterioration and Title V Greenhouse Gas Tailoring Rule, 40 CFR pts 51, 52, 70, and 71 (2009).

A proposal to amend the Clean Air Act best demonstrates this phenomenon. On 14 February 2002, President Bush announced The Clear Skies Act of 2003, formally entitled: 'A bill to amend the Clean Air Act to reduce air pollution through expansion of cap and trade programs, to provide an alternative regulatory classification for units subject to the cap and trade program, and for other purposes'.[103] The proposed bill originated in the White House and was then introduced in the House and the Senate first in July 2002, and then again in February 2003. After the bill failed to move, Senators Inhofe and Voinovich introduced a new version of the bill on 24 January 2005.[104]

The Clear Skies Act proposed potentially positive changes, such as capping mercury emissions from coal-fired power plants starting in 2010, phasing in tighter caps on sulfur dioxide and nitrogen oxides for all fossil fuel-driven power plants by 2010, setting new caps for key pollutants by 2018, and initiating a new market-based cap-and-trade program for nitrogen oxide and mercury. Despite initial positive appearances, the bill posed numerous challenges and was filled with loopholes that would have undermined existing provisions in the CAA.[105]

At the top of the list of problems associated with the Clear Skies Act was the postponement of the deadline to clean up air in highly populated regions from 2010 to 2018. Even the 2018 deadline was misleading because it was based on credits that would not have been 'cashed in' until seven years later. Effectively, this meant that the EPA would not be able to achieve the bill's 70% reduction goal before 2025—and probably not even by that late date. In a more blatant backtrack, the bill would have potentially negated findings made during the Clinton administration that designated mercury as a toxic pollutant. Instead, the bill would have permitted up to seven times more mercury pollution before triggering regulatory response and would have delayed the implementation of any substantive measures to create a comprehensive mercury regulatory regime. The bill would also have repealed existing new source review standards under the CAA, delayed deadlines for meeting public health standards, and weakened the rights and ability of states to sue their neighboring states for transboundary pollution.

Basic air pollution issues aside, the proposed bill failed to provide any measures or means for regulating carbon dioxide emissions from any source. The proposal did not include a single reference or measure to reduce, limit, or even slow the growth of carbon dioxide emissions.

---

[103] Clear Skies Act, S 485, 108th Cong (2003).
[104] EPA, 'Clear Skies: Legislative Information' (EPA's analysis of the Clear Skies Act 2003) <http://www.epa.gov/air/clearskies/legis.html> accessed 1 June 2010.
[105] See David W Rugh, 'Clearer, But Still Toxic Skies: A Comparison of the Clear Skies Act, Congressional Bills and the Proposed Rule to Control Mercury Emissions from Coal-Fired Power Plants' (2003) 28 *Vermont L Rev* 201.

Despite executive support, the Clear Skies Act again floundered after its reintroduction to Congress in 2005. Competing proposals to refine air pollution control and to regulate carbon dioxide emissions gained momentum in Congress, ultimately overtaking the Clear Skies Act. But the bill is important, nevertheless, because it exemplified continuing efforts, not merely to avoid implementing legislation to regulate carbon dioxide, but to actually propose new legislation that would weaken existing environmental laws.

Efforts to block legislative and administrative measures to control climate change continue. In the summer of 2009, for example, a scandal erupted in Congress over the discovery that lobbyists working on behalf of the American Coalition for Clean Coal Electricity sent forged letters to House Democratic representatives Tom Perriello of Virginia and Kathy Dahlkemper and Chris Carney of Pennsylvania under the guise of environmental organizations, urging them to vote against the Waxman–Markey climate bill. Similarly, in August 2009, Utah Governor Gary Herbert stated that he intended to host the first 'legitimate' debate about anthropogenic climate change, while the US Chamber of Commerce, the country's largest business group, threatened to sue EPA if it did not agree to hold a public debate enabling a 'credible weighing' of the scientific evidence that global warming endangers human health. Additionally, the National Association of Manufacturers and the National Federation of Independent Businesses—two of the nation's most prominent business associations—began a massive ad campaign opposing climate change legislation in states that are home to key swing-vote senators.

Internal wrangling over the realities of human induced climate change and the vagaries of political response to climate change thus continued to characterize US domestic climate change politics leading up to and following the 2009 global climate change negotiations in Copenhagen.

International Mechanisms for Domestic Climate Change Action

The Inter-American Commission for Human Rights

Measures to influence the course of US federal climate change policy extend beyond internal players and institutions. Citizens of the US, as well as those of foreign countries, look to international institutions, such as the Inter-American Commission on Human Rights (IACHR) and the World Heritage Convention, as mechanisms for holding the US government accountable for alleged violations of international responsibilities and, potentially, for liabilities resulting from a failure to implement effective climate change policies.

In the first instance, on 7 December 2005, the Center for International Environmental Law (CIEL) filed petitions with the IACHR against the US on behalf of 63 Inuit petitioners, representing both American and Canadian citizens.[106] The petitions concerned the 'impact of global warming on the Inuit and other vulnerable communities in the Americas, and the implication of these impacts for human rights'.[107]

The Inuit petitions were based on the US's alleged contribution to and its failure to address global warming.[108] Specifically, the petitions emphasized that the US, with only 5% of the world's population, is responsible for 25% of the world's greenhouse gas emissions, and that the US government is not only refusing to participate in the international climate change regime, but is 'actively impeding the ability of the global community to take collective action'.[109]

The IACHR rejected CIEL's petition on 16 November 2006, without prejudice. Although the petition was dismissed, in February 2007, the IACHR invited the petitioners to appear before the Commission to provide testimony on the links between climate change and human rights. On 5 March 2007, Sheila Watt-Cloutier, an Inuit petitioner and former Chair of the Inuit Circumpolar Conference (as well as a Nobel Prize nominee), CIEL Senior Attorney Donald Goldberg, and Earthjustice Managing Attorney Martin Wagner gave testimony before the Commission.[110]

In her testimony, Ms Watt-Cloutier began by stating that '[c]limate change brings into question the basic survival of indigenous people and indigenous cultures throughout the Americas'.[111] She continued by emphasizing that '[d]eteriorating ice conditions imperil Inuit in many ways' and that the impacts of climate change are 'destroying [the Intuit's] rights to life, health, property and means of subsistence'.[112] Ms Cloutier emphasized both the physical and non-physical impacts of climate change for Inuit communities, and repeated the statement of an Inuit hunter in Alaska who said that '[t]here's a lot of anxieties and angers that are being felt by some of the hunters that no longer can go and hunt. We see the change, but we can't stop it, we can't explain why it's changing it... our way of life

---

[106] See Center for International Environmental Law, 'Inuit File Petition with Inter-American Commission on Human Rights, Claiming Global Warming Caused by United States Is Destroying Their Culture and Livelihoods' (7 December 2005) <http://www.ciel.org/Climate/ICC_Petition_7Dec05.html> accessed 29 December 2009.

[107] CIEL, 'Global Warming and Human Rights Gets Hearing on the World Stage' (5 March 2007) <http://www.ciel.org/Climate/IACHR_Inuit_5Mar07.html> accessed 29 December 2009 (Global Warming on the World Stage); see Sara C Aminzadeh, Note, 'A Moral Imperative: The Human Rights Implications of Climate Change' (2007) 30 *Hastings Intl & Comp L Rev* 231 (discussing the relationship between global climate change and international human rights).

[108] See Global Warming on the World Stage (n 107).

[109] Yuill Herbert, 'President Bush, See You in Court Judging the Cost of Climate Change' (25 August 2004) The Dominion <http://www.tuvaluislands.com/news/archived/2004/2004-08-25.htm> accessed 1 June 2010.   [110] Global Warming on the World Stage (n 107).

[111] Sheila Watt–Cloutier, 'Nobel Prize Nominee Testifies About Global Warming: Inuit leader Sheila Watt–Cloutier's Testimony Before the Inter-American Commission on Human Rights Put Spotlight on Climate Change and Indigenous Peoples' (Testimony before Inter-American Commission on Human Rights, 1 March 2007) <http://www.ciel.org/Publications/IACHR_WC_Mar07.pdf> accessed 1 June 2010.   [112] Ibid 2.

is changing up here, our ocean is changing.'[113] She summed up her testimony by referring to the threat that climate change poses to the individual and collective rights of 'peoples to their culture' and by encouraging the Commission to:

> protect the sentinels of climate change—the indigenous people. By protecting the rights of those living sustainably in the Amazon Basin or the rights of the Inuit hunter on the snow and ice, this commission will also be preserving the world's environmental early-warning system.[114]

Following Sheila Watt-Cloutier, Donald Goldberg and Martin Wagner provided supplemental testimony on the links between climate change and human rights from an institutional and legal perspective. Goldberg emphasized the existence of unique links between indigenous and deprived communities and their natural environment, and the past record of the Inter-American and International human rights system in responding to and protecting these distinctive land–resource–culture links.[115] These past threats, he suggested, pale in comparison to the threats that climate change represents to indigenous communities. Pointing to specific habitats that stand to be destroyed by climate change, Goldberg drew highly specific links between the impacts of global warming and the physical and cultural well-being of indigenous communities.[116] On this basis, he urged the Commission to issue a 'clear statement that global warming has implications for human rights'[117] as a way of motivating states to address their international responsibilities. In so doing, he implored the Commission 'to be the institutional conscience of the Americas: to assist, prod, and compel countries to meet their obligations under human rights law to protect our communities from the ravages of global warming'.[118]

Mr Wagner focused his testimony on explaining the precise relationship between global warming and three specific human rights: the rights of indigenous peoples, the right to enjoy and use property without undue interference, and the rights to life, physical integrity, and security.[119] He further discussed the right of people to enjoy the benefits of culture and the relevance of customary

---

[113] Ibid 4. [114] Ibid 5.
[115] Donald Goldberg, 'Global Warming & Human Rights Gets Hearing on the World Stage: CIEL Senior Attorney Donald Goldberg, Inuit leader and Nobel Prize nominee, Sheila Watt-Cloutier, and Earthjustice Managing Attorney Martin Wagner Provide Testimony before the Inter-American Commission on Human Rights' (Testimony before Inter-American Commission on Human Rights, 1 March 2007) <http://www.ciel.org/Publications/IACHR_Goldberg_Mar07.pdf> accessed 1 June 2010.
[116] Ibid 2. Goldberg suggested that 'The forests protected in the Awas Tingni and Belize Maya cases could be destroyed altogether as a result of the increased risk of fires, pests, and other adverse conditions caused by global warming'. Ibid.   [117] Ibid 2.
[118] Ibid 2–3.
[119] Martin Wagner, 'Testimony of Earthjustice Managing Attorney Martin Wagner before the Inter-American Commission on Human Rights' (Testimony before Inter-American Commission on Human Rights, 1 March 2007) <http://www.ciel.org/Publications/IACHR_Wagner_Mar07.pdf> accessed 1 June 2010.

international law to the topic of climate change and human rights.[120] In his testimony, Wagner stressed that customary international law obligates every State 'not to allow knowingly its territory to be used for acts contrary to the rights of other States'.[121] Based on this principle, Wagner reasons that international law provides a mechanism for accountability in the case at hand, stating that '[b]ecause the emission of greenhouse gases in one State causes harm in other States, this norm provides context for assessing States' human rights obligations with respect to global warming'.[122] Wagner draws the links between international human rights law and customary international law to implore the IACHR to play a central role in creating jurisprudence on the links between human rights and global climate change.

Together, the three witnesses used their testimony to create a full picture of the physical, cultural, and legal links between climate change and human rights in the hopes of creating tangible and enforceable links between international human rights law, international environmental law, and global climate change. In the wake of this requested testimony, the IACHR has the opportunity to be one of the first international institutions to address the question of how far existing norms of international law can go towards addressing the far-reaching physical, social, cultural, and economic problems posed by global climate change.

While the IACHR is one of the first international institutions to confront the links between climate change and international law, it will not be the last. The Inuit petitions signal a trend whereby sovereign states and members of civil society seek redress for the harms posed by climate change through international, rather than, or in addition to, state mechanisms. This trend is evidenced by statements made by the government of the island nation of Tuvalu that it plans to lodge similar complaints against either the US and/or Australia with the International Court of Justice (ICJ).

Tuvalu is a low-lying island state and is, thus, especially vulnerable to climate change. The government of Tuvalu alleges that rising sea levels brought about by climate change threaten the continued survival of the entire island and its inhabitants. Tuvalu first raised the issue of taking a claim to the ICJ in 2002.[123] Since that time, the country has neither filed a case in the ICJ nor pushed forward with its claims in other international venues. Even if Tuvalu subsequently brings its complaints before the ICJ, the island state will have a difficult time making its case. Neither the US nor Australia is likely to agree to the jurisdiction of the ICJ. Because ICJ jurisdiction in contentious state–state cases is based on consent,

---

[120] Ibid 3–4.
[121] Ibid 3, quoting the Trail Smelter Arbitration, 3 RIAA 1905 (1938) (requiring the Canadian company operating the Trail smelter to cease causing further damage to the State of Washington).
[122] Ibid.
[123] See Tom Price, 'The Canary is Drowning: Tiny Tuvalu' (Global Policy Forum, 3 December 2002) <http://www.globalpolicy.org/nations/micro/2002/1203canary.htm> accessed 29 December 2009.

absent the US and Australia consenting to the court's jurisdiction the ICJ would be limited to submitting an advisory opinion on the matter, if such an opinion was requested by a UN agency or body. Advisory opinions do not possess any binding effect but they are highly regarded as statements of policy by the international community.

The Inuit petitions to the IACHR and Tuvalu's threatened complaint have attracted considerable public attention. However, the Inuit petitions were dismissed and the Tuvalu case has yet to materialize. Nevertheless, the IACHR request for testimony from CIEL and Inuit leader Sheila Watt-Cloutier, and the simple decision by affected groups to challenge the US's climate change strategy in international tribunals, suggests that, both within the US and extra-jurisdictionally, nations, citizens, and international institutions are increasingly willing and able to contest the US's current legal and political stance on climate change.

The World Heritage Convention

Another international venue confronting the connections between climate change and international law is the UNESCO Convention Concerning the Protection of the World Cultural and Natural Heritage (hereinafter 'World Heritage Convention').[124] Adopted in 1972, the World Heritage Convention came into force in 1975 and boasts extensive state participation, currently having 186 States Parties.[125]

The underlying goal of the World Heritage Convention is to recognize that certain sites of 'cultural or natural heritage are of outstanding interest and therefore need to be preserved as part of the world heritage of mankind as a whole'.[126] The World Heritage Convention reflects increasing acceptance of the concept of 'cultural internationalism', which 'views cultural property as belonging to the world's peoples and not limited to the citizens of the state where the property is located'.[127]

The central feature of the World Heritage Convention is the creation of a list of critically important sites of cultural and natural heritage, known as the World Heritage List. The List creates a 'means of recognizing that some sites,

---

[124] As early as 21 September 2004, the Sydney Centre for International and Global Law at the Faculty of Law released a report finding that the World Heritage Convention creates legal obligations for States Parties to cut greenhouse gas emissions because of their legal duty as members of the Convention and to protect important listed heritage sites from the threats posed by climate change. Sydney Centre for International and Global Law, 'Global Climate Change and the Great Barrier Reef: Australia's Obligations under the World Heritage Convention' (21 September 2004) 20 para 47.

[125] There are currently 186 States Parties to the World Heritage Convention.

[126] Convention Concerning the Protection of the World Cultural and Natural Heritage (adopted 16 November 1972, entered into force 1975) 1037 UNTS 151 (World Heritage Convention).

[127] See Mehmet Komurcu, 'Cultural Heritage Endangered by Large Dams and Its Protection under International Law' (2002) 20 *Wisconsin Intl LJ* 233, 284.

both cultural and natural, are important enough to be recognized by and be the responsibility of the International Community, and as sites to be the target of preservation/conservation efforts'.[128] The World Heritage Convention focuses on identifying and creating mechanisms for protecting cultural, natural, and/or mixed cultural and natural heritage sites worldwide. In total, the World Heritage List includes 890 sites: 689 Cultural Sites, 176 Natural Sites, and 25 Mixed Sites, located in 148 states.[129]

Participation in the World Heritage Convention commits States Parties first to identify/nominate potential sites, and second, to care for any World Heritage Sites designated within their sovereign territories.[130] As members of the World Heritage Convention, States Parties also commit to 'adopt policies, set up services, undertake scientific and technical research; take appropriate legal, scientific, technical, administrative and financial measures necessary to identify, protect, conserve, present and rehabilitate heritage sites; and foster establishment of regional training centers'.[131] The designation of World Heritage Sites thus carries prestige and economic potential—eg, cultural tourism—as well as on-going caretaking responsibilities for nominated States Parties.[132]

The World Heritage Convention is singular in its creation of an international listing mechanism for protecting areas of both cultural and natural heritage, and for creating a regime that allows for the sharing of power and responsibility between national governments and the international community. In addition to creating a formal international protection regime, the World Heritage Convention has done much to raise civil society's awareness of the importance of protecting cultural and natural heritage. To this end, World Heritage Sites have become 'high profile' and when listed sites are threatened, civil society responds. Civil society has recently focused attention on the threats climate change poses to World Heritage Sites worldwide. Specifically, environmental groups have petitioned for several World Heritage Sites to be designated as 'in danger' due to the threats of global climate change.

The 'in danger' characterization is central to the World Heritage listing process. The Convention created this category for identifying listed sites where conditions threaten 'the very characteristics for which a property was inscribed on the World Heritage List, and to encourage corrective action'.[133] The 'List of

[128] See UNESCO, 'World Heritage List' (List of World Heritage Properties) <http://whc.unesco.org/pg.cfm?CID=31&l=EN> accessed 20 May 2010.
[129] World Heritage Convention (n 126) arts 1 and 2.
[130] Ibid arts 3 and 4 (obligating each State party to 'identify and delineate the different properties situated on its territory', and creating the 'duty of ensuring the identification, protection, conservation, presentation and transmission to future generations of the cultural and natural heritage...').   [131] Ibid art 5.
[132] Designation of a site as a World Heritage Site requires consent by the State party. Ibid art 11.
[133] UNESCO, 'World Heritage in Danger' (List of World Heritage Sites in danger) <http://whc.unesco.org/en/158/> accessed 29 December 2009.

World Heritage in Danger' is created and maintained by the World Heritage Committee and includes properties for which conservation will require 'major operations' and 'for which assistance has been requested under this Convention'.[134] If a site is inscribed on the List of World Heritage in Danger, the World Heritage Committee must work in conjunction with the State Party where the site is located to create a program for rehabilitating and monitoring the condition of the site with the end goal of removing the site from the list of properties in 'danger'.

On 16 February 2006, 12 conservation groups from the US and Canada lodged a petition with the World Heritage Committee to list Waterton-Glacier International Peace Park—located across the US and Canadian borders—on the List of World Heritage Sites in Danger as a consequence of the threats that climate change poses to the natural environment of the Site. Specifically, the petition alleged that 'less than one fifth of the park's glaciers still exist—and those precious few that remain are melting rapidly due to human-induced climate change'.[135] Based on the risk posed to the Site by climate change, the petitioners requested that the World Heritage Committee list the Site as in danger and adopt a management plan with a set of corrective measures that should 'focus on reductions in U.S. greenhouse gas emissions because the glaciers, which are so rapidly melting, are within the U.S.'s territory, implicating the obligation of the World Heritage Convention to conserve and protect natural and cultural heritage within a Party's boundaries'.[136] Professor Erica Thorson prefaced the petition to the World Heritage Committee with the argument that:

The effects of climate change are well-documented and clearly visible in Glacier National Park, and yet the US has not taken action to protect the world heritage of the park by reducing its greenhouse gas emissions pursuant to its obligations under the World Heritage Convention.[137]

The Waterton-Glacier International Peace Park petition has the twofold goal of protecting the Site from further degradation and finding a legal foothold for forcing the US to regulate greenhouse gas emissions.

Four other petitions have been filed by conservation organizations worldwide to add Mount Everest, the Peruvian Andes, the Great Barrier Reef, and the Belize Barrier Reef to the List of World Heritage in Danger due to threats posed by

---

[134] World Heritage Convention (n 126) art 2(4).
[135] International Environmental Law Project, 'Petition to the World Heritage Committee Requesting Inclusion of the Waterton-Glacier International Peace Park on the List of World Heritage in Danger as a Result of Climate Change and for Protective Measures and Actions' (Executive Summary, 16 February 2006) vii <http://www.lclark.edu/livewhale/download/?id=166> accessed 1 June 2010 (World Heritage Petition). [136] Ibid viii.
[137] International Environmental Law Project, 'IELP Petitions International Committee to List Waterton-Glacier International Peace Park as a World Heritage Site in Danger due to Climate Change' (16 February 2006) <http://law.lclark.edu/clinics/international_environmental_law_project/news/story/?id=4473> accessed 1 June 2010.

climate change.[138] Together, the five climate-related petitions put the issue of climate change high on the agenda of the World Heritage Committee. In response to mounting concerns, in 2005, the World Heritage Committee commissioned an expert report on the effects of climate change on heritage.[139] The report found that the effects of climate change may jeopardize World Heritage natural and cultural sites and determined that the 'fact that Climate Change poses a threat to the outstanding universal values of some World Heritage sites' requires the Committee to, among other things, 'design appropriate measures for monitoring the impacts of Climate Change and adapting to the adverse consequences'.[140] After establishing the validity of climate change as a threat to heritage, the report analyses the key issues and options for the World Heritage Committee to consider when deciding how to respond to climate change.

At its annual meeting in July 2006, the World Heritage Committee took the issue into consideration. After consultation, the Committee issued a decision on the question of climate change and heritage.[141] The Committee's decision acknowledged concerns about the impact of climate change on heritage; endorsed the expert strategy on 'Predicting and Managing the Effects of Climate Change on World Heritage'; requested that State Parties adopt this strategy as a way to protect properties from the threats of climate change; encouraged the creation of pilot projects at World Heritage Sites with the goal of developing best practices for implementing the strategy; recommended that State Parties work with the IPCC to include a chapter on the impacts of climate change on heritage in the next IPCC report; requested that 'the World Heritage Centre...prepare a policy document on the impacts of climate change on World Heritage properties'; and, decided that 'the decisions to include properties on the List of World Heritage in Danger because of threats resulting from climate change are to be made by the World Heritage Committee, on a case-by-case basis'.[142]

The Committee's decision formally acknowledges the links between climate change and heritage and the importance of creating an institutional strategy for

---

[138] See Climate Justice Programme, 'Advance Notice: World Heritage Group must Protect Top Sites from Climate Change' (World Heritage Committee Annual Meeting, July 2006) <http://www.climatelaw.org/media/2006Jul11/> accessed 1 June 2010; Climate Justice Programme, 'UNESCO Danger-Listing Petitions Presented' (17 November 2004) <http://www.climatelaw.org/media/2004Nov17/> accessed 20 May 2010.
[139] May Cassar, *Predicting and Managing the Effects of Climate Change on World Heritage: A Joint Report from the World Heritage Centre, its Advisory Bodies, and a Broad Group of Experts to the 30th session of the World Heritage Committee* (Report for the World Heritage Centre in Vilnius 2006) <http://whc.unesco.org/en/news/262> accessed 20 May 2010.   [140] Ibid 2–3 paras 3 and 5.
[141] UNESCO World Heritage Committee, 'Decision 30 COM 7.1' (Decision adopted at the 30th session of the World Heritage Committee in Vilnius 8–16 July 2006) WHC-06/30.COM/19. The Committee also adopted a strategy on heritage and climate change. UNESCO World Heritage Committee, 'A Strategy to Assist States Parties to Implement Appropriate Management Responses' (10 July 2006) <http://whc.unesco.org/en/news/262> accessed 1 June 2010; see also UNESCO World Heritage Committee, 'World Heritage Committee Adopts Strategy on Heritage and Climate Change' (10 July 2006) <http://whc.unesco.org/uploads/news/documents/news-262-2.doc> accessed 20 May 2010.   [142] Cassar (n 139) paras 13 and 14.

responding to this new challenge. The decision, however, fell short of the hopes and expectations of many individuals, organizations, and climate and heritage campaigners. With this decision, the Committee created an institutional framework for beginning to respond to the impacts of climate change on world heritage. It did not, however, create any mechanisms for addressing the causes of global climate change. The Committee rejected a call by campaigners to cut greenhouse gas emissions and took no direct action on the pending petitions to place properties on the 'in danger' list. In addition, the final decision reflected concessions to State Parties, including the US, by deleting references to the Kyoto Protocol and IPCC scientific findings.

The World Heritage Committee's Decision on climate change and heritage advances international dialogue on inter-institutional climate linkages by creating an official institutional record of the link between climate change and heritage and institutionalizing efforts to address this link. However, it does not create the foundation for new principles of international environmental law as it does not draw any formal institutional links between the World Heritage Committee and other international environmental agreements, eg, the UNFCCC and the Kyoto Protocol. Although the decision takes international environmental lawyers one step closer to creating recognized links between climate change and other areas of international environmental law, eg, heritage, it fails to create the basis for legally binding requirements for States to address global climate change.

The Committee's decision on climate change and heritage also fails to achieve the specific desired goal that is of relevance to our discussion here—that is, listing the compromised Sites on the 'in danger' list and, consequently, requiring the US and other State Parties to cut greenhouse gas emissions so as to protect the Park and other threatened sites from further degradation.

Complementary International Activity

Domestic and international climate change litigation and alternative modes of legal actions cluster around the activities—or omissions—of the US. This is, however, not to suggest that climate change litigation is not increasing elsewhere or that international legal actions are directed solely at the US. Climate based litigation is increasing in developed and developing countries. In Australia, for example, the result of recent litigation mandates that government agencies consider the impacts of greenhouse gas emissions associated with new building projects and land development when making planning decisions.[143]

---

[143] *Gray v The Minister for Planning* (2006) NSWLEC 720 (Environment Court of New South Wales); see also *Australian Conservation Foundation v Minister for Planning* (2004) VCAT 2029 (Victorian Civil and Administrative Tribunal). See generally Michael Kerr, 'Tort Based Climate Change Litigation in Australia' (Discussion Paper Prepared for the Climate Change Litigation Forum, London, March 2002) <http://www.acfonline.org.au/uploads/res_climate_change_litigation.pdf> accessed 20 May 2010.

Nearby, in 2005, the New Zealand Environmental Court permitted an appeal against a District Council for refusing permission for the construction of a wind farm.[144] Citing the New Zealand Resource Management Act 1991, the petitioners argued that the Council Commissioners must consider the positive impacts of reducing greenhouse gas emissions—even if they are *de minimis*—when determining whether a 'proposal promotes the sustainable management of natural and physical resources'.[145] The Court further noted that 'the effects of climate change and the benefits to be derived from the use and development of renewable energy' are explicitly included in the Resource Management Act and that 'this is a clear recognition by Parliament of both the importance of the use and development of renewable energy and the need to address climate change, both of which are key elements in the proposed wind farm'.[146] Based on analysis of the positive and negative consequences of the proposed wind farm, the Court found that the positive attributes—particularly the provision of renewable energy and its benefits—outweighed the negative aspects of the project. The Court, thus, overturned the decision of the Council and granted the application for resource consent.[147]

In reaching its decision in *Genesis Power Limited*, the Environmental Court of New Zealand relied on its earlier decision in the 2002 case of *Environmental Defence Society v Auckland Regional Council and Contact Energy Limited*.[148] Here, the Environmental Defence Society sought to impose a condition on the mandatory air consent permit that Contact Energy Limited needed to build a new gas-fired power plant—namely, that Contact Energy Limited be required to offset all of its carbon dioxide emissions, thus minimizing the plant's contribution to global warming.[149] In the decision, the Court offered a detailed analysis of the greenhouse effect and the international legal regime for addressing climate change.[150] In analyzing the UNFCCC, the Kyoto Protocol, and New Zealand governmental documents on climate change, the Court concluded that 'there is a general commitment by New Zealand, and developed countries, to limit emissions and enhance carbon sinks' and that it is up to national governments—as a matter of public policy—to decide how to achieve the desired ends.[151] Based on principles of international law, the Court found that the provisions of both the UNFCCC and the Kyoto Protocol were relevant to its decision-making process in the case at hand under the terms and spirit of the New Zealand Resource Management Act 1991.[152] The Court found that, based on the evidence presented, 'the greenhouse effect and the possibility of climate change are a matter of serious concern', therefore concluding that 'the greenhouse effect is likely to result in significant changes to the global environment, including New Zealand and

---

[144] *Genesis Power Limited v Franklin District Council*, Environment Court Decision A148 [2005] NZRMA 241 (New Zealand Environment Court).
[145] Ibid para 51.   [146] Ibid para 222.   [147] Ibid para 232.
[148] *Environmental Defence Society v Auckland Regional Council* [2002] NZRMA 492 (New Zealand Environment Court).
[149] Ibid para 4.   [150] Ibid.   [151] Ibid para 20.   [152] Ibid para 64.

the Auckland region'.[153] The Court further confirmed the importance of climate change as an issue to consider in the decision-making process by 'accept[ing] that the present scientific consensus is that the cumulative anthropogenic emissions of carbon dioxide on a global basis contribute to climate change'.[154] In the end, however, the Court denied the Environmental Defence Society's request for an appeal of the decision granting Contact Energy Limited the necessary consents to construct the gas-fired power plant. The Court held that the question of how the State should respond to climate change 'quintessentially [involved] policy decisions, to be arrived at after much research, discussion, and consultation'; that the Court was unable to assess the reasonableness, consequences, and efficacy of the proposed conditions; and, that the question raised was best dealt with at the national level in the political context.[155]

Despite denying the appeal, the decision in *Environmental Defence Society* was significant, both nationally and internationally. The case represented one of the first times that any judiciary formally recognized, accepted, and relied on climate change science and confirmed the importance of considering human impacts on the environment, in the context of climate change.

Across the globe, on 14 November 2005, the Federal High Court of Nigeria ordered that gas flaring must stop in a Niger Delta community when it decided *Gbemre v Shell Petroleum Development Company of Nigeria Ltd et al.* The Court found that the gas flaring violated citizens' constitutional rights to life and dignity. The prohibition affects the activities of major oil and gas companies, including ExxonMobil, ChevronTexaco, TotalFinaElf, Agip, and Shell. While the holdings in the case did not rest on the legal relationship between flaring and climate change, the decision is significant for purposes of addressing climate change because gas flaring in Nigeria is a primary source of greenhouse gas emissions in sub-Saharan Africa.[156] After Shell failed to comply with the order of the Court, the company requested and received a 'conditional stay of execution' in April 2006 that allowed it to continue flaring, on the grounds that it immediately begin to phase-out flaring and was able to show technical and management plans for stopping all flaring by 2007. Shell has not met the requirements of the Court's order and legal and political negotiations are on-going.

Climate change litigation is also increasingly prevalent in Europe. In Germany, for example, an environmental and development group, Germanwatch, has led legal actions challenging both the State and private industry on their contributions to climate change. In 2004, in the case of *Bundes für Umwelt und Naturschutz Deutschland eV & Germanwatch eV v Bundesrepublik Deutschland*,

---

[153] Ibid para 65.   [154] Ibid para 88.   [155] Ibid paras 86 and 88.
[156] World Bank Group Climate Justice Program and Environmental Rights Action, *Global Gas Flaring Reduction Initiative: Report No.1: Report on Consultations with Stakeholders* (1 January 2004).

vertreten durch Bundesminister für Wirtschaft und Arbeit,[157] Germanwatch and Friends of the Earth Germany initiated litigation to force the German Economic Ministry to reveal how decisions made by its export credit agency, Euler Hermes AG, contributed to climate change, with particular reference to the energy production projects funded by the agency. In January 2006, the case ended in a court settlement with a legal judgment rejecting the German government's claims that it was immune from the freedom of environmental information laws and that the agencies' actions did not affect climate change. As a consequence of this case, in the future, the Ministry will have to disclose the impacts of its export credits on global climate change.

Following on from this victory, on 7 May 2007, Germanwatch launched another challenge, this time against a private actor. Here, Germanwatch filed a complaint with the National Organisation for Economic Co-operation and Development (OECD) Contact Point in the German Federal Ministry of Economics and Technology claiming that Volkswagen Corporation had violated OECD Guidelines for Multinational Enterprises in 15 specific cases.[158] In particular, Germanwatch claimed that Volkswagen voluntarily committed itself to the OECD Guidelines, which in turn committed the company to creating an institutional strategy for minimizing its contribution to climate change.

Germanwatch has stated that it intends to lodge further complaints against other members of the auto industry. In the German lawsuits and petitions, unlike those in the US, the plaintiffs are seeking to pressure states and private actors that are already bound by international and domestic commitments, thus creating an extra layer of compliance pressure.

The increase in climate change litigation worldwide targets perceived 'rogue' States and private actors, such as the US and Shell, as well as outwardly Kyoto compliant States, such as New Zealand and Germany. Litigation is being used to re-interpret existing legislation, force the creation of new laws and regulations, and ensure compliance with active laws, policies, and agreements. Climate change jurisprudence is evolving rapidly and via diverse avenues. The rise in litigation as a mechanism for compelling public and private actions is emblematic of the larger trend within both US society and the global community of accessing diverse pressure points as a means of influencing climate change policy. This trend signals a

---

[157] *Bundes für Umwelt und Naturschutz Deutschland & Germanwatch v Bundesrepublik Deutschland, vertreten durch Bundesminister für Wirtschaft und Arbeit*, VG 10 A 215.04 (FRG 10 January 2006) (German Administrative Court).

[158] See Organisation for Economic Co-operation and Development, 'Text of the OECD Guidelines for Multinational Enterprises' (27 June 2000) <http://www.oecd.org/document/28/0,2340,en_2649_34889_2397532_1_1_1_1,00.html> accessed 1 June 2010. The OECD Guidelines seek to promote 'the highest sustainable economic growth and employment and a rising standard of living in member countries, while maintaining financial stability, and thus to contribute to the development of the world economy; to contribute to sound economic expansion in member as well as non-member countries in the process of economic development; and to contribute to the expansion of world trade on a multilateral, nondiscriminatory basis in accordance with international obligations'. Ibid.

new era in environmental policy where domestic and international responsibilities intermingle and national, regional, and local actors display unprecedented levels of cooperation and vision in their efforts to influence the actions and policies of the public and private sector. The State remains the dominant actor in international climate change law and the principal architect of federal policies, but international and domestic laws and policies alike are increasingly shaped via local and regional policies, private actions, litigation, and recourse to international mechanisms. The State, thus, remains the primary actor, but the insulated nature of national and international decision making is perforated, molded, and constrained by emergent modes of influence.

Overview of Sub-Federal Climate Change Initiatives

The changing role of states, regional partnerships, local governments, NGOs, businesses, and even tools of law—eg, litigation, regulation, and petitions—in US climate change law and policymaking could be reduced to a 'super grassroots' movement. This would be a mistake. The phenomenon of widespread efforts to create and influence climate change policy at all levels of government is not a mere continuation of a legacy of citizen activism left over from the environmental movement of the 1970s. Nor can the activities of states in creating independent and regional climate change laws, policies, and partnerships be categorized as a normal pattern of policymaking in the process of federalism. While it is true that NGOs have played a consistent role in initiating grassroots campaigns to influence environmental policy and states have, on occasion, adopted environmental standards more stringent than federal standards, the level, breadth, and power of sub-federal efforts to create shared climate change policies in the absence of a strong federal framework represents a new development and, arguably, a new era in environmental policymaking in the US.

Facing an environmental issue of global and local significance, and confronting the federal government's failure—arguably, refusal—to discharge its customary role as the principal architect of environmental policy, sub-federal entities have stepped in to fill the policy void. As one commentator notes, 'there has emerged a fairly stunning proliferation and diversification of state efforts to create elements of a policy architecture to reduce greenhouse gases'.[159]

The question that plagues this movement, however, is whether the agreements, laws and policies evidenced are merely 'symbolic' or whether they are substantive enough to bring about real change.

The review of sub-federal climate change initiatives provided in this book confirms that many efforts are, in large part, 'symbolic'. That is, the programs create

---

[159] Barry G Rabe, 'North American Federalism and Climate Change Policy: American State and Canadian Provincial Policy Development' (2004) 14 *Widener LJ* 121, 152.

ambitious policy goals but often omit specific obligations or discrete measures for implementation and compliance.[160] The 'symbolic' nature of sub-federal schemes, however, should not be underestimated or written off as efforts by local politicians to gain cost-free political goodwill. The central issue here is the relationship between symbolic and substantive policies, and what—if any—value these symbolic policies provide.

In environmental law, the line between procedural and substantive measures is fine, and often difficult to discern. Many laws and policies that are largely procedural in nature create substantive rights and value in practice, eg, environmental impact assessment measures in NEPA, judicial standing requirements, listing procedures in the ESA, regulatory notice and review provisions in the Federal Administrative Procedure Act. In the context of climate change laws and policies, the difference between symbolic and substantive policies perhaps turns on whether the law or policy creates enforceable rights and/or obligations. The creation of substantive and enforceable rights is critical to securing positive action to address climate change. In the context of US federalism—where states are constrained by federal pre-emption—even 'symbolic' actions can take on substantive meaning. Efforts by sub-federal governmental entities to create interstate research, regulatory and political links, and to adopt statements of intent, executive orders, resolutions, and formal recommendations on climate change have both symbolic and substantive impact.

The initiatives offer varied symbolic values. For local citizens, the initiatives signal concern for citizen welfare, recognition, and validation of the problem, and willingness to begin to act. For the private sector, the initiatives signal a forthcoming regulatory change that it can plan for and that will offer regulatory certainty. Critically, for the federal government, the initiatives carry multiple symbolic values. First, independently and collectively, climate change initiatives signal to the federal government that widespread sectors of the populace are taking climate change seriously, that they are concerned about the social and economic consequences of climate change, and that they are willing to act to address the problem. Second, sub-federal initiatives demonstrate that certain policies are more effective and more socially acceptable, ie, the case of states serving as the laboratories for federal policies.[161] Third, the proliferation of sub-federal climate

---

[160] Eg, Engel and Saleska have noted that:

With several important exceptions, few state and local climate initiatives currently impose mandatory emissions reductions from existing sources. Instead, many constitute preliminary steps toward eventual regulation, affording state and local politicians notoriety on an issue being largely ignored by the federal government without committing the locality to costly controls on in-state greenhouse gas sources. While the federal government is clearly not out ahead of the states on the issue of climate change, nor are the states quite as far out ahead of the federal government as one might believe as a result of the recent attention garnered by state and local climate change initiatives. Engel and Saleska (n 2) 222–23.

[161] Barry Rabe notes that: '[i]n many respects, this pattern is in keeping with the traditions of American federalism. Even before the most recent era of state resurgence, states have long been

change policies signals to the federal government that, should it choose to adopt climate change laws, a wide sector of the public is likely to support these laws. Fourth, sub-federal actions demonstrate that many cities and states are willing to act 'unilaterally', if necessary. In this regard, state actions mirror federal actions in the international context.

The US government has frequently acted unilaterally to achieve desired ends in international law. From banning tuna and shrimp imports, to threatening sanctions against commercial whalers, the US has gained a reputation as a country that is willing to act independently to achieve its desired goals—even when that unilateral action threatens international relations. In these circumstances, the State does not act with the mistaken belief that it can unilaterally generate a new substantive international policy. Instead, unilateral actions are taken with the dual purpose of prompting a desired act from specific State(s) and encouraging multilateral dialogue that leads to the creation or enforcement of a substantive international obligation. In this regard, US states have learned from their own Federal State that willingness to act—whether symbolically or substantively—often prompts legal and political change at a higher level. Of course, in the international context, the US has the confidence to act unilaterally because of its economic and political prowess in the global community. Nevertheless, US actions frequently step on the toes of the primary international organization, the United Nations.

Within the domestic context, the states have varying degrees of economic and political power. California has both. This is one of the primary reasons why California can and has led domestic efforts to re-shape regional—and, ultimately, federal—climate change policies, particularly, by adopting laws that exceed federal standards. The states have not been so foolhardy as to adopt laws that directly or intentionally encroach upon federal authority, but multiple states are demonstrating a willingness to fill in the federal policy void.

While it is true that many of the measures adopted by sub-federal governmental entities are largely symbolic, cast against the backdrop of a federal climate change policy regime that is almost entirely symbolic in nature, it is worth comparing the content of sub-federal and federal 'symbolism' to highlight the fact that many sub-federal initiatives are significantly more robust in method and objective.

---

incubators of policy ideas that ultimately sweep across regions and, in some instances, would be embraced in some later form at the federal level. That pattern, of course, has only intensified in recent decades as state policymaking capacity has risen steadily alongside an attendant decline at the federal level'. Rabe (n 159) 130. I agree with Rabe's assessment that current state activity is 'in keeping' with the American federalist tradition, but I argue that it goes above and beyond previous levels of state activism, especially in regard to the proliferation of inter-state partnerships and initiatives and the progressive, and, often, aggressive ways that, individually and cooperatively, states are trying to both fill the policy void and chip away at federal resistance to force a more comprehensive federal climate change policy architecture.

Further, there are 'clear exceptions to the "symbolic" diagnosis'.[162] California, for example, has adopted concrete, enforceable emission standards for passenger automobiles—which have subsequently been adopted by 15 other states—and is poised to begin regulating greenhouse gases economy-wide in 2011. The Northeast states are operating a regional greenhouse gas emissions trading program, while numerous states nationwide have begun regulating carbon dioxide emissions from new and existing power plants.

The combination of symbolic and substantive policies employed by US states lays the foundation for active inter-state dialogue and cooperation. It also creates momentum for states to develop additional and improved climate change policies. Individual US states have a long history of and demonstrated capacity for policy innovation. In the context of climate change, the states are only beginning to reveal their capacity for innovation, determination, and courage.

Conclusion

Any in-depth analysis of climate change policymaking in the US reveals a complicated picture of pushes and pulls—of stagnation and resistance to change at the top, meeting innovation and pressure for progress from below. The phenomenon of grassroots pressure driving environmental change is not new. In the case of climate change, however, the profile and ingenuity of these so-called 'grassroots' efforts to influence policymaking in the US suggest a new type of environmental problem and a new style of political change. Grassroots efforts to force the federal government to adopt more progressive climate change policies reveal the unheralded urgency of the issue, the diversity of interested and potentially affected parties, and the striking alterations in the nature of US environmental politics.

Much has been written about sub-federal efforts to adopt climate change policies, but this is just the tip of the iceberg. From adopting local policies, to employing common law, criminal law, and tort-based litigation, to using existing federal environmental laws and regulations, to invoking the jurisdiction of international institutions, public bodies and civil society alike are using every possible mechanism to overcome a history of stagnation and resistance at the national level and thereby drive a more progressive climate change policy agenda from the bottom up.

[162] Engel and Saleska (n 2) 220–21.

PART III

CLIMATE CHANGE LAW AND POLICY IN THE EUROPEAN UNION

5
Law and Policy in the European Union

As compared to the United States, the European Union comprises a group of member states possessing relatively long and complex legal and political histories. In contrast to the histories of its member states, however, the overarching regional governance regime is young, distinctive, and evolving.[1] The marriage of old and young governance systems creates significant challenges for coordinating decision making amongst its diverse members but it also offers great opportunities for developing innovative systems of governance, especially in new and evolving legal fields, such as environmental law. The EU has taken advantage of its unique position to advance a progressive strategy to address climate change.

Introduction to the European Union

The EU is a *sui generis* organization. Often described as supranational in nature, the EU functions as both a conventional regional organization, eg, providing coordinating functions for sovereign states and helping harmonize EU policy through a centralized governance system, as well as taking on many of the traditional roles and responsibilities of sovereign states, eg, enacting laws that have direct effect within sovereign states and harmonizing domestic laws and standards to create compatible legal architectures for the Community as a whole.[2]

EU efforts to harmonize regional policy are framed and constrained by the principle of subsidiarity, which dictates that decisions should be taken at the most decentralized level possible and that Community actions be limited to those issues best addressed at the regional level. The dueling principles of harmonization and subsidiarity define the parameters of European law and policymaking and create a constant system of checks and balances on Community policymakers.

---

[1] The evolving nature of the EU is exemplified by the governance shift that occurred on 1 December 2009 upon the entering into force of the Treaty of Lisbon, which replaced the European Community with the European Union and dictated certain changes in capacity and functioning of the various EU institutions. Treaty of Lisbon amending the Treaty on European Union and the Treaty establishing the European Community [2007] OJ C306/01, 17 December 2007.

[2] Paul Craig and Gráinne de Búrca, *EU Law: Texts, Cases & Materials* (4th edn, OUP, Oxford, 2007); Nuno S Lacasta, Surage Dessai, and Eva Powroslo, 'Consensus Among Many Voices: Articulating the European Union's Position on Climate Change' (2002) 32 *Golden Gate U L Rev* 351, 356 (Consensus Among Many Voices).

Unlike other regional organizations where States surrender specific competences but continue to operate as discrete decision-making institutions in domestic and international matters, the EU possesses a legal personality that is separate from the legal personalities of its member states. As a distinct legal institution, the EU possesses the competence to adopt supranational rules and enter into international agreements, with binding effect on its member states. In creating and/or joining the EU, member states delegate both shared and exclusive competences to the Community's principal governing institutions, which then execute these functions on behalf of the member states.

The EU was created following World War II as an instrument for promoting economic cooperation and preventing further conflict in Europe. The foundation for European integration developed largely as a security mechanism in Europe, but has since expanded to address a variety of social functions as well.

In brief, the 1951 Treaty of Paris, the 1957 European Atomic Energy Community Agreement, and the 1957 Treaty Establishing the European Community (Treaty of Rome) created the foundations for the European Economic Community (EEC), which has been known as the European Union (EU) since the Treaty on European Union (TEU) (or Treaty of Maastricht) in 1992. Thus, while the EU arose as a mechanism for addressing post-war economic and security concerns common to the countries of Western Europe, it has since evolved to become one of the more influential entities in regional and international politics.[3] The EU has evolved from its post-War mandate to become an increasingly important forum for regional harmonization and governance across a full range of policy arenas, including environmental law.

European Union Decision-Making Institutions

The EU operates a unique governance system that shapes EU climate change law and policy. It consists of four primary decision-making institutions, including the Council of the European Union (the Council), the European Commission (the Commission), the European Parliament (the EP), and the Court of Justice of the European Union (the Court).[4]

The Council

Generally regarded as the principal legislative or decision-making institution in the EU, the Council is the central institution involved in ratifying EU law

---

[3] Maurice Sunkin, D Ong, and R Wight, *Sourcebook on Environmental Law* (2nd edn, Cavendish Publishing, London, 2002) 6.
[4] See M Zander, *The Law-Making Process* (6th edn, CUP, New York, 2004) ch 7.

based on policy proposals from the Commission. While the Council was once supreme in its legislative duties, the TEU and the Lisbon Treaty have elevated the status of the EP in this process such that the Council and the EP now share decision making responsibilities via assent, consultation and co-decision procedures,[5] with co-decision now being termed the ordinary legislative procedure.[6]

The Council represents the EU's member governments, consisting of one ministerial representative from each member state. The Council shares lawmaking and budgetary powers with the EP while possessing central authority in EU common foreign and security policy and economic coordination. In its role as lawmaker, the Council acts primarily on proposals put forward by the Commission—and, increasingly, only in concert with the EP through co-decision procedures. In the area of environmental regulation, the Council shares decision making authority with the EP according to the co-decision procedure, as outlined in articles 251 and 252 of the EC Treaty.

The Commission

Within the EU, the Commission is the institution that designs policy proposals and then implements ratified Community legislation; it is the heart of the Community's executive functioning and is tasked with promoting the European public interest. The Commission is responsible for the following:

- safeguarding treaties,
- proposing legislation on which the Council and the European Parliament will act,
- implementing common policies,
- managing the budget, and
- managing the programs of the EU.

Each member state nominates a single commissioner to serve, but the Commission is much broader than the individual commissioners. The Commission is 'divided into some 40 directorates-general (DGs) and services, which are subdivided in turn into directorates, and directorates into units.'[7]

---

[5] The co-decision procedure was initiated by the Maastricht Treaty and later reinforced by the Treaty of Amsterdam and the Treaty of Lisbon. Co-decision making divides legislative power between the European Parliament and the Council. European Union, 'Consolidated Versions of the Treaty on European Union and of the Treaty Establishing the European Community' (Consolidated versions with amendments, 29 December 2006) <http://eur-lex.europa.eu/LexUriServ/LexUriServ.do?uri=OJ:C:2006:321E:0001:0331:EN:pdf> accessed 30 December 2009 (Maastricht Treaty and EC Treaty) art 251.   [6] Treaty of Lisbon (n 1) art 249A(1).
[7] European Commission, 'The European Commission at Work: Basic Facts' <http://ec.europa.eu/atwork/basicfacts/index_en.htm> accessed 1 June 2010.

The DGs, which are responsible for designing, implementing, and monitoring legislation, constitute an extensive network of scientific, economic, social, political, and legal expertise that informs the European policymaking process. Together, the Commission and the DGs have an integral role in European policymaking, especially with regard to environmental and climate change laws and policies—where the DG Environment and DG Enterprise, for example, are charged with undertaking policy research and formulation.

The European Parliament

The third governance institution of the EU is the EP, which consists of 736 members.[8] The Members of the EP (MEPs) are directly elected by national populations every five years based on a system of degressive proportionality. The EP is the only directly-elected decision making body of the EU, and provides a critical, open, democratic forum for debating Community policies.

The EP has legislative and budgetary powers, both of which it shares with the Council. Although its powers have historically been modest compared to those of a traditional domestic parliament, the EP's authority has expanded over time. Since its establishment in the 1950s, the EP has evolved from a purely advisory body to an institution with substantial legislative competence, placing it on near equal footing with the Council on most matters of EU lawmaking. The EP's authority has similarly been extending in regards to international agreements, wherein the EP is now required to consent to all international agreements in fields covered by the ordinary legislative procedure.[9]

The EP is also invested with critical power-balancing tools: it has powers to approve the European Commission and to adopt a motion of censure against the Commission as a whole. Once a motion of censure is tabled, the EP can vote to dismiss the entire Commission by a two-thirds majority of the votes cast, making the EP a powerful instrument for balancing institutional power in the Community. The motion of censure is an extreme measure but one that invests the EP with power in clearly specified cases.

The Court of Justice of the European Union

As the EU governance institution equipped with judicial competence, the Court[10] is the primary judicial institution for the EU. The Court possesses the

---

[8] The Treaty of Lisbon capped the number of MEPs at 751, which allows for 750 MEPs plus the President of Parliament. Each member state will be represented by at least six but no more than 96 MEPs. Treaty of Lisbon (n 1) Title III 9A(2).   [9] Treaty of Lisbon (n 1) 188 N6(a).
[10] The Treaty of Lisbon modified the EU Court of Justice, renaming it the 'Court of Justice of the European Union' and merging the Court of Justice with the previously distinct General Court and the specialized courts to create one single judicial institution for the Union. Treaty of Lisbon (n 1) arts 9F(1) and 204.

legal competence to assess the validity of EU law and implementation measures. Consisting of one judge from each of the member states, the Court has the authority to make rulings on disputes about EU law that arise in national litigation, and to make decisions on actions raised by the EU Institutions, by member states, or by individuals that assert that a breach of obligation under Community law has occurred.

The Court also has the authority to settle disputes between member states if they have exhausted other administrative procedures through the Commission. Although interpretations by the Court regarding Community law have direct effect in the member states, the Court does not have enforcement rights, a reality that leaves member states with the responsibility to enforce Court rulings.

Governance Overview

Together, the Council, the Commission, the EP, and the Court form a novel and essential system of governing institutions. The Court may be characterized as the Community's judiciary. Describing the Council and the EP as legislative bodies and the Commission as an executive body, however, is an oversimplification of EU governance, where lawmaking, implementation, and enforcement capacities are more complexly shared not only between the EU institutions but also among the EU institutions and the member states.

Unlike governance in the US, where the legislative, executive, and judicial branches have distinct, constitutionally defined roles in lawmaking, implementation, and enforcement—even if these responsibilities overlap in practice— the roles and responsibilities of the decision-making institutions of the EU are more intricate and less transparent to the public. The separation of powers in the EU is functional rather than formal; it is not defined in a constitution or in a treaty.

Because decision making processes differ according to the policy arena in question, the respective roles of the Council, the Commission, and the EP differ from decision to decision. The triadic relationship between the three institutions creates a system of checks and balances that is perpetually evolving. This complexity does not correlate with better decision-making; it produces a multi-layered and intricate system of governance that is procedurally and conceptually difficult for the public to comprehend.

The complexity amongst the EU institutions is further complicated by the relationship between the EU institutions and the member states. For example, one key area where tensions continue to complicate governance questions relating to the roles and responsibilities of the EU and the member state is on the question of harmonization. While the harmonization of regional policy is one of the central objectives of the EU, fundamental differences continue to exist over when, where, and how EU law establishes minimum rather than exhaustive requirements. That is, when does EU law establish a floor beneath which no member state may fall but which any member state can choose to exceed so long as it is in compliance with the EU Treaties; and when does EU law establish both the floor and the ceiling, thus precluding member states' ability to adopt more stringent standards? These questions are especially pertinent in the area of environmental policy, where member states may often wish to adopt stricter standards at the domestic level.

As a general matter, EU governance is premised on the principle of minimum harmonization but exhaustive harmonization is used as a tool to ensure basic rights and standards in such areas as health and safety. Pursuant to the EU Treaty, environmental measures automatically follow the principle of minimum harmonization because the Treaty explicitly recognizes the power of member states to adopt stricter measures.[11] In other areas of policy, however, the legal basis for determining whether EU law creates the floor or the ceiling is less determinative, requiring judicial construction.[12] Even within the field of environmental policy, however, degrees of harmonization vary. For example, in the more narrow field of climate policy, degrees of harmonization vary from phase to phase in the EU Emissions Trading Scheme—the cornerstone of EU climate policy—with the Commission exercising increasing levels of control in the later phases of the scheme. Similarly, as discussed further below, in a tangential area of environmental policy—the Integrated Pollution Prevention and Control Directive—member states' ability to establish more stringent limits in pollution permits has been partially circumscribed.

Questions of harmonization reveal the extent to which the EU is increasingly confronting governance questions which, while discussed in the language of multi-level governance, are similar to the federalism questions that the US confronts in balancing state and federal climate change lawmaking. Thus, while variations between the supranational and federal decision-making structures of the EU and the US are to be expected, there are emerging similarities as well.

The respective domain and jurisdiction of the US federal government is defined by the US Constitution. And, while broad authorities are left to the states, the US federal government has specific enumerated and implied powers over its states. These powers have expanded over time, giving the federal government extensive lawmaking, implementation, and enforcement powers. The powers of the EU have similarly expanded over time, with the Treaty of Lisbon being the most recent example. Yet, the reach and extent of 'federal' power in the US and the EU continue to vary significantly in degree. While member states have gradually

---

[11] European Union, 'Consolidated Versions of the Treaty on European Union and of the Treaty Establishing the European Community' (Consolidated versions with amendments, 29 December 2006) <http://eur-lex.europa.eu/LexUriServ/LexUriServ.do?uri=OJ:C:2006:321E:0001:0331:EN:pdf> accessed 30 December 2009 (Maastricht Treaty and EC Treaty), art 176 (now TFEU 193).

[12] M Dougan, 'Vive la Différence? Exploring the Legal Framework for Reflexive Harmonisation within the Single European Market', in R Miller and P Zumbansen (eds), *Annual of German and European Law 2003* (Berghahn Books) 113–65.

ceded increasing levels of authority to the EU, they have resisted the creation of a traditional federalist system of governance at the regional level and have continued to emphasize the importance of subsidiarity and minimum harmonization as defining characteristics of EU competence. Member states continue to maintain significant sovereignty over domestic matters as well as important rights of legal deviation. Thus, while both US and EU governance systems contain key components of a federalist system of governance, the mechanisms for power sharing and the degrees to which power is shared continue to differ.

Subsidiarity

The principle of subsidiarity defines the EU system of governance. It limits the EU's ability to act in policy domains where local or state governments are competent to act and dictates that, when possible, decisions should be made 'as closely as possible to the citizen and that constant checks are made as to whether action at Community level is justified in the light of the possibilities available at national, regional or local level'.[13] Subsidiarity, thus, constrains the EU from taking actions unless it is more effective to do so at the Community level than at the member state or local level.

Subsidiarity allows for the effective allocation of responsibility for decisions related to environmental policymaking by constraining the legal competences of the Council and the Commission in policy areas where local regulation is more appropriate. Similarly, notions of subsidiarity empower EU efforts to facilitate coordinated action in certain areas, such as climate change, that are best addressed through harmonized policy at the Community level.

The principle of subsidiarity is also correlated with the principle of proportionality, which establishes that Community actions should not exceed the point necessary to meet Treaty goals. Together, subsidiarity and proportionality limit the actions of the EU by determining that supranational decision-making is appropriate only in limited circumstances.

Thus, in contrast with the US, where the federal government has historically focused on creating a clearly defined policy framework that delimits the actions of the states, the EU has traditionally been 'less concerned to tell Member states what to do, than how to do it. It is less concerned to prescribe concrete outcomes than it is to develop devices intended to promote critical reflection on the nature of the problem, its causes, and possible solutions, and to promote the pooling of knowledge and experiences by the member states.'[14] Yet, as will be explored later in this chapter and elsewhere in the book, both the US and the EU have diverged from their normal governance approach in regards to climate change revealing

---

[13] Europa, 'Europa Glossary: Subsidiarity' <http://europa.eu/scadplus/glossary/subsidiarity_en.htm> accessed 30 December 2009.
[14] Eileen Denza, 'Law and Environmental Governance in the EU' (2002) 51 *Intl & Comp LQ* 996, 1005.

the shifting dynamics of environmental policy in both contexts. In the US, for example, climate policy has been driven from the bottom up with the federal government exhibiting great reluctance to enact a clearly defined framework at the national level, whereas the EU has displayed greater willingness to impose binding targets that, in fact, prescribe specific member state actions.

The Legislative Process

EU governance functions through a multi-layered legal system including the founding treaties, legislative instruments, and delegated legislation, as well as membership in international agreements. The Treaties of the EU enumerate the parameters of the EU system of governance; they create legally enforceable rights for individuals and States and obligations that require States to act before the measure is given full effect.[15] EU legislative instruments, in contrast, are more diverse and more common, taking the form of regulations, directives, and decisions. Within the fields of environmental and climate change policy, legislative instruments are the primary rulemaking instruments.

In contrast to the US system, in which federal legislation is automatically enforceable in state systems, EU law can be distinguished according to whether it is directly applicable to or has direct effect in member states. The Council and the Commission have the power to enact and implement legislation that will immediately impact member state law and be directly applicable in domestic legal systems; alternatively, the Council and the Parliament can pass legislation that imposes obligations on member states, but which requires member states to take legally binding acts in order to give full effect to the law.[16] This gradation is exemplified by the three categories of secondary legislation already mentioned—regulations, directives, and decisions. Differentiation among the categories of secondary law allows Community governance institutions to function while continuing to respect principles of subsidiarity and proportionality.

Regulations are binding in their entirety and have direct effect in the member states. Immediately upon their enactment, regulations bind member states and create rights that are legally enforceable at the national and Community levels.

Directives, in contrast, normally do not have direct effect in the member states.[17] They are binding only insofar as the 'result to be achieved'. Directives

---

[15] R Baldwin, *Rules and Government* (Clarendon Press, Oxford, 1995) ch 8 ss I and II.

[16] The Council and Commission must ground the authority for regulations and directives in specific Treaty provisions. The role of Parliament depends on the Treaty articles used to justify the adoption of certain regulations and directives; a legal presumption exists in favor of choosing the legal basis that confers the greatest opportunities for democratic participation through an increased role for Parliament.

[17] The Court may, however, find that certain directives have direct effect where the provisions of the directive are precise and unconditional, eg, the Environmental Impact Assessment Directive. Council Directive (EC) 85/337 of 27 June 1985 on the assessment of the effects of certain public and private projects on the environment [1985] OJ L175/40. In addition, directives may 'produce

require member states to adopt their own legally binding measures to give full effect to the directive within their domestic legal system. Unlike regulations, directives give member states flexibility to determine how they will fulfil the measure in terms of the form and method of implementation. More specifically, directives accommodate regional diversity by allowing States to design solutions that are culturally, economically, and environmentally appropriate for their respective domestic conditions. Directives, thus, facilitate the harmonization of regional law while acknowledging differences in social and legal culture; they are the most common form of legislation in the field of environmental law and in the context of climate change.

Finally, decisions may or may not be directly applicable to member states—the effect of each is specified in its content. Decisions differ from regulations and directives in that they may be directed at individuals or at companies, rather than just at member states. Like regulations, decisions are binding in their entirety against those to whom they are addressed and do not leave any discretion as to mode of implementation.

Regulations, directives, and decisions are the main tools through which the EU creates uniform or harmonized laws.

Although the Community uses all three types of secondary legislation in its environmental policy, as previously mentioned, directives are the most common form of environmental rule. They enable the Community to harmonize environmental standards while preserving member state flexibility to implement the most effective regulatory strategy for specific local circumstances.

Additional Tools

The Community also uses additional tools to inform State and regional policy, including recommendations, opinions, and guidelines. These policy instruments generally lack legal force, but are persuasive in influencing national policy choices and can serve as indicators for trends in Community law.

There is, for example, a growing body of soft law that expands and refines the influence of the EU, and especially the Commission's influence in the field of environmental policy. In the specific context of climate change, for example, the Commission is required by to adopt guidelines for the monitoring and reporting of greenhouse gas emissions under the EU Emissions Trading Scheme (ETS).[18] While these guidelines do not possess the same normative force of legislative instruments, they are legally binding pursuant to the overarching EU ETS Directive and compliance with the guidelines is vital to the effective

---

'similar effects' to regulations after the time limit for their implementation has expired and the State has not properly implemented them'. Craig and De Búrca (n 2) at 282.

[18] Directive 2003/87/EC of the European Parliament and of the Council and Commission Decision C(2004) 130 final, art 14.

implementation and enforcement of the EU ETS.[19] The EU increasingly relies on the use of guidelines and other binding and non-binding policy tools, such as programs of action and structural funding, to improve the 'quality, effectiveness of simplicity'[20] of legislative instruments, particularly in the area of climate policy.

In this way, the EU operates a multi-tiered system of primary and secondary legislation that provides the Council, the Commission, and the EP with varying levels of legislative control over member states, depending on the respective policy issue. As defined in the TEU, the EU's competence to legislate in particular policy areas varies significantly. In the field of environmental law, the EU does not have exclusive competence, but rather, has shared competence with the member states.[21] In cases of direct conflicts between Community law and member state law, however, Community law takes precedence. The body of EU law is referred to as the *acquis* and is an increasingly comprehensive body of common rights and obligations, especially in the area of environmental law.

European Union Competences in Environmental Law

Until the 1970s, environmental considerations were absent from EU law due to the fact the EU lacked the competency to act in the field of environmental policy. Further, the EU Treaties did not specifically address the Community's competence in the environmental arena until the Single European Act of 1986, when the EU's competencies were expanded to include environmental protection.[22]

Community Programme of Action on the Environment 1973

During the dawning of the global environmental era of the 1970s, Western Europe became increasingly concerned about the scope of environmental problems and the inconsistency between the legal and political instruments that various EU countries employed to counter environmental degradation. Responding to growing concerns, in 1973, the Council issued the first Community Programme of Action on the Environment.[23] The Programme included the Council's declaration of intent to harmonize Community efforts to reduce growing environmental

---

[19] 2007/589/EC: Commission Decision of 18 July 2007 establishing guidelines for the monitoring and reporting of greenhouse gas emissions pursuant to Directive 2003/87/EC of the European Parliament and of the Council.

[20] Craig and De Búrca (n 2) 161.   [21] See Consensus Among Many Voices (n 2) 358–60.

[22] European Union, 'Consolidated Versions of the Treaty on European Union and of the Treaty Establishing the European Community' (Consolidated versions with amendments, 29 December 2006) <http://eur-lex.europa.eu/LexUriServ/LexUriServ.do?uri=OJ:C:2006:321E:0001:0331:EN:pdf> accessed 30 December 2009 (Maastricht Treaty and EC Treaty).

[23] Alexandre Kiss and Dinah Shelton, *Manual of European Environmental Law*, (2nd edn, CUP, New York, 1997) 24.

problems, and stated that economic development and expansion:

> cannot now be imagined in the absence of an effective campaign to combat pollution or nuisance or an improvement in the quality of life and the protection of the environment; the improvement in the quality of life and the protection of the natural environment are among the fundamental tasks of the Community.[24]

In the absence of a clear Treaty-based authorization to address environmental problems, the Council drew its legal competences to address environmental issues from articles 94 (now 95), 95 (now 94(9)), and 235 (now 308) of the TEU.[25] Articles 94 and 235 enabled the Council to harmonize laws and administrative processes and to take appropriate measures that would enable the EU to fulfil the broader Treaty objectives,[26] while article 95 was at times used to confer exclusive rather than shared environmental competence.

---

[24] Ibid, quoting the First Action Programme.   [25] Maastricht Treaty and EC Treaty (n 22).
[26] Article 94 (now 95) of the EC Treaty provides:

> The Council shall, acting unanimously on a proposal from the Commission and after consulting the European Parliament and the Economic and Social Committee, issue directives for the approximation of such laws, regulations or administrative provisions of the Member States as directly affect the establishment or functioning of the common market. Treaty Establishing the European Community (EC Treaty) art 94.

Article 235 of the EC Treaty (now 308) provides:

> If action by the Community should prove necessary to attain, in the course of the operation of the common market, one of the objectives of the Community, and this Treaty has not provided the necessary powers, the Council shall, acting unanimously on a proposal from the Commission and after consulting the European Parliament, take the appropriate measures. Ibid art 308.

Article 95 of the EC Treaty (now 94(9)) provides:

> ...The Council shall, acting in accordance with the procedure referred to in Article 251 and after consulting the Economic and Social Committee, adopt the measures for the approximation of the provisions laid down by law, regulation
>
> ...
>
> The Commission, in its proposals envisaged in paragraph 1 concerning health, safety, environmental protection and consumer protection, will take as a base a high level of protection, taking account in particular of any new development based on scientific facts. Within their respective powers, the European Parliament and the Council will also seek to achieve this objective.
>
> If, after the adoption by the Council or by the Commission of a harmonisation measure, a Member State deems it necessary to maintain national provisions on grounds of major needs referred to in Article 30, or relating to the protection of the environment or the working environment, it shall notify the Commission of these provisions as well as the grounds for maintaining them.
>
> Moreover, without prejudice to paragraph 4, if, after the adoption by the Council or by the Commission of a harmonisation measure, a Member State deems it necessary to introduce national provisions based on new scientific evidence relating to the protection of the environment or the working environment on grounds of a problem specific to that Member State arising after the adoption of the harmonisation measure, it shall notify the Commission of the envisaged provisions as well as the grounds for introducing them.
>
> The Commission shall, within six months of the notifications as referred to in paragraphs 4 and 5, approve or reject the national provisions involved after having verified whether or not they are a means of arbitrary discrimination or a disguised restriction on

During the 1970s, with the promulgation of the first Programme of Action on the Environment, EU Treaty provisions were interpreted so as to justify effort to protect and preserve the environment.[27]

Single European Act of 1986

Recognizing the need for a clear mandate to address environmental issues, the Community ratified the Single European Act of 1986 (SEA) and thereby provided direct authority for the EU 'to preserve, protect and improve the quality of the environment, to contribute towards protecting human health, and to ensure a prudent and rational utilization of natural resources' in cases where Community action would be more effective than individual State action.[28]

The SEA was the first formal legal basis for Community environmental policy, and it relieved the Council and the Commission from using economic grounds to justify its environmental interventions. The SEA was a transformative Treaty with respect to environmental policy. It established the foundation for formal environmental governance in Europe and provided the Commission with valuable legal instruments and competences to propose and implement policy in this sector.

Maastricht Treaty of 1992, Amsterdam Treaty of 1999, Nice Treaty of 2001

The TEU, or Maastricht Treaty of 1992 establishing the European Union, furthered the environmental provisions of the SEA by affirming that environmental protection be integrated into all areas of EU law.[29] Following the TEU, the Treaty of Amsterdam of 1997 more firmly embedded environmental protection into formal EU competences by further extending the legislative authority of the EP into environmental policy. In key part, the Treaty of Amsterdam codified the principle of sustainable development in the environmental integration rule, found in article 6 of the Treaty.[30] This rule establishes the obligation to integrate the environment into Community policies.[31] In that same year, at the European Council's Luxembourg Summit, the process

---

trade between Member States and whether or not they shall constitute an obstacle to the functioning of the internal market....

When, pursuant to paragraph 6, a Member State is authorised to maintain or introduce national provisions derogating from a harmonisation measure, the Commission shall immediately examine whether to propose an adaptation to that measure.

When a Member State raises a specific problem on public health in a field which has been the subject of prior harmonisation measures, it shall bring it to the attention of the Commission which shall immediately examine whether to propose appropriate measures to the Council.

[27] See *Sourcebook on Environmental Law* (n 3) 6–17.
[28] Single European Act of 1986 art 130.   [29] Maastricht Treaty (n 22) art 130 r2.
[30] Treaty of Amsterdam Amending the Treaty on European Union, the Treaties Establishing the European Communities and Certain Related Acts (Amsterdam Treaty).
[31] See Martin Unfried, 'The Cardiff Process: The Institutional and Political Challenges of Environmental Integration in the EU', *RECIEL* 9(2) 2000.

of environmental integration was started, premised on the notion that environmental protection must be integrated throughout EU policies and activities with the fundamental goal of promoting sustainable development. In June of 2008, at the European Council's Cardiff Summit, the Commission presented an integration strategy entitled 'Partnership for Integration' that created guidelines for integrating environmental protection into specific policy areas.[32] These guidelines prompted government-wide efforts to develop environmental integration strategies. Finally, in the Treaty of Nice of 2001, the Conference issued a declaration confirming the contracting parties' commitment to article 175 of the TEU declaring that the High Contracting Parties are determined to 'see the European Union play a leading role in promoting environmental protection in the Union and in international efforts pursuing the same objective at global level'.[33] Together, these instruments and the ensuing process of environmental integration created foundations for modern EU environmental law.

The Consolidated EU Treaty

Until December 2009, the fundamental basis for contemporary EU environmental law rested in article 2 of the Consolidated EU Treaty, which provided that:

The Community shall have as its task...to promote throughout the Community a harmonious, balanced and *sustainable development* of economic activities, a high level of employment and of social protection, equality between men and women, *sustainable* and non-inflationary growth, a high degree of competitiveness and convergence of economic performance, *a high level of protection and improvement of the quality of the environment*, the raising of the standard of living and quality of life, and economic and social cohesion and solidarity among Member States.[34]

The Community's competences to act in the area of environmental policy were further enumerated in articles 6 (now TFEU 11), 95 (now TFEU 114), and 174–176 (now TFEU 191–193) of the EU Treaty. In relevant part, article 6 provided that '[e]nvironmental protection requirements must be integrated into the definition and implementation of the Community policies and activities...with a view to promoting sustainable development'. Article 95 then details the procedures surrounding the harmonization of national laws,

---

[32] EU Commission, *The Cologne Report on Environmental Integration: Mainstreaming of Environmental Policy*, <http://ec.europa.eu/environment/docum/pdf/cologne.pdf>, accessed 9 March 2010.

[33] Treaty of Nice Amending the Treaty on European Union, the Treaties Establishing the European Communities and Certain Related Acts (Nice Treaty).

[34] Maastricht Treaty and EC Treaty (n 22) art 2 (emphasis added). In addition, both the Treaty on European Union and the Treaty Establishing the European Community provide for considerations of sustainable development and environmental protection as integral to the creation of the EU. Ibid; Maastricht Treaty (n 22).

providing that harmonized environmental standards will be based on a 'high level of protection'.[35]

The core objectives of the Community's environmental policy, however, were set out in articles 174–176 (TFEU 191–193) of the Treaty. Article 174 (TFEU 191) enumerated the specific Community goals of: 'Preserving, protecting and improving quality of environment; Protecting human health; Prudent and rational utilisation of natural resources; Promoting measures at international level to deal with regional or worldwide environmental problems.'[36] It further grounds Community environmental policies in the precautionary principle and on principles of prevention and that the polluter pays.[37]

Of particular significance, the Treaty established environmental protection as being on an equal footing with economic development by providing a safeguard clause allowing member states 'to take provisional measures, for non-economic environmental reasons, subject to a Community inspection procedure'.[38] Article 175 then defined the substantive and procedural guidelines for legislating in the area of environmental protection, while article 176 provided the foundation for allowing member states to adopt 'more stringent protective measures'.[39]

The Treaty of Lisbon

The Treaty of Lisbon was signed on 13 December 2007 and came into force on 1 December 2009. The Treaty replaces the European Community with the European Union, which succeeds the EC and takes over all its rights and obligations. With the adoption of the Treaty of Lisbon, the TEU maintained the same name but the Treaty establishing the European Community (Treaty of Rome) became the Treaty on the Functioning of the European Union (TFEU).

The Treaty of Lisbon maintains but expands upon the core environmental provision of the EU Treaties while consolidating the provisions of the TEU relevant to environmental policy in the TFEU. In key part, the Lisbon Treaty embeds sustainable development as a fundamental principle of EU governance[40] and identifies environmental policy as an area of shared competence,[41] while the TFEU strengthens EU environmental capacity, including an express reference to

---

[35] Maastricht Treaty and EC Treaty (n 22) art 95 ss 3–5. [36] Ibid art 174 s 1.
[37] Ibid art 174 s 2.
[38] Ibid art 174 also details what factors the Community should take into account when determining environmental policies and provides for cooperation with third countries and international organizations on matters of environmental protection, with relative competency dependent on the type of negotiations at hand. Ibid.
[39] Ibid arts 175 and 176. The Council is largely responsible for (1) provisions of a fiscal nature; (2) measures concerning town and country planning, land use with the exception of waste management and measures of general nature and management of water resources; and (3) measures significantly affecting member state choice between different energy sources. Ibid art 175.
[40] Treaty of Lisbon (n 1) art 2(3) and (5).   [41] Treaty of Lisbon (n 1) art 2C(2)(e).

climate change in the provision on environmental competence in the TFEU.[42] In addition, the TFEU expands EU capacity in the area of energy policy by adding a new 'solidarity clause' that addresses the possibility of adopting regional economic measures to respond to energy supply crises, as well as a more general reference to solidarity in energy policy.[43]

More specifically, in relation to climate change, the TFEU amends articles 174 (now 191)—the core environmental provision of the EU Treaty—to explicitly address climate change. In relevant part, the provision is modified to read that 'Community policy on the environment shall contribute to pursuit of the following objectives', the statement 'promoting measures at international level to deal with regional or worldwide environmental problems', was replaced with the climate change-specific statement 'promoting measures at international level to deal with regional or worldwide environmental problems, *and in particular combating climate change*'.[44]

Thus, while the Treaty of Lisbon does not fundamentally alter the EU's institutional set-up, it reinforces EU action in the critical areas of environmental and energy policy by supplementing existing environmental policy provisions via specific reference to climate change and enhancing common energy policy with regards to security, efficiency, and interconnectivity of supply and solidarity.

European Environment Agency 1993

Although the legal mandate for environmental protection has continuously evolved alongside the EU treaty system, throughout this process Community institutions have acted to develop secondary institutions and measures for environmental protection. For example, amidst the evolving treaty framework and pursuant to the evolving mandate for Community action in environmental policy, the EU formed the European Environment Agency (EEA) in 1990.[45] The mandate of the EEA is to 'help the Community and member countries make informed decisions about improving the environment, integrating environmental considerations into economic policies and moving towards sustainability', and to 'coordinate the European environment information and observation network'.[46] In furtherance of its mandate to coordinate environmental and climate change policies, the Agency disseminates environmental information and provides scientific and technical support to member states. This support has proved important in implementing and improving EU climate change policies.

---

[42] Consolidated Version of the Treaty on the Functioning of the European Union [2008] OJ C 115/47, 9 May 2008 <http://eur-lex.europa.eu/LexUriServ/LexUriServ.do?uri=OJ:C:2008:115:0047:0199:EN:pdf>, accessed 8 March 2009, art 174(1) (TFEU).
[43] TFEU (n 42) arts 100(1) and 176A(1).   [44] TFEU (n 42) art 191(1) (emphasis added).
[45] Council Regulation (EEC) No 1210/90 of 7 May 1990, on establishment of the European Environment Agency and the European environment information and observation network [1990] OJ L120/1.
[46] European Environment Agency, 'Who Are We' <http://www.eea.europa.eu/about-us/who> accessed 30 December 2009.

The scope of the EEA, however, is not confined to information provision. The Agency also plays a critical role in monitoring and implementing Community environmental legislation, setting criteria for evaluating the environmental impacts of projects, and advising member states on environmental policy development. Furthermore, the EEA works in eight priority areas, the first of which is 'air quality and atmospheric emissions'. Within this priority area, there is specific focus on 'transfrontier, plurinational and global phenomena', giving the EEA a central role in helping formulate EU climate change laws and policies.[47]

Community governance on climate change is carried out according to two principal methods. Primarily, the EU employs a wide range of secondary legislation—regulations, directives, and decisions—to create Community-wide legal obligations. Additionally, the EU adopts Community Environmental Action Programmes (EAPs). Through its contributions to the EAPs, the EEA plays a pivotal role in furthering the development and implementation of Community climate policy. To date, there have been six EAPs, each providing a framework for Community environmental policies during a specified time period, and thereby shaping the environmental agenda for the EU. The EAPs are similar to international framework agreements, such as the UNFCCC, in that they take the form of decisions that do not create enumerated legally binding obligations, but are binding on member states in so far as their goals and objectives are concerned. The sixth and most recent EAP categorizes climate change as a main priority of EU environmental concern. The EEA has been one of the main contributors to the climate change component of the sixth EAP.

Climate Change Law and Policy in the European Union

The substantive climate governance systems in the EU contrast sharply with the US model. The primary difference between the two models is that the EU and its member states have ratified and/or approved both the UNFCCC and the Kyoto Protocol, and thus, are legally bound to meet specific international climate change obligations, while the US is bound only by the UNFCCC.[48] The EU and the US, thus, draft climate change laws and policies with different points of origin.

The EU can enter negotiations, sign, and ratify treaties with other States and international organizations. In the context of climate change, the EU formally shares competence with the member states. Consequently, during the negotiations for the UNFCCC and the Kyoto Protocol, the EU negotiated on behalf of

---

[47] Council Regulation (EC) 401/2009 of 23 April 2009 on the European Environment Agency and the European environment information and observation network [2009] OJ L126/13 art 3 s 2.

[48] Council Decision (EC) 2002/358 on 25 April 2002 concerning the approval, on behalf of the European Community, of the Kyoto Protocol to the United Nations Framework Convention on Climate Change and the joint fulfillment of commitments thereunder (Kyoto Protocol Approval Decision) [2002] OJ L130; see also UNFCCC, 'Kyoto Protocol Status of Ratification' <http://unfccc.int/files/essential_background/kyoto_protocol/application/pdf/kpstats.pdf> accessed 1 June 2010.

its member states. The EU approved the UNFCCC on 15 December 1993,[49] and the Kyoto Protocol on 25 April 2002.[50] For purposes of the Kyoto Protocol, the EU operates in partnership as a 'bubble'.

The EU is the only party to the UNFCCC and the Kyoto Protocol that participates as a regional economic organization.[51] As a party to the Kyoto Protocol, the EU—consisting of the 15 member states that constituted the EU at the time of the Protocol's ratification (the EU-15)—is jointly responsible for meeting a Community-wide obligation to reduce greenhouse gas emissions to 8% below 1990 levels by the first Kyoto compliance period, 2008–2012.[52]

EU-15

The member states constituting the EU-15 include: the Kingdom of Belgium, the Kingdom of Denmark, the Federal Republic of Germany, the Hellenic Republic, the Kingdom of Spain, the French Republic, Ireland, the Italian Republic, the Grand Duchy of Luxembourg, the Kingdom of the Netherlands, the Republic of Austria, the Portuguese Republic, the Republic of Finland, the Kingdom of Sweden, and the United Kingdom of Great Britain and Northern Ireland. Each EU-15 nation has also independently ratified or accepted the Kyoto Protocol. Thus, the EU-15 States have legal obligations under the Kyoto Protocol both as members of the EU 'bubble' and as sovereign State Parties to the Treaty.

To satisfy its dual obligations as a regional party to the UNFCCC and the Kyoto Protocol, the European Commission—on behalf of the EU—is required to assess whether the actual and projected progress of the individual EU-15 nations in reducing greenhouse gases will enable the EU to fulfil its Community-wide commitments.[53] As part of this assessment, the Commission prepares annual progress reports that are reviewed by the Council and by the EP to evaluate Community progress in fulfilling its international commitments and regional goals. In 2012, the end of the first Kyoto

---

[49] Council Decision (EC) 94/69 of 15 December 1993 concerning the conclusion of the United Nations Framework Convention on Climate Change [1993] OJ L033.

[50] Kyoto Protocol Approval Decision 2002/358.

[51] European Environment Agency, *The European Community's Initial Report under the Kyoto Protocol* (EEA Technical Report No 10/2006) <http://unfccc.int/files/national_reports/initial_reports_under_the_kyoto_protocol/application/pdf/ec_initial_report_new.pdf> accessed 30 December 2009 (EC's Initial Report).

[52] Under the Kyoto Protocol, the 15 member states that were part of the EU prior to its enlargement on May 1, 2004—when 10 new member states joined, raising the overall number of member states to 25—are obligated to reduce their collective greenhouse gas emissions by 8% below 1990 levels during the first Kyoto Compliance period, 2008–2012. This target is shared among the 15 member states under a legally binding burden-sharing agreement. Kyoto Protocol Approval Decision 2002/358. See European Climate Change Programme Commission, *Second ECCP Progress Report: Executive Summary* (April 2003) <http://ec.europa.eu/environment/climat/pdf/second_eccp_report_xsum.pdf> accessed 1 June 2010 (Second ECCP Progress Report).

[53] Council Decision (EC) 280/2004 of 11 February 2004 concerning a mechanism for monitoring Community greenhouse gas emissions and for implementing the Kyoto Protocol [2004] OJ L49.

compliance period, the emission reductions of the EU-15 will be aggregated to determine if the EU has met its international obligations under the Kyoto Protocol.

Beyond the EU-15

Subsequent to the EU's ratification of the Kyoto Protocol, 12 additional nations have joined the EU. The accession of these 12 nations does not alter the EU's Kyoto Commitments—as a regional Party to the Kyoto Protocol, the EU only legally represents and binds the EU-15.

Of the 12 new member states, ten have independent Kyoto emission limitations or reduction commitments. The Czech Republic, Estonia, Hungary, Latvia, Lithuania, Poland, the Slovak Republic, Slovenia, Bulgaria, and Romania have ratified the Kyoto Protocol as Annex I parties, with concomitant emission limitation or reduction obligations. Cyprus and Malta are also parties to the Protocol; however, they were both initially classified as non-Annex I parties and, thus, do not have binding emission reduction obligations for the first Kyoto Protocol compliance period.[54]

Independent of their participation in the EU-15 or their status as Annex I or Annex II parties, all 27 EU member states are required by the provisions of the UNFCCC to submit individual reports to the European Commission detailing national greenhouse gas inventories and steps taken to meet the objectives of the Convention.[55] Thus, while the EU's commitments under the Kyoto Protocol are limited to the collective actions of the EU-15, the Community retains authority to create Community-wide climate change policies and to oversee the progress of the entire EU-27 in complying with the UNFCCC and the Kyoto Protocol.

Allocation of Responsibilities within the European Union

In 2002, the EU adopted an internal 'burden sharing' agreement to define how the 8% reduction goal under the Kyoto Protocol would be balanced among the EU-15 member states.[56] The 'burden sharing' agreement takes the form of a Council Decision, which is legally binding on the member states. Within the EU Kyoto 'bubble,' the 'burden sharing' agreement determines that each of the EU-15 member states is allocated a quantified emission limitation and reduction commitment, based on historical contribution to greenhouse gas emissions and national capacity to reduce emissions.[57]

---

[54] During the Copenhagen Climate Change Conference, pursuant to a proposal by Malta the COP amended Annex I to add Malta as an Annex I party to the Convention. UNFCCC, Draft decision -/CP.15, 'Amendment to Annex I of the Convention' <http://unfccc.int/files/meetings/cop_15/application/pdf/cop15_mal_auv.pdf> accessed 31 December 2009.

[55] United Nations Framework Convention on Climate Change (adopted 9 May 1992, entered into force 21 March 1994) 31 ILM 849 (UNFCCC) art 12.

[56] See Kyoto Protocol Approval Decision 2002/358.

[57] Kyoto Protocol to the United Nations Framework Convention on Climate Change (adopted 11 December 1997, entered into force 16 February 2005) 37 ILM 32 (Kyoto Protocol) art 4;

To further clarify this point, the UK is obligated to cut its emissions by 12.5% below 1990 levels by 2008–2012, while Germany is required to cut its emissions by 21% below 1990 levels.[58] Greece and Ireland, however, are permitted to increase their emissions from 1990 levels in the first compliance period—Greece by 25% and Ireland by 13%.[59] Each EU-15 country is responsible for ensuring that its aggregate greenhouse gas emissions do not exceed the amount assigned according to the EU 'burden sharing' agreement. National governments are also legally required to establish and enforce national climate change policies.[60] The burden sharing distribution is outlined in Table 5.1 below.

Table 5.1. EU and member states' greenhouse gas reduction targets

EU-15 member state	QELRC* (% reduction of base year/period emissions)
Austria	-13
Belgium	-7.5
Denmark	-21
Finland	0
France	0
Germany	-21
Greece	+25
Ireland	+13
Italy	-6.5
Luxembourg	-28
The Netherlands	-6
Portugal	+27
Spain	+15
Sweden	+4
United Kingdom	-12.5
Total EU-15 Commitment: -8%	

* QELRC=Quantified Emission Limitation and Reduction Commitment
Source: European Commission: DG Environment[61]

Council Decision (EEC) 93/389 of 24 June 1993 for a monitoring mechanism of Community CO2 and other greenhouse gas emissions [1993] OJ L167.

[58] Department of the Environment, Transport and the Regions, 'Climate Change: The UK Programme Summary' (2003) <http://www.ieadsm.org/Files/Exco%20File%20Library/Country%20Publications/cclsumm.pdf> accessed 1 June 2010 (DETR).

[59] Commission of the European Union, 'UN Conference on Climate Change: EU Set to Keep Momentum in the Global Fight Against Climate Change' (Press Release, 3 December 2004) <http://europa.eu/rapid/pressReleasesAction.do?reference=IP/04/1437&format=HTML&aged=1&language=EN&guiLanguage=en> accessed 30 December 2009.

[60] See Joseph A Kruger and William A Pizer, 'Greenhouse gas trading in Europe: the New Grand Policy Experiment' [October 2004] 46[8] *Environment* 14; DETR (n 58); 'UN Conference on Climate Change: EU Set to Keep Momentum in the Global Fight Against Climate Change' (n 59).

[61] Europa, 'The Kyoto Protocol and climate change—background information' (31 May 2002) <http://europa.eu/rapid/pressReleasesAction.do?reference=MEMO/02/120&format=HTML&aged=0&language=EN&guiLanguage=en> accessed 30 December 2009.

Beyond these commitments, member states individually determine how they will meet their assigned emissions reduction obligations. In contrast with the federal boundaries that limit US state and local policies, within the EU 'bubble', member states—as sovereign entities—are free to establish more ambitious reduction goals.[62] For example, the UK has set a domestic goal of cutting emissions by 20% below 1990 levels by 2010.[63] At a minimum, however, each EU-15 nation is bound to meet its designated emissions reduction obligations in order to ensure that the EU meets its combined emissions reduction targets.

European Union Climate Change Policy Framework

The EU and its member states, thus, share emission reduction targets under the Kyoto Protocol. Beyond this more general target, however, the EU and the EU-15 member states work collectively and independently to establish climate change laws and policies.

European Climate Change Programme

EU climate change laws and policies are delivered through the European Climate Change Programme (ECCP), which was created in June 2000 to 'identify and develop all the necessary elements of an EU strategy to implement the Kyoto Protocol.'[64] The ECCP provides a framework for developing climate change laws and policies and for helping the Community and its member states identify the most cost-effective measures for reducing greenhouse gas emissions. The ECCP embodies and manages the core elements of the EU's climate change strategy, and serves as an umbrella for integrating diverse, sector-based efforts that address the causes and consequences of climate change.[65] The development and implementation of the ECCP does not occur entirely within the public sphere; rather, it involves numerous public and private stakeholders, including the governing bodies of the EU, member states, industry, and environmental groups.[66]

The structure of the ECCP has facilitated the adoption of numerous cross-cutting legal instruments, including directives on the following:

- creating the CO2 Emissions Trading Scheme;[67]

- linking Kyoto flexibility mechanisms with the EU Emissions Trading Scheme;[68]

---

[62] Kyoto Protocol Approval Decision 2002/358.   [63] See DETR (n 58).
[64] Europa, Commission of the European Union, 'European Climate Change Programme' <http://ec.europa.eu/environment/climat/eccp.htm> accessed 15 April 2006 (Climate Change Homepage).
[65] See ibid.   [66] See Second ECCP Progress Report (n 52).
[67] Council Directive (EC) 2003/87 of 13 October 2003 establishing a scheme for greenhouse gas emission allowance trading within the Community and amending Council Directive 96/61/EC [2003] OJ L275/32.
[68] Council Directive (EC) 2004/101 of 27 October 2004 amending Directive 2003/87/EC establishing a scheme for greenhouse gas emission allowance trading within the Community,

Law and Policy in the EU 163

- promoting renewable energy;[69]
- promoting combined heat and power to promote high efficiency co-generation;[70]
- promoting biofuels;[71]
- improving energy efficiency in buildings;[72]
- restructuring the taxation of energy products and electricity;[73]
- establishing a framework for setting eco-design requirements for energy-using products;[74]
- using biofuels in transport;[75]
- controlling emissions from air conditioning systems in motor vehicles;[76] and
- recovering methane from biodegradable waste in landfills.[77]

The ECCP has also facilitated the passage of Council decisions on Monitoring and Reporting Guidelines, which require all member states to monitor and report to the Commission annually on national greenhouse gas emissions, and Commission regulations on standardizing national greenhouse gas registry systems, establishing common rules for direct support schemes under the common agricultural policy, and establishing certain support schemes for farmers (carbon credit for energy crops).[78] These directives, regulations, and decisions

---

in respect of the Kyoto Protocol's project mechanisms. Text with EEA relevance [2004] OJ L338/18.

[69] Council Directive (EC) 2001/77 of 27 September 2001 on the promotion of electricity produced from renewable energy sources in the internal electricity market [2001] OJ L283/33.

[70] Council Directive (EC) 2004/8 of 11 February 2004 on the promotion of cogeneration based on a useful heat demand in the internal energy market and amending Directive 92/42/EEC [2004] OJ L52/50.

[71] Council Directive (EC) 2003/30 of 8 May 2003 on the promotion of the use of biofuels or other renewable fuels for transport [2003] OJ L123/42.

[72] Council Directive (EC) 2001/30 of 2 May 2001 amending Directive 96/77/EC laying down specific purity criteria on food additives other than colours and sweeteners [2001] OJ L146/1.

[73] Council Directive (EC) 2003/96 of 27 October 2003 restructuring the Community framework for the taxation of energy products and electricity [2003] OJ L283/51.

[74] Council Directive (EC) 2005/32 of 6 July 2005 establishing a framework for the setting of ecodesign requirements for energy-using products and amending Council Directive 92/42/EEC and Directives 96/57/EC and 2000/55/EC of the European Parliament and of the Council [2005] OJ L191/29.

[75] Council Directive (EC) 2003/30 of 8 May 2003 on the promotion of the use of biofuels or other renewable fuels for transport [2003] OJ L123/42.

[76] Council Directive (EC) 2006/40 of 17 May 2006 relating to emissions from air-conditioning systems in motor vehicles and amending Council Directive 70/156/EEC [2006] OJ L161/12.

[77] Council Directive (EC) 1999/31 of 26 April 1999 on the landfill of waste [1999] OJ L182/1.

[78] Commission Decision 280/2004 of 11 February 2004 concerning a mechanism for monitoring Community greenhouse gas emissions and for implementing the Kyoto Protocol [2004] OJ L49. The Environment Directorate General maintains the EU emission registry. Council Regulation (EC) 1782/2003 of 29 September 2003 establishing common rules for direct support

cumulatively provide a foundation for Community action with respect to climate change.

The First European Union Climate Change Programme

The EU has a multifaceted climate change policy program, including the aforementioned policy goals and specified supranational legal obligations designed with the ambitious goal of exceeding its Kyoto Protocol obligations.[79] In contrast with the US model, which still struggles to create comprehensive federal climate change laws and policies, the EU began formulating climate change policies as early as 1991, when it developed its first Community-wide strategy for limiting carbon dioxide emissions and improving energy efficiency.[80] In order to effectuate these Community-wide strategies, the EU creates a system of common coordinated policies and measures (CCPMs) that provide a structure for member state policies and measures.

The ECCP provides the foundation for this policy umbrella, which sets minimum emission reduction requirements for the member states and provides a forum for coordinating Community-wide action, for facilitating voluntary programs and public–private partnerships, and for administering emissions trading schemes. To date, the ECCP has made possible the implementation and coordination of key EU-level measures, to include:

- Environmental Action Programmes, which provide for the establishment of the Community-wide Emission Trading System;

- Voluntary agreements between the Commission and key European, Japanese, and Korean car manufacturers to reach a 25% reduction of CO_2 emissions from new passenger cars by 2008;

- Measures under the Community-wide SAVE program, including proposals on the taxation of energy products, that aim to improve energy efficiency and reduce the environmental impact of energy use in the transport, industry, commerce, and domestic sectors;

- European Best Practice Initiative, which demonstrates the scope of coordinated action on energy efficiency and best technology deployment across the EU, and is based on the UK Energy Efficiency Best Practice

---

schemes under the common agricultural policy and establishing certain support schemes for farmers and amending Regulations (EEC) No 2019/93, (EC) No 1452/2001, (EC) No 1453/2001, (EC) No 1454/2001, (EC) 1868/94, (EC) No 1251/1999, (EC) No 1254/1999, (EC) No 1673/2000, (EEC) No 2358/71, and (EC) No 2529/2001 [2003] OJ L270/1.

[79] See Second ECCP Progress Report (n 52).
[80] Europa, 'European Union Greenhouse Gas Emission Trading System' (2005) <http://ec.europa.eu/environment/climat/emission/index_en.htm> accessed 30 December 2009 (EU ETS).

Programme—revealing the benefit of information sharing between the EU and its member states.[81]

In addition, the EU has put in place numerous directives, regulations, and decisions that focus on improving energy efficiency, reducing carbon intensity, and encouraging best practices in the transport, waste, building, and industrial sectors.

During the first phase of the ECCP, 2000–2001, the focus was on formulating policies aimed at the economic sectors most closely associated with greenhouse gas emissions, including the energy, transport, and industry sectors.[82] To achieve this goal, the EU established working groups to analyze the problem of climate change and to make policy recommendations.[83] The result of the first round of working groups was the 2001 ECCP Report identifying 42 measures which, if implemented, had the potential to reduce emissions by double the amount actually required of the EU during the first Kyoto Compliance Period.[84]

The strategy was designed to achieve these emission reductions in a cost-effective manner, ie, less than €20 per tonne of carbon dioxide,[85] and identified the following six most cost-effective ways for the EU to meet its Kyoto commitments:

1. Decarbonization of energy supplies;
2. Improvement of energy efficiency—especially in industry, households, and the services sector;
3. Reduction of nitrous oxide in specified industries;
4. Reduction of methane emissions from coal mining, oil and natural gas, and the waste and agricultural sectors;
5. Reduction of fluorinated gases in specified applications;
6. Energy efficiency improvement measures in the transport system.[86]

The EU used the 2001 Report to develop a package of three measures to combat climate change, all of which have since been employed.[87] First, the EU issued the 'Communication from the Commission,' outlining the implementation strategy for the first phase of the ECCP.[88] Second, it proposed a plan for the EU to ratify the Kyoto Protocol, which it did in April 2002.[89] Third, the EU recommended adopting a Directive on Greenhouse Gas Emissions Trading, which would allow specific businesses and industries to trade their allocations of CO2 emissions, and thus, reduce emissions in a cost-efficient manner.[90] The trading system is now

---

[81] See Second ECCP Progress Report (n 52).
[82] See ibid. [83] See ibid. [84] See ibid.
[85] Europa, European Commission on Environment, 'EU Emissions Trading: An Open Scheme Promoting Global Innovation to Combat Climate Change' (Brochure discussing the EU Emissions Trading Scheme 2009) <http://ec.europa.eu/environment/climat/pdf/brochures/ets_en.pdf> accessed 30 December 2009 (ETS Brochure).
[86] Consensus Among Many Voices (n 2) 389–90.
[87] Second ECCP Progress Report (n 52); see also U Steiner Brandt and G Tinggaard Svendsen, 'Rent-Seeking and Grandfathering: the Case of GHG Trade in the EU' (2004) 15 *Energy & Environment* 69.
[88] See Climate Change Homepage (n 64). [89] See ibid. [90] See ibid.

one of the central features of the EU umbrella program and is discussed in more detail below.[91] The EU has followed through with all three proposed measures—implementing the first phase of the ECCP, ratifying the Kyoto Protocol, and creating the underlying framework for a Community-wide emissions trading scheme.

The Second European Climate Change Programme

Drawing upon the lessons of the First ECCP, the EU launched the Second ECCP (ECCP II) on 24 October 2005.

The ECCP II prioritizes progress in the following areas: emissions trading, energy efficiency, renewable energy, the transport sector (including aviation and maritime transport), carbon capture and storage, and adaptation.[92] It also expands work on Community efforts to identify the impacts of climate change and on efforts to begin developing climate change adaptation strategies. Furthermore, the ECCP II broadens the objectives of the first ECCP—which was confined to meeting Kyoto Protocol emission reduction targets—and is more multifaceted than the original, given that it expands the mandate beyond halting climate change to also include adapting to climate change and improving international cooperation, technology transfer, research, and education.

The primary goal of the second phase of the ECCP is to 'facilitate and support the actual implementation of the priorities identified in the first phase'.[93] Thus, during the second phase of the ECCP, the goal is to make concrete and measurable progress towards mitigating and adapting to climate change. The Commission has primary responsibility for coordinating and facilitating the implementation of the goals of the ECCP II. As early as March 2005, the Commission had made substantial progress in implementing the ECCP II by initiating the following plans and programs:

- The proposal for an EU framework for emissions trading (see discussion below);
- An action plan to improve energy efficiency in the EU;
- A proposal for a directive on the promotion of electricity from renewable energy sources in the internal electricity market;
- A communication and proposal for a directive on the promotion of biofuels;
- A proposal for a directive to promote combined heat and power biofuels;
- A communication regarding vehicle taxation; and, in 2006,
- A proposal to include aviation in the EU ETS.

Further progress has since been made toward developing policies and measures for implementing the Kyoto Protocol's flexibility mechanisms, for reducing

---

[91] See Commission (EC), 'Greenhouse Gas Emissions Trading within the European Union' (Green Paper) COM (2000) 87 final, 15 September 2000; ETS Brochure (n 85).
[92] Commission (EC), 'Winning the Battle Against Climate Change' COM (2005) 35 final, 2 September 2005.   [93] Climate Change Homepage (n 64).

greenhouse gas emissions from agriculture, and for understanding the potential value of using carbon sinks in both the agricultural sector and in forests.[94] In addition to working toward the goals of the first phase, the second phase of the ECCP also focuses heavily on promoting the use of renewable energy sources.[95]

As will be discussed in the sections below, renewable energy is a central element of the ECCP II. However, it is only one of the Programme's many components. The ECCP II also facilitates measures to review and improve the emissions trading scheme, efforts to promote energy efficiency, programs to improve the carbon footprint of the transport sector, plans to promote carbon capture and storage, and policies to support climate adaptation measures.

The ECCP II creates the policy framework for the EU's ongoing legal and political efforts to address climate change. The ECCPs—I and II—have both substantive and symbolic value. They 'provide strong, long-term signals to industry, member state governments and the wider international community that the EU is committed to tackling climate change and expects all of its institutions, businesses and citizens to play their part'.[96] Furthermore, they provide the framework within which the Commission, the Council, and the EP can work to institute legally binding mechanisms and long-term policy strategies designed to meet Community goals and obligations.

Air Quality Framework Directives and IPPC Directive

In addition to the legal measures and voluntary agreements adopted through the ECCP, the EU has other measures in place to control air quality, including directives limiting emissions that contribute to climate change.[97] EU air quality policy is delivered through the European Air Quality Framework Directive, as expanded by numerous Daughter Directives and the 2008 New Air Quality Directive.[98] Similar to the effect of the CAA on air quality control in the US, the European Air Quality Control Directives identify and establish threshold limits or targets for key air pollutants and provide mechanisms for harmonizing monitoring strategies, measuring methods, and quality assessment procedures in order to create a uniform set of standards.

Air quality directives, combined with the ECCP, set minimum environmental standards and goals for member states in controlling climate change. Member states

---

[94] See ibid. [95] See ibid.
[96] Commission (EC), 'Fourth National Communication from the European Union under the Framework Convention on Climate Change (UNFCCC)' (Communication) COM (2006) 40 final, 2 August 2006.
[97] See Council Directive (EC) 2001/80 of 23 October 2001 on the limitation of emissions of certain pollutants into the air from large combustion plants [2001] OJ L301/9; Council Directive (EC) 2001/81 of 23 October 2001 on national emission ceilings for certain atmospheric pollutants [2001] OJ L309/22.
[98] Council Directive (EC) 96/62 of 27 September 1996 on ambient air quality assessment and management [1996] OJ L296/55; Council Directive (EC) 2008/50 of 21 May 2008 on ambient air quality and cleaner air for Europe [2008] OJ L152/1.

are then autonomous in determining how they will conform to the EU standards and whether they will establish national rules that exceed basic EU requirements.

Beyond the Air Quality Directive, there also exists an interesting and controversial dynamic between EU climate policy and the EU directive concerning integrated pollution prevention and control (IPCC)[99] (soon to be Industrial Emissions Directive (96/61)).[100] The IPCC Directive mandates that all industrial and agricultural operations with high pollution potential must obtain a permit, which can only be issued if certain environmental conditions are met. The objective of the directive is to internalize the costs of pollution prevention, to maximize pollution prevention, and to ensure that the parties responsible for creating pollution bear the burden of preventing and reducing pollution at its source. While the IPCC is institutionally distinct from EU climate policy, the relationship between the IPCC and the EU ETS is pertinent because the EU ETS precludes member states from including in IPPC permits emission limit values for direct emissions of greenhouse gases for installations covered by the trading scheme unless it is necessary to ensure that no significant *local* pollution is caused. Similarly, for activities covered by the EU ETS, under the IPCC directive, member states may choose not to impose energy efficiency requirements for covered units.[101] Implementation of the EU ETS, thus, intrudes upon the functioning of the IPPC, which is heralded as one of the EU's more progressive environmental protection regimes. This relationship has prompted questions about whether EU efforts to reduce greenhouse gases under the ETS undermine what could potentially be more effective emission controls under the IPPC as well as creating secondary pollution problems.

Promoting Renewable Energy

Returning to climate-specific actions, the evolution of renewable energy policy in the EU offers a helpful tool for charting the larger development of Community climate policy. The EU established initial measures to develop renewable energy as early as the late 1990s. In 1997, after recognizing the role of renewable energy in reducing CO2 emissions, enhancing sustainability, and ensuring the security of energy supplies, the EU issued a White Paper for a Community strategy on renewable energy.[102] The White Paper proposed an ambitious target of 'doubl[ing] the share of renewable energies in gross domestic energy consumption in the EU by 2010 (from the [then] 6% to 12%), including a timetable of actions to achieve this objective in the form of an Action Plan'.[103]

---

[99] Council Directive (EC) 2008/1/EC of 15 January 2008 concerning integrated pollution prevention and control [1996] OJ L24/8 (IPCC).

[100] Commission Proposal COM (2007) 844 final, Proposal for a directive of the European Parliament and of the Council on industrial emissions (integrated pollution prevention and control), 21 December 2007.   [101] IPCC (n 99) art 9(3).

[102] Commission (EC), 'Renewable energy: White Paper laying down a Community strategy and action plan' COM (97) 599 final, 26 November 1997.

[103] Europa, European Commission of Energy, 'Renewable Energy: What do we want to achieve?' (Webpage) <http://ec.europa.eu/energy/renewables/index_en.htm> accessed 30 December 2009.

The central feature of the Action Plan was the 'Campaign for Take-Off' of renewables, which established a framework for 'action to highlight investment opportunities and attract the necessary private funding which is expected to make up the lion's share of the capital required'.[104] One of the goals of the Campaign was to facilitate both public and private investment in renewable energy. During its implementation period, 2000 through 2003, the Campaign maintained a high profile and focused considerable attention on key renewable sectors, including solar, wind energy, and biomass. Despite modest gains, the EU continued to lag behind its renewable energy goals.[105]

Seeking to overcome the slow pace of progress in developing renewable capacity, the Commission issued a Renewable Energy Road Map in January 2007.[106]

Renewable Energy Road Map 2007

The Road Map describes the progress made, as of 2007, towards increasing the market shares of renewable energy as 'patchy and highly uneven' and categorizes much of the EU regulatory framework for renewable energies as 'relatively weak'.[107] The Commission emphasized that the only area in which substantial progress had been achieved was the electricity sector, resulting from the 2001 directive on renewable electricity. For this reason, the Commission proposed in its Road Map a *legally binding* 20% commitment by 2020 that is based upon comprehensive, long-term mandatory targets.[108] However, the Commission recognized that without binding new Community and national implementation measures, the EU would struggle to meet the 20% goal.[109] The Road Map therefore provided a series of proposed policies and legal measures, including:

- legal provisions that remove barriers to the integration of renewable energy sources in the EU energy system;
- legislation that addresses barriers to growth in the use of renewable energies in the heating and cooling sectors;
- measures for assessing member state support systems for renewable energy, with a view toward harmonizing schemes for renewables in the EU internal electricity market;
- support for a proposal for an incentive/support system for biofuels;

---

[104] See Renewable Energy for Europe, 'Renewable Energy Sources for Europe: The Campaign for Take-Off' (2001) <http://ec.europa.eu/energy/idae_site/catalog_text.html> accessed 30 December 2009.

[105] Between 1990 and 2003, the share of renewable energy in the EU-15's electricity consumption grew very slightly from 13.4% to 13.7%. The market share of renewable electricity has increased a modest amount since 2003. European Environment Agency, *Greenhouse Gas Emissions Trends and Projection in Europe 2005* (Report 2005) <http://www.eea.europa.eu/publications/eea_report_2005_8> accessed 1 June 2010.

[106] Commission (EC), 'Renewable Energy Road Map—Renewable energies in the 21st century: building a more sustainable future' (Communication) COM (2006) 848 final, 10 January 2007.

[107] Ibid. [108] Ibid 9–10. [109] Ibid 35.

- plans for promoting renewable energies sources in public procurement;
- plans for promoting the exchange of best practices on renewable energy sources;
- measures for internalizing the external costs of conventional fossil fuel energy.[110]

The Commission used the Road Map to promote major modifications to Community energy policy. Of particular importance, the Road Map states that 'the Commission is convinced that a legally binding target for the overall contribution of renewables to the EU's energy mix plus mandatory minimum targets for biofuels are now called for'.[111] In this way, the Commission laid the groundwork for negotiating a comprehensive and binding regulatory regime for renewable energy in the EU.

Renewable Energy Directive 2009

The culmination of the Roadmap was the adoption in 2009 of a new directive for the promotion of energy from renewable sources mandating that the market share of renewable energy be 20% by 2020.[112] Directive 2009/28/EC is the legal cornerstone of EU efforts regarding renewable energy. The directive establishes individual targets for the proportion of renewables in member states' final energy consumption. These binding national targets vary widely across the EU, ranging from a 10% target in Malta to a 49% target in Sweden, but together create a pathway to ensuring that the EU reaches its 20% goal by 2020. The directive further stipulates that each member state must ensure a 10% renewable share in domestic *transport* sectors and lays down criteria for the sustainable use of biofuels.

The objectives of the directive are ambitious given that the current market share for renewable energy in the EU wavered around 16% in 2006 and will need to double to over 30% to allow the EU to reach its 20% renewable target by 2020 but the sectoral and geographic differentiation is designed to create a viable pathway for gradual increases in market share.[113] In addition, existing and proposed EU directives for improving the market share of biofuels in transport and for increasing electricity generation from combined heat and power will help achieve the increased target for renewable energy.

---

[110] Ibid 12–13. [111] Ibid 18.
[112] See Council Directive (EC) 2009/28 of 23 April 2009 on the promotion of the use of energy from renewable sources and amending and subsequently repealing Directives 2001/77/EC and 2003/30/EC [2009] OJ L140/16.
[113] Europa, European Commission of Energy, 'The promotion of electricity from renewable energy sources' <http://ec.europa.eu/energy/renewables/electricity/electricity_en.htm> accessed 30 December 2009.

European Union Emissions Trading System

The ability of the EU to create such an ambitious renewable energy target was enabled, at least in part, by the existence of its emissions trading system. As early as 1999, the European Commission proposed the creation of a Community-wide emissions trading scheme.[114] On 13 October 2003, the EU finalized a scheme for a GHG Emission Trading System (ETS) within the region.[115] The ETS, which began operating on 1 January 2005, is both the first and the largest multi-country, multi-sector greenhouse gas emission trading scheme to exist worldwide.[116] It was designed to operate in three phases to allow for growth and improvement. The first phase lasted from 2005–2007; the second phase coincides with the first Kyoto Compliance period, 2008–2012; the third phase will run from 2013–2020.

The ETS Directive promotes the use of more energy efficient technologies, including combined heat and power.[117] Initially, the EU scheme was limited to trading CO_2 from large stationary sources (such as power plants or industrial facilities); but during phase three, coverage is being extended to other sectors and to greenhouse gases other than CO_2.[118] The ultimate goal of the ETS is to assist member states in meeting their Kyoto Protocol commitments by facilitating cost-effective compliance programs.[119]

National Allocation Plans

Under the EU ETS system of tradable allowances, member states allocate allowances among the public and private sectors and establish electronic registries where allowances will be held. Each member state is responsible for creating a 'National Allocation Plan,' (NAP) using EU guidelines and advice.[120] NAPs

---

[114] Commission (EC), 'Preparing for Implementation of the Kyoto Protocol' (Communication) COM (99) 230 final, 19 May 1999.

[115] Commission Decision (EC) of 29/01/2004 establishing guidelines for the monitoring and reporting of greenhouse gas emissions pursuant to Directive 2003/87/EC of the European Parliament and of the Council C(2004) 130 final; Council Directive (EC) 2004/101 of 27 October 2004 amending Directive 2003/87/EC establishing a scheme for greenhouse gas emission allowance trading within the Community, in respect of the Kyoto Protocol's project mechanisms [2004] OJ L338/18.

[116] ETS Brochure (n 85). [117] Ibid.

[118] Stationary sources include heat generators, oil refineries, ferrous metals, cement, lime, glass and ceramic materials and pulp and paper facilities. See ETS Brochure (n 85) 15. In addition, as concerns the greenhouse gases covered by the trading scheme, member states also have the option of opting in other greenhouse gases during phase II, eg, from 2008 onwards the EU ETS was broadened to cover emissions of nitrous oxide from the production of nitric acid in the Netherlands and in Norway. [119] See ETS Brochure (n 85); see also Kruger and Pizer (n 60).

[120] See Commission Regulation (EC) 2216/2004 of 21 December 2004 for a standardised and secured system of registries pursuant to Directive 2003/87/EC of the European Parliament and of the Council and Decision No 280/2004/EC of the European Parliament and of the Council [2004] OJ L386/1; see also Council Directive (EC) 2003/87 of 13 October 2003 establishing a scheme for greenhouse gas emission allowance trading within the Community and amending Council Directive 96/61/EC arts 9–11 and Annex III.

indicate the level of CO2 emission allowances that each member state proposes to allocate during each specified trading period, and the method for distributing these allowances among the companies that are responsible for CO2 emissions.[121] The European Commission is charged with analyzing and approving the NAPs and it controls the distribution of total allowances and allocations, almost 80% of which it allocates to sources in the EU-15 member states, because of their additional commitment under the Kyoto Protocol.

The Commission analyzes each NAP using 11 allocation criteria, including 'consistency with the country's overall strategy to reach its Kyoto target and emissions developments, non-discrimination, respect for European competition and state aid rules, and certain technical aspects'.[122] By 20 June 2005, the Commission had reviewed and approved—often with revisions and modifications—NAPs for 25 countries, encompassing all of the member states participating in the 2005–2007 trading period.[123] Together, the NAPs accounted for 6.57 billion allowances from more than 11,400 installations, representing more than half (52%) of all of Europe's CO2 emissions.[124]

The EU Central Administrator oversees the registries using the 'Community independent transaction log' to monitor transactions for irregularities.[125] The system resembles bank operations except that instead of monitoring the ownership and movement of money, it monitors the ownership and movement of emission allowances. During the first two years of existence, McKinsey & Company and Ecofys monitored the program in order to analyze the impact and success of the system.[126]

The ETS operates in addition to and supplementary to the respective member states' national programs, as well as member state participation in Joint Implementation and Clean Development Mechanism projects under Kyoto's flexibility mechanisms, to which it is also linked.[127] The linking directive joins up the EU ETS with the Joint Implementation and Clean Development Mechanisms under the Kyoto Protocol, thus allowing credits generated via these two project based mechanisms to be traded in the EU market. The objective of the linking directive was to increase participation in the Kyoto flexibility mechanisms by creating new incentives for businesses to participate in these projects—thus bolstering the creation of a stable carbon market—while simultaneously increasing the cost-effectiveness of member state efforts to meet their commitments under the EU emissions allowance trading scheme. The

---

[121] See ETS Brochure (n 85) 11.
[122] Europa, 'Emissions Trading: Commission Approves Last Allocation Plan Ending NAP Marathon' (Press Release 20 June 2005) <http://ec.europa.eu/rapid/pressReleasesAction.do?reference=IP/05/762&format=HTML&aged=0&language=EN&guiLanguage=en> accessed 30 December 2009.
[123] Ibid.   [124] Ibid.   [125] Ibid.   [126] ETS Brochure (n 85).
[127] Council Directive (EC) 2004/101 of 27 October 2004 amending Directive 2003/87/EC establishing a scheme for greenhouse gas emission allowance trading within the Community, in respect of the Kyoto Protocol's project mechanisms [2004] OJ L338/18; Kyoto Protocol (n 57) arts 3, 6, 12, and 16.

linking directive has proved critical to the functioning of the ETS as well as to overall member state efforts to meet their national targets, with healthy trading of flexibility mechanism credits being traded in the European market.

During its first phase, 2005–2007, the EU ETS was, at best, a qualified success; while phase one did not witness the creation of a stable market leading to significant emission reductions, it put in place the necessary market institutions and served its purpose as a test phase by allowing the EU to identify strengths and weaknesses of existing strategies and institutions. In the first year of the project, over 362 million tonnes of CO2 were traded on the market and market prices remained fairly healthy—reaching a peak of €30 a tonne in April 2006. However, shortly thereafter, the market crashed following an unintentional leak of verified emissions data revealing that individual countries had established such high emissions caps that there was little need for many industries to reduce their emissions.[128] Due to resulting market insecurities, market prices declined to less than €1 a tonne in spring 2007. Furthermore, market price declines were attributable to over-allocations, as well as the inability to bank unused emission allowances for phase two. As a result, 2007 was an unsuccessful year for the EU ETS. This challenged the Commission to overhaul and revive the system for more successful subsequent phases.[129]

Market-based trading problems are inevitably difficult to get right in the first phase due to information deficits and economic and programmatic complexities, but can improve over time, as the keystone US sulfur dioxide trading program demonstrates. The first phase of the EU ETS provided valuable insight into the changes that are needed to improve market functioning. In particular, the Commission identified the need for the stricter review of NAPs to deter over-allocations that undermine market prices.

From a structural perspective, the second phase of the EU ETS also differs in that it expands the scope of the scheme to allow the inclusion of non-CO2 greenhouse gases, to provide mechanisms for linking Clean Development Mechanism and Joint Implementation credits to the European scheme through the 2004 linking directive, and to include non-EU members.[130] Responding to problems of complexity and lack of transparency in phase one, the Commission made significant modifications to the parameters of participation between phases one

---

[128] Karan Capoor and Philippe Ambrosi, *State and Trends of the Carbon Market 2007* (World Bank Report, Washington DC, May 2007) <http://www.ieta.org/ieta/www/pages/getfile.php?docID=2281> accessed 30 December 2009.

[129] Climate Action Network Europe, *National Allocation Plans 2005–2007: Do they Deliver?—Key Lessons for Phase II of the EU ETS* (Summary Report for Policymakers) <http://qualenergia.it/UserFiles/Files/CAN_NAPsReport_Summary0306.pdf> accessed 30 December 2009.

[130] Council Directive (EC) 2004/101 of 27 October 2004 amending Directive 2003/87/EC establishing a scheme for greenhouse gas emission allowance trading within the Community, in respect of the Kyoto Protocol's project mechanisms [2004] OJ L338/18. See also Commission Decision (EC) 2006/780 of 13 November 2006 on avoiding double counting of greenhouse gas emission reductions under the Community emissions trading scheme for project activities under the Kyoto Protocol pursuant to Directive 2003/87/EC of the European Parliament and of the Council [2006] OJ L316/12.

and two that increased the stringency of review while decreasing complexity and increasing transparency. In particular, the Commission identified new criteria for more rigorously assessing NAPs.[131] As of 11 December 2009, the Commission had reviewed and accepted 25 NAPs for phase two of the EU ETS. At this point, mid-way through the second phase, Estonia and Poland still lacked approved NAPS, having had their plans rejected by the Commission on multiple occasions due to the over allocation of allowances.[132] As the repeated rejections of Estonia and Poland's plans revealed, the Commission applied a heightened level of scrutiny in reviewing member state NAPs for the second phase of the ETS. In particular, during the review process the Commission required substantive changes on multiple occasions where:

- the proposed total of allowances ('cap') for the 2008–2012 trading period is not consistent with meeting the member state's Kyoto target;
- the proposed total of allowances is not consistent with expected emissions or the technological potential to reduce emissions, taking into account independently verified emissions in 2005, anticipated changes in economic growth, and carbon intensity;
- the proposed limit on the use by companies of credits from emission-reduction projects in third countries carried out under the Kyoto Protocol's flexible mechanisms is not consistent with the rule that the use of these mechanisms should be supplementary to domestic action to address emissions.[133]

The Commission approved most of the plans on the condition that they include a reduction in the total number of the proposed emission allowances. In the case of the first nine NAPs approved for Greece, Ireland, Latvia, Lithuania, Luxembourg, Malta, Slovakia, Sweden, and the UK, for example, the Commission reduced the proposed allowances by almost 7%, resulting in a 7% cut from 2005 emission allowances.

Implementation of the second phase of the ETS—and the Commission's decisions therein—has also given rise to an important body of case law concerning the parameters of the trading scheme.[134] By 2009, the ETS had been

[131] Commission (EC), 'Assessment of national allocation plans for the allocation of greenhouse gas emission allowances in the second period of the EU Emissions Trading Scheme' (Communication) COM (2006) 725 final, 29 November 2006.

[132] Europa, 'Emissions trading: Commission takes new decisions on Estonian and Polish national allocation plans for 2008-2012' (Press Release, 11 December 2009) <http://europa.eu/rapid/pressReleasesAction.do?reference=IP/09/1907&format=HTML&aged=0&language=EN&guiLanguage=en> accessed 30 December 2009.

[133] Europa, 'Emissions trading: Commission decides on first set of national allocation plans for the 2008–2012 trading period' (Press Release, 29 November 2006).

[134] See Josephine AW vanZeben, 'The European Emissions Trading Scheme Case Law' Amsterdam Center for Law & Economics Working Paper No 2009-12, 27 August 2009, <http://papers.ssrn.com/sol3/papers.cfm?abstract_id=1462651> accessed 10 March 2010; Navrah Singh Ghaleigh, 'Emissions Trading Before the European Court of Justice: Market Making in

the subject of an impressive 40 proceedings before the Courts on a variety of grounds, including challenges to the validity of the directive and challenges to Commission decisions on NAPs.[135] The majority of challenges to the legality, functioning, and enforcement of the ETS have failed and the emerging case law has largely had the effect of 'bullet proofing' the Commission and its decisions, which are seen to strengthen the stability of the emerging carbon market.[136]

With the second phase on-going, the Commission is already tasked with undertaking major revisions to the program during phase three. In April 2009, the EU adopted a new directive to improve and extend the EU ETS.[137] Phase three will run from 2013 to 2020, following the conclusion of the first Kyoto Protocol commitment period in 2012. During this phase, the Commission is requiring emission allocations to be cut by 21% from 2005 allocations. In addition, greenhouse gases will be reduced by 20% from 1990 levels and the Commission will require 20% of power to be derived from renewable energy sources. NAPs will no longer be necessary because there will be one EU-wide emissions allocation.[138] Auctioning will be prevalent—phase three will provide for 60% of EU allowances to be auctioned, a significant increase from the 10% in phase two. The 60% provision will continue to be increased later in the phase, with the goal of full auctioning by 2027. Further, the scope of the scheme will be enlarged to include new sectors, not just large industrial installments. Petrochemical, ammonia, aluminum, and aviation (as of 2012) sectors are included, covering approximately 50% of all EU emissions. Additional plans are in place to expand the EU ETS to include the aviation sector following an European Econimic and Social Committee (EESC) recommendation suggesting that inclusion of air transport in the EU ETS offered the most cost-effective and environmentally preferable solution for beginning to regulate greenhouse gas emissions from aviation.[139] After long deliberations, in 2006, the Commission decided to include emission from the aviation sector starting in 2012. Sectors that are not included are road transport, shipping, agriculture, and forestry, mainly due to the difficulty in measuring emissions viably.[140] Smaller installations have

---

Luxembourg', in *Legal Aspects of Carbon Trading: Kyoto, Copenhagen and Beyond,* D Freestone and C Streck (eds) (Oxford, OUP, 2009).

[135] Ghaleigh (n 134) 12. [136] Ibid 32.
[137] Council Directive (EC) 2009/29 of 23 April 2009 amending Directive 2003/87/EC so as to improve and extend the greenhouse gas emission allowance trading scheme of the Community [2009] OJ L140/63.
[138] Phase three exhibits significantly higher degrees of harmonization, with the Commission replacing member state caps with an EU-wide cap and establishing harmonized rules for allocating emissions allowances.
[139] Commission (EC), 'Opinion of the European Economic and Social Committee on the Communication from the Commission to the Council, the European Parliament, the European Economic and Social Committee and the Committee of the Regions—Reducing the Climate Change Impact of Aviation' COM (2005) 459 final, 21 April 2006.
[140] Euractiv, 'EU Emissions Trading Scheme' (22 January 2007) <http://www.euractiv.com/en/climate-change/eu-emissions-trading-scheme/article-133629> accessed 30 December 2009.

the option to be excluded from the ETS if they emit under 25,000 tons of CO2 per year if alternative reduction measures are enacted.

In April 2009, at the same time that the EU expanded the ETS, it also took steps to improve emission reduction efforts in sectors excluded from the emissions trading scheme. Looking to reduce emissions from non-EU ETS sectors, the EU adopted an 'Effort Sharing Decision' that aims to reduce emissions from key sectors such as transport, buildings, agriculture and waste.[141] Similar to the 2009 Renewables Directive, under the Effort Sharing Decision each member state has agreed to a binding national emissions limitation target for 2020 determined on the basis of the State's relative wealth; the targets range from emissions reduction commitments of 20% to allowed emission increases of 20%. Together, the member state targets are designed to cut the EU's collective emissions from non-ETS sectors by 10% by 2020 compared with 2005 levels.

Although supplemental to, and institutionally separate from the ETS, the Decision is important in the context of the ETS because it helps to even out EU climate policy and create a more comprehensive policy regime for promoting a region-wide move to a low-carbon society.

It is still too early to analyze the overall success of the EU ETS. Although the first phase of the EU ETS saw ambiguous results, and a tough end to the second year, the Commission is careful to point out that the first phase of the EU ETS was always envisioned as a learning period.[142] The second phase of the EU ETS, 2008–2012, corresponding with the first Kyoto compliance period, is the critical phase for determining whether the Commission has crafted a scheme that successfully internalizes climate change concerns into private sector decision making with the end result of reducing member state—and, thus, Community—greenhouse gas emissions. Mid-way through the second phase many of the allocation problems of the first phase have been addressed and the price of carbon has begun to stabilize, allowing greater predictability and functioning.[143] As the price of carbon continues to stabilize and the market becomes more established, it is anticipated that the ETS will become increasingly effective. Mid-way through the second phase, however, the price of carbon—while relatively stable—remained low, raising fundamental questions about whether

---

[141] Council Decision No 406/2009 (EC) of 23 April 2009, on the effort of Member States to reduce greenhouse gas emissions to meet the Community's greenhouse gas emission reduction commitments up to 2020 [2009] OJ L140/136.

[142] Commission (EC), 'Assessment of national allocation plans for the allocation of greenhouse gas emission allowances in the second period of the EU Emissions Trading Scheme' (Communication) COM (2006) 725 final, 29 November 2006.

[143] Eg, A Denny Ellerman and Barbara K Buchner, 'The European Union Emissions Trading Scheme: Origins, Allocation, and Early Results' (2007) 1(1) *Rev Envtl Econ & Policy* 66; Frank J Convery and Luke Redmond, 'Market and Price Developments in the European Union Emissions Trading Scheme', (2007) 1(1) *Rev Envtl Econ & Policy* 88, Joseph Kruger, Wallace E Oates, and William A Pizer, 'Decentralization in the EU Emissions Trading Scheme and Lessons for Global Policy' (2007) 1(1) *Rev Envtl Econ & Policy* 112.

the EU ETS can effectively induce the shifts necessary to move the EU towards a low-carbon economy in the long-term.[144]

The EU ETS is a globally significant initiative. It is the first in the history of the EU and the largest effort to date globally to create a comprehensive greenhouse gas emissions trading scheme with the objective of incrementally reducing emissions. The ETS creates opportunities for substantive emissions reductions, for changing business practices, and for global dispersal of emissions reduction techniques. Both the successes and the failures of the EU ETS offer useful lessons for global efforts to commodify carbon and other greenhouse gases.

The 2°C Goal

The European Council reaffirms that absolute emission reductions are the backbone of a global carbon market. Developed countries should continue to take the lead by committing to collectively reducing their emissions of greenhouse gases in the order of 30% by 2020 compared to 1990. They should do so also with a view to collectively reducing their emissions by 60% to 80% by 2050 compared to 1990.[145]

Bolstering the evolving renewable energy and emissions trading initiatives, in 2007, the European Commission issued a Communication to the Council suggesting that the 'EU must adopt the necessary domestic measures and take the lead internationally to ensure that global average temperature increases do not exceed pre-industrial levels by more than 2°C'.[146] The Commission emphasized the economic and technical feasibility of the 2°C goal and the importance of bringing the major emitters together to act swiftly. In making this proposal, the Commission set a new standard for Europe and for the rest of the world.

To achieve its target, the EU must reduce its greenhouse gas emissions by 30% by 2020 (as compared to 1990 levels). Even absent global participation and the negotiation of a new international agreement, the Commission proposed that the EU 'now take on a firm independent commitment to achieve at least a 20% reduction of GHG emissions by 2020, ... [via] the EU emission trading scheme (EU ETS), [and] other climate change policies and actions in the context of energy policy'.[147] Such a commitment, the Commission argued, would limit the extent of climate change, demonstrate international leadership on climate change, and provide a positive signal to industry that the ETS would continue and would, therefore, encourage the creation of a well-functioning market.

---

[144] The price of carbon in Europe dipped following the economic recession that began in 2008, suggesting that problems of over-allocation once again plagued its success and raising concerns about whether the ETS lacks the necessary flexibility to respond to social and economic changes that modify market conditions.

[145] Council of the European Union, 'Council Conclusions on climate change' (Conclusions from the 2785th Environment Council Meeting, 20 February 2007) <http://ec.europa.eu/environment/climat/pdf/future_action/feb20draft.pdf> accessed 30 December 2009.

[146] Commission (EC), 'Limiting Global Climate Change to 2 degrees Celsius: The way ahead for 2020 and beyond' (Communication) COM (2007) 2 final, 10 January 2007.

[147] Ibid.

In addition to using the Communication as a vehicle for contrasting the costs and benefits of action over inaction, the Commission also created a list of concrete actions that the EU should take to effectively meet the 2ºC goal. The underlying foundation for the Commission's proposal is the EU's reduction of regional greenhouse gas emissions by 20–30% by 2020. To reach this goal, the Commission proposed:

- improving the EU energy efficiency by 20% by 2020;
- improving the share of renewable energy to 20% by 2020;
- adopting an environmentally safe carbon capture and geological storage policy;
- strengthening the EU ETS by: making allocations for more than five years, providing predictability, extending the scheme to other gases and sectors, recognizing carbon capture and geological storage, harmonizing allocation process among member states, and linking the EU ETS to compatible mandatory schemes (eg, in California and Australia);
- limiting transport emissions by: including aviation in the ETS, linking car taxes to CO2 emissions, adopting further measures to tackle CO2 emissions from cars, strengthening demand-oriented measures, limiting GHG emissions from road freight transport and shipping; and by reducing life-cycle emissions of CO2 in transport fuels;
- reducing GHG emissions in other sectors, to include residential and commercial buildings and from non-CO2 gases;
- increasing resources dedicated to research and technological development;
- improving cohesion in climate change programs;
- examining all the possible ways to reduce GHG emissions and to ensure 'the environmental and economic consistency of the measures to be adopted'.[148]

The 20/20/20 plan—ie, 20% GHG reduction below 1990 levels, 20% improvement in energy efficiency, 20% share of renewables in the total energy mix (up from 8.5 % today) by 2020—redefined the existing climate strategy and prompted subsequent efforts in fall 2008 and spring 2009 to revise EU climate strategy, as discussed later in this chapter.

The Communication also offered a series of suggestions for improving international action to combat climate change. In the Communication, the Commission reiterated the importance of globally cooperative and coherent action. In particular, the Communication emphasized the importance of including major emitters in global negotiations and creating a post-2012 framework that 'contain[s] binding and effective rules for monitoring and enforcing commitments'.[149]

---

[148] Ibid 5–7.   [149] Ibid 9.

Further, the Communication recommended ways to improve options to encourage developing countries to take further actions to curb GHG emissions, including: streamlining and expanding the Clean Development Mechanism, improving developing countries' access to finance, and introducing sector-wide company-level emissions trading at the national and/or global levels. The Commission also suggested that countries adopt emission reduction commitments as their development improves, but recommended that least developed countries should not be asked to adopt any commitments for legally binding emission reductions.

Additionally, the Commission considered climate change in its wider context and stressed the importance of improving international research and technological cooperation, supporting actions to halt deforestation, and encouraging adaptation measures, as well as creating an international agreement on energy efficiency standards to harmonize international standards and promote market consistency.

Ultimately, the Communication instituted an extensive series of measures and suggestions—the Commission used the Communication to promote the new Community goal of limiting the global temperature rise to 2ºC. Furthermore, the Commission used the Communication to present a blueprint for future Community actions to address climate change and, in doing so, to create a series of short-term and long-term objectives.

The Commission's Communication is not legally binding. Nevertheless, it is a politically important document because it establishes Commission priorities, recommends legal and political actions, and lays the foundation for future legislative proposals. In this way, the Communication provides a positive indication of the future direction of climate change governance in the EU.

The emissions reductions proposed by the Commission far outpace current commitments under the Kyoto Protocol and establish the foundations for an aggressive post-Kyoto agreement. As discussed in Chapter 1, during the 2007 G8 meeting, President Angela Merkel argued for the 2ºC goal as the cornerstone of a proposed G8 Heads of State statement on climate change. President George W. Bush rejected this target, but President Merkel used the G8 to endorse and promote the 2ºC goal on a global scale. As will be discussed in Chapter 9, the 2ºC goal was later endorsed by President Obama and has now been embedded as the cornerstone of global climate negotiations via the terms of the 2009 'Copenhagen Accord'.

Increasing the Role of Civil Society and Regional and Local Authorities

Efforts to address climate change in Europe and around the world—including in the US—focus primarily on the dual roles of the State and of the private sector, mostly at the expense of attempts to increase the role and responsibility of civil society and local government in addressing climate change. In 2006, however, the EESC and the Committee of the Regions took important steps to reduce this gap by issuing opinions on: (1) The role of civil society in meeting the challenges of climate change and (2) The contribution of local and regional authorities to combating climate change.[150]

In its opinion of 14 September 2006, on 'The role of civil society in meeting the challenges of climate change', the EESC suggested modifying the framework of climate change discussions. The EESC has taken an aggressive stance on the threat posed by climate change, characterizing it as 'the biggest challenge facing the European Union and the entire planet over the coming years and decades.'[151] Here, the EESC advocates widening the scope of the European climate change debate to make it more inclusive.

The opinion laments the fact that climate change discussions focus 'excessively on the macro level and events in the distant future' and suggests that '[t]here is a clear need for a debate on how climate change affects—and will affect citizens in their everyday life'.[152] The EESC argues that 'ownership of the climate change debate must be extended so as to involve actively the social partners and organised civil society at large'.[153] The EESC suggests that focusing on long-term and macro level discussions results in an overly-narrow debate that ignores both the micro level impacts of climate change for civil society and the potential contributions civil society can make to help mitigate the effects of climate change in the short and long-term.

Using the opinion to describe the scope of the challenges presented to civil society by climate change, the EESC emphasizes the ways in which climate change will affect everyday life, and concludes by identifying ten specific sectors that offer opportunities for civil society to address climate change, including: urban and community planning, building/construction, road transport, travel and leisure sector, agriculture, workplace issues, and disaster management. Further, the opinion suggests that engaging civil society in issues that directly affect citizens will increase understanding and involvement in addressing the causes of climate change. With this in mind, the EESC makes a series of recommendations, including the following:

- that EU member states set up offices to advise on issues of adaptation;
- that the Community launch an EU-wide dialogue on ways to deter further climate deterioration and to begin to take steps towards adapting to climate change; and,

---

[150] European Economic and Social Committee, 'Opinion of the European Economic and Social Committee on Meeting the challenges of climate change—The role of civil society' (14 September 2006) 2006/C 318/17 [2006] OJ C318/102.

[151] European Economic and Social Committee, 'Opinion of the European Economic and Social Committee on Sustainable development in agriculture, forestry and fisheries and the challenges of climate change' (18 January 2006) 2006/C 69/02, [2006] OJ C69/5 at 5.1.

[152] Ibid s A2.   [153] Ibid s 1.5.

- that the Community instigate an on-going dialogue with citizens with 'clear and continuous participation of stakeholders at the local and regional level'.[154]

The EESC advocates promoting citizen awareness on issues of climate change using a bottom-up approach that engages local and regional organizations, consumer organizations, industry, and enterprises, and advises using tools such as the Eco-Management and Audit Scheme—a voluntary environmental management scheme—to raise awareness and to reward organizations that take extra steps to improve environmental performance.

Finally, the EESC issues a challenge for civil society to play an active part in the European Dialogue on Climate Change by helping raise awareness of climate change, mobilizing consumers to make sound choices, and encouraging new programs for urban planning.[155] Furthermore, the EESC challenges civil society organizations to serve as conduits for information between citizens and government and for cooperating with global civil society to 'take constructive action to mitigate the effects of climate change'.[156]

The EESC opinion regarding the role of civil society in climate change comes at a critical time—there is increasing recognition that climate change will have widespread impact on aspects of everyday human life such as health, transport, work, and general social and economic well-being. Furthermore, as the EESC opinion highlights, government laws and policies in the EU and elsewhere focus primarily on the role of government and industry in addressing climate change, thereby creating a gap in efforts to halt and adapt to climate change.

Civil society awareness and participation in initiatives to address and adapt to climate change will be essential to any long-term strategy to meet the challenges that climate change poses.[157] Thus, while the EESC opinion is not binding, it establishes critical groundwork for extending climate change discussions to the people who will be affected by climate change and whose actions, choices, and political support are integral to addressing climate change in the short-term and long-term.

In a complementary opinion issued the previous year on 17 November 2005, the Committee of the Regions recommended that the Commission take a 'portfolio approach' to climate change that encouraged 'cooperation between all spheres of government, in partnership with the private sector, healthcare, community and education groups, and with energy efficiency organizations'.[158] While calling for further aggressive emission reduction targets and improvements in the

---

[154] Ibid s 4.7.   [155] Ibid s 6.2.   [156] Ibid.
[157] See European Economic and Social Committee, 'Involvement of civil society organisations essential for success of EU energy and climate change policy' (Press Release, 1 February 2008) <http://www.eesc.europa.eu/activities/press/cp/index_en.asp?year=2008> accessed 30 December 2009.
[158] Committee of the Regions, 'Opinion of the Committee of the Regions on the contribution of local and regional authorities to combating climate change' (17 November 2005) 2006/C 115/20, [2006] OJ C115/88.

energy sector, the opinion emphasizes the importance of local and regional governments in providing resources and demonstration projects. Here, again, the opinion highlights the importance of extending the current debate to include a wider group of stakeholders.

Before continuing to analyze the most recent developments in EU climate policy, it is worth pausing to juxtapose US and EU efforts to address climate change in order to highlight key distinctions.

Progress and Challenges in European Union Climate Change Policy

The EU is currently on track to meet its climate change goals, having made significant progress over the past decade. In the first half of the decade it seemed unlikely that the EU would meet its regional or international obligations. For example, the Second ECCP Progress Report (2003), the European Environment Agency's report, 'Greenhouse Gas Emission Trends and Projections in Europe 2005', and the European Communities' Fourth National Communication under the UNFCCC (2006) all indicated that the EU was unlikely to meet its Kyoto Protocol target unless it expanded domestic policies and measures.[159] These reports suggested that EU-15 and the EU-27 could, however, each meet and exceed their individual and collective targets by implementing additional policies and measures, in particular by drawing heavily upon the Kyoto Protocol flexibility mechanisms.[160]

The EEA 2005 report also revealed the extent to which member state progress towards meeting national emission reduction commitments varied. According to the 2005 report, among the EU-15, the UK and Sweden, and possibly France and Germany, were on track to meet their emission reduction obligations under the EU bubble.[161] Greece, Luxembourg, Austria, Belgium, Finland, and the Netherlands were not on track to meet their targets but still stood a chance of meeting their individual emission reduction obligations using a combination of domestic policies and measures in conjunction with heavy reliance on the Kyoto Protocol's flexibility mechanisms.[162] Meanwhile, in 2005, States such as Denmark, Spain, Portugal, Italy, and Ireland actually reported increases in their levels of greenhouse gas emissions and seemed unlikely to be able to achieve

---

[159] Commission (EC), 'Fourth National Communication from the European Union under the Framework Convention on Climate Change (UNFCCC)' (Communication) COM (2006) 40 final, 2 August 2006 at 3; Second ECCP Progress Report (n 52) 4.   [160] Ibid.
[161] European Environment Agency, *Greenhouse Gas Emission Trends and Projections in Europe 2005* (Report No 8, 2005) <http://www.eea.europa.eu/publications/eea_report_2005_8/GHG2005.pdf> accessed 1 June 2010.   [162] Ibid.

their emission reduction targets using any combination of domestic and Kyoto measures.[163]

Projections based on the 2005 emissions statistics, thus, suggested that the EU-15 would not meet its Kyoto 'bubble' target of reducing emissions by 8% below the 1990 baseline.[164] The European Environment Agency, however, suggested that if countries increased their use of the Kyoto Protocol flexibility mechanisms, and if certain countries—including the UK, Germany, and Sweden—exceeded their targets, the EU might still meet its overall Kyoto reduction target.[165] A majority of EU-15 and EU-27 member states expressed intent to use the flexible mechanisms—including Joint Implementation, Clean Development Mechanisms, and Emission Trading—to meet their individual Kyoto targets.

The findings in the 2005 report were reaffirmed in 2006 when the EU submitted its first report under the Kyoto Protocol 'to facilitate the calculation of the assigned amount...and to demonstrate its capacity to account for its emissions and assigned amount' for the first Kyoto compliance period.[166] In this report, the EU provided information on Community-wide greenhouse gas inventories, detailed information on how emissions reduction obligations are calculated, and gave an account of the progress made towards meeting EU commitments under the UNFCCC and the Kyoto Protocol.

The report indicated that, in 2004, EU-15 greenhouse gas emissions were 1.1% lower than emissions in the base year.[167] The report also confirmed the findings of the ECCP Progress Report and the EEA 2005 Report. Despite the apparent shortcomings in the EU's progress, when read together, even these three early reports revealed the extent to which the EU, in direct contrast to many of its developed country counterparts, was taking considerable steps towards putting in place a comprehensive climate change regime. Nevertheless, the reports revealed significant shortcomings in EU climate policy.

Following this rather discouraging report, just a few years later, in 2009, the EEA released an updated report suggesting that EU member states were progressing more rapidly than expected in their efforts to combine domestic and flexibility-based measures to achieve their emissions targets but that progress was mixed across sectors.[168] The EEA 2009 report estimates EU emission reduction progress based upon greenhouse gas emissions inventories for 1990 to 2007,

---

[163] Ibid.   [164] Ibid.   [165] Ibid.   [166] EC's Initial Report (n 51) 6.
[167] Three EU-15 member states established 1990 as the base year, while 12 member states set 1995 as the base year. The EU inventory aggregates the total findings based on the base year each member state has chosen. This calculation excludes emissions and removals associated with land use, land use change, and forestry.
[168] European Environment Agency, *Greenhouse gas emission trends and projections in Europe 2009: Tracking progress towards Kyoto targets* (Report No 9, 2009) <http://www.eea.europa.eu/publications/eea_report_2009_9/ghg-trends-and-projections-2009-summary.pdf> accessed 30 December 2009.

available estimates of 2008 emissions and greenhouse gas emission projections for 2010, 2015, and 2020, based on data provided by the member states.[169]

Overall, this report confirms that EU greenhouse gas emissions are decreasing and are projected to continue decreasing alongside the continued implementation of member state measures. Noting that the year 2008 marked the fourth consecutive year that EU emissions have decreased to reach their lowest level since 1990, the EEA lauds recent efforts of member states to decouple emissions from economic growth while simultaneously cautioning that emissions continued to increase in certain sectors, including electricity and heat production, industry, and transport.[170] Noting that the transport sector and growth in hydrofluorocarbons pose two of the most significant problem areas, the report suggests that for sectors, such as transport, not currently covered by the EU ETS, additional measures are being employed to facilitate new efforts to begin curbing emissions through, for example, modal shift, biofuels, and new efficiency standards.

The EEA also suggests that a greater number of EU-15 and EU-27 member states are on-track to meet their individual obligations using a combination of domestic policies and measures and credits generated through the Kyoto flexibility mechanisms. Based on current efforts the EEA determines that five EU-15 (France, Germany, Greece, Sweden, and the UK) and nine EU-12 member states (Bulgaria, Czech Republic, Estonia, Hungary, Latvia, Lithuania, Poland, Romania, Slovak Republic), and Croatia have already achieved GHG emission levels below their Kyoto target during 2003–2007 or 2004–2008 and that the EU-15 as a whole is capable of reducing its greenhouse gas emissions levels to 8.5% below the Kyoto base year.[171] The ability of the EU-15 to reach its target, however, is contingent on the anticipated combined emission reductions that exceed burden-sharing targets in the main emitting member states, to include, France, Germany, Spain, and the UK. These five States expect to meet their burden-sharing targets using existing measures. Additional domestic policies and measures, sinks, and flexibility mechanism projects will create additional emissions reductions.

The remaining ten states in the EU-15, however, will rely on emissions reductions generated through carbon sink activities and flexibility mechanism projects to meet their targets. Drawing upon these supplemental measures, all but one EU member state, Austria, anticipate meeting their commitments under the Kyoto Protocol. This projection differs significantly from analysis of just one year before that suggested that at least four EU-15 member states—Denmark, Italy, Spain, and Austria—would fail to reach their burden-sharing targets. This progress is attributable to existing EU polices, in particular the EU ETS, renewable energy policies and policies aimed at improving energy efficiency in buildings as well as internal energy market policies and improved uptake of flexibility mechanism projects. Existing measures will be complemented by planned initiatives in the

[169] Ibid 8. [170] Ibid 8. [171] Ibid 9.

areas of renewable energy, energy end-use efficiency; energy services might also provide additional savings. Emissions are also expected to experience situational decreases due to the recent economic downturn.[172]

Beyond its ability to achieve its 8.5% Kyoto target, the EEA determines that the EU is capable of achieving additional emissions reductions of 5.1% based on additional existing and planned measures and the implementation of Kyoto flexibility mechanism projects and carbon sinks. The EU expects ten member states to draw heavily on the flexibility mechanisms to meet their individual assigned amounts. These efforts, in combination with plans for new carbon sinks and further expansion and implementation of the EU ETS will boost projected emissions reduction levels. While utilization of the flexibility mechanisms and carbon sinks offer opportunities to boost emissions reduction numbers, heavy reliance on emissions reduction units generated through the flexibility mechanisms is not ideal. Relying on Joint Implementation and Clean Development Mechanism projects to generate emissions reduction units means that states are not implementing necessary domestic policies. Furthermore, validating Joint Implementation and Clean Development Mechanism projects is complex and plagued with ethical dilemmas and substantive and procedural challenges, challenging the authenticity and appropriateness of many of the alleged GHG emissions reduction units.[173]

In relation to the more ambitious regional target of reducing EU emissions 20% by 2020 as against 1990 levels, the report suggests that the EU is 'making good progress', noting that estimates indicate that the EU-27 reduced domestic greenhouse gas emissions by roughly 10.7% between 1990 and 2008.[174] Already more than halfway towards meeting its target, continued implementation of existing and planned regional measures offers opportunities to further reduce regional emissions by 14% below 1990 levels by 2020, enabling the EU to achieve three-quarters of its 20% target using domestic policies and measures alone.

The EEA projections are optimistic and contingent on multiple oscillating factors, including effective implementation and enforcement of existing and planned domestic measures, over-performance by the main polluters, effective utilization of carbon sink activities, and heavy reliance on the flexibility mechanisms, but they are not unrealistic. Existing EU policies and measures fall far short of creating a comprehensive climate policy that enables the EU to shift smoothly towards

---

[172] Beyond the EU-15 in the EU-12, all member states bound by commitments under the Kyoto Protocol are expected to meet or over-achieve their Kyoto targets, creating a surplus of Kyoto units. Only one state, Slovenia, anticipates that it will need to utilize the Kyoto Protocol flexibility mechanisms to meet its commitment. Ibid 12.

[173] Eg, Cinnamon Carlarne, 'Risky Business: The Ups and Downs of Mixing Economics, Security and Climate Change' (2009) 10(2) *Melbourne J Intl L* (2009); Michael W Wara and David G Victor, 'A Realistic Policy on International Carbon Offsets' (2008) Program on Energy and Sustainable Development Working Paper #74 <http://pesd.stanford.edu/publications/a_realistic_policy_on_international_carbon_offsets/> accessed 30 December 2009.

[174] 'Greenhouse gas emission trends and projections in Europe 2005' (n 161) 12.

a low-carbon economy. Yet, the measures provide firm foundations and significant advances in this regard.

Prioritizing Climate Change

Despite existing obstacles, the EU continues to prioritize actions to address climate change. In fact, the EU has made combating climate change the first priority of the 6th Environmental Action Programme and a key component of the EU Sustainable Development Strategy.

In making climate change a priority area for regional policy, the EU is working to create an integrated and comprehensive climate change policy framework.[175] It is actively developing new mechanisms for meeting Kyoto Protocol targets—including new legislative frameworks for limiting carbon dioxide emissions from passenger automobiles and for energy policy.[176]

The EU is also compelling member states to meet their respective regional obligations. For example, the European Commission recently confirmed that it will 'open or continue infringement proceedings against 6 Member States for failing to provide important information required as part of the EU's efforts to combat climate change'.[177] Additionally, the Commission plans to take Luxembourg to the Court for failing to present the Commission with adequate information on its domestic policies and measures to reduce emissions, and its projected future emissions. In total, the Commission will issue warnings to 14 other member states that have not yet submitted information as required by the Emissions Trading Directive.[178] Furthermore, in a move that is essential to any long-term efforts

---

[175] Europa, 'Climate Change and the EU's Response' (Press Release, 15 February 2007) <http://europa.eu/rapid/pressReleasesAction.do?reference=MEMO/07/58> accessed 1 June 2010.

[176] The proposed energy package made world news by creating ambitious targets to be met by 2020, including improving energy efficiency by 20%, increasing the market share of renewable energy source to 20%, and increasing the share of biofuels in transport fuels to 10%. Ibid.

[177] Europa, 'Climate Change: Commission Takes Legal Action Against Six Member States Over Missing Information' (Press Release, 22 March 2007) <http://europa.eu/rapid/pressReleasesAction.do?reference=IP/07/386> accessed 1 June 2010; see also Council Decision (EC) 280/2004 of 11 February 2004 concerning a mechanism for monitoring Community greenhouse gas emissions and for implementing the Kyoto Protocol [2004] OJ L49.

[178] France, Estonia, Greece, Lithuania, and Poland are due to receive their first warning letters, while Germany and Luxembourg are due to receive their second and final warning letters for not providing the requisite information on the allocation of their Assigned Amount Units. Austria, Czech Republic, Denmark, Hungary, Italy, Portugal, Slovenia, and Spain are due to receive warning letters for failing to produce National Allocation Plans. Europa, 'Commission Asks Member States to Provide Important Information in the Fight Against Climate Change' (Press Release, 12 October 2006) <http://europa.eu/rapid/pressReleasesAction.do?reference=IP/06/1364&format=HTML&aged=0&language=EN&guiLanguage=en> accessed 1 June 2010. The Commission is also taking action against Bulgaria, Estonia, Greece, Italy, and Malta for failing to provide adequate information to the Commission under the ETS. Europa, 'Climate Change: Commission Takes Legal Action Against Six Member States Over Missing Information' (Press Release, 22 March 2007) <http://europa.eu/rapid/pressReleasesAction.do?reference=IP/07/386&format=HTML&aged=0&language=EN&guiLanguage=en> accessed 1 June 2010.

to address climate change, the EU argues that it has made significant progress towards decoupling emissions from economic growth.[179]

Miles to Go: Revision and Recession

Despite the optimistic report released by the EEA in 2009 revealing the progress the EU has made over the last five years in structuring a more effective climate regime, the EU continues to face significant regional and international challenges. The accession of ten new member states to the EU in 2004, the accession of two further member states in 2006, and the slow pace of policy implementation in existing EU member states continues to thwart EU efforts at the regional level, while—as will be discussed in detail in Chapter 9—the changing dynamics of global politics at the international level increasingly sideline EU efforts to push for a more aggressive global climate regime.

Responding to perceived shortcomings in existing policy initiatives, between December 2008 and spring 2009, the EU embarked on a mission to revitalize and extend its climate program. Unfortunately, these efforts coincided with the global economic downturn of fall 2008 and the ascendancy of the Czech Republic to the Presidency of the Council of the European Union for the first half of 2009, chaired by Czech President Vaclav Klaus—an open climate and environmental skeptic. The combination of economic stress and political division threatened to throw EU climate policy off course.

However, as discussed further in Chapter 7, through extensive negotiations, EU leaders successfully overcame very real divisions in member state opinion to negotiate in December 2008, and subsequently adopt in April 2009, the final legal texts of an energy and climate change package of legislation intended to achieve the previously agreed to 20/20/20 goals. The final package of legislation embodied six key measures, most of which were discussed in context in the earlier parts of this chapter:

- Revision of the EU ETS to cut allowances, reduce greenhouse gas emissions, mandate full auctioning of allowances for the power sector post-2013 and phase in auctioning for other sectors between 2013 and 2027;
- An 'effort-sharing' agreement setting out binding emission-reduction targets by member states for sectors outside the ETS, eg, transport and agriculture, in order to reach an overall cut of 10% by 2020;
- A regulatory framework for carbon capture and sequestration to support the new technologies before they are commercially viable;

---

[179] 'Climate Change and the EU's Response (n 175) 7. Decoupling refers to breaking the link between 'environmental bads' and 'economic goods'. OEUD, 'Indicators to Measure Decoupling of Environmental Pressure from Economic Growth' (16 May 2002) <http://www.oecd.org/dataoecd/0/52/1933638.pdf> accessed 1 June 2010.

- A directive for the promotion of energy from renewable sources that establishes individual targets for the proportion of renewables in member states' final energy consumption, to reach 20% EU-wide, that stipulates that each member state must ensure a 10% renewable share in domestic transport sectors, and that establishes criteria on the sustainable use of biofuels;
- CO2 limits for new automobiles;
- Standards for fuel quality.[180]

The new legislation regulating greenhouse gas emissions from automobiles is particularly important and has not been discussed previously in this chapter. The decision by the EU to create a new regulatory regime for automobiles was prompted by the failure of voluntary agreements with the auto industry to create tangible results. As a result of these failed agreements, emissions from the EU transport sector proved to be one of the greatest Achilles heel in the evolving EU climate regime. In an effort to respond to the regulatory gaps in the transport sector, in 2007 the Commission adopted a proposal for legislation to reduce the average carbon dioxide emissions of new passenger cars. Following multiple years of contentious debate, on 23 April 2009, the EU adopted a regulation setting emissions performance standards for new passenger cars.[181] In relevant part, the regulation establishes limits on carbon dioxide emissions for new cars beginning in 2012, creates a strategy for how emissions reductions should be achieved, establishes a penalty regime for missed targets and sets a long-term emissions target.[182] The regulation is the first of its kind in the EU and it begins to round out existing sectoral imbalances in EU climate policy.

In adopting the final texts of this comprehensive energy and climate change package – despite economic recession and Klaus's skeptical presence at the helm – the EU's Council of Ministers demonstrated that while the EU faces intrinsic challenges in coordinating policy in the broad fields of energy and climate change, it has laid the groundwork for continuing internal coordination and global leadership.[183]

Basic Differences

Overall, the EU is leading the international community in prioritizing climate change as a critical local, regional, and international issue. The EU has created a comprehensive and constantly evolving framework for regional climate change policy.

---

[180] Euractive, 'EU wraps up climate and energy policy' (News Release, 7 April 2009) <http://www.euractiv.com/en/climate-change/eu-wraps-climate-energy-policy/article-181068> accessed 30 December 2009.

[181] Regulation (EC) No 443/2009 of 23 April 2009 setting emission performance standards for new passenger cars as part of the Community's integrated approach to reduce CO2 emissions from light-duty vehicles [2009] OJ L140/1.

[182] The specific carbon dioxide limits for new cars are as follows: 120 g/km for 65% of new cars in 2012, 75% in 2013, 80% in 2014, and 100% in 2015.

[183] Ibid 66.

Furthermore, EU member states demonstrate a wide range of successes and failures in their implementation of climate change laws and policies to meet or surpass their respective international and supranational obligations. However, the EU and its member states are investing substantial effort in devising effective legal and political strategies to mitigate and adapt to climate change. The breadth of supranational and national laws and policies specifically targeting climate change is remarkable.

In contrast, the US federal government leads neither international nor national efforts to address climate change, as climate change does not figure prominently on the US international or domestic agendas. At the federal level, the executive branch points to diverse research and voluntary programs as creating the backbone for a US climate change strategy. However, the federal government has yet to enact any legislation specifically targeting greenhouse gas emissions or to otherwise create a robust set of laws and policies to address global climate change. At the sub-federal levels, US states are beginning to prioritize climate change, to implement climate change laws and policies, and to exert pressure on the federal government for substantive action. Thus far they have not been successful.

The critical difference between the respective approaches of the US and the EU to climate change policy at the federal and supranational levels, respectively, is the EU's willingness to make legally enforceable, absolute and measurable commitments to reduce greenhouse gas emissions. For example, the EU and its member states have committed to international obligations, to developing and implementing robust policy frameworks for climate change, to implementing enforceable legislation in conjunction with voluntary programs to achieve reductions in greenhouse gas emissions, and to encouraging innovative political and technological strategies to address climate change. The EU and its member states have made climate change an important component of regional and national environmental governance.

Conclusion

EU environmental law, once a second-class partner to regional economic laws and policies, is now itself a thriving area of European policy. It has 'grown rapidly in size and complexity in recent years so that it now consists of at least 300 environmental directives and regulations.'[184] Once a modest counterpart to the US as a global environmental leader, the EU in many areas is now surpassing the US both in creating new regulatory regimes for complex environmental problems, and in leading

---

[184] Clifford Rechtschaffen, 'Shining the Spotlight on European Union Environmental Compliance' (2007) 24 *Pace Envtl L Rev* 161, 161.

global negotiations to address trans-boundary environmental problems, such as global climate change, biodiversity protection, and chemicals regulation.[185]

EU environmental law, however, has not escaped the problems of implementation and enforcement that plague environmental law worldwide. The proliferation of environmental law in the EU has created a comprehensive but, at times, complex and confusing legal system. Normal problems of implementation and enforcement are further complicated and constrained in the EU by the shared competence between the EU and its member states. The division of authority between the supranational and national entities produces a web of law that is difficult to manage, monitor, and enforce.

EU climate change laws and policies are similarly complex. EU climate change law involves a collection of regional players: the Council, the Commission, multiple Directorates-Generals within the Commission, including DG Environment, DG Enterprise, the Joint Research Centre, the EP, and the European Environment Agency. Beyond these regional governmental actors, EU climate policy involves 27 member states with 27 different political systems and political agendas, incalculable business lobbies, local governments, and civil society organizations, among others. The complexities of climate change policy combined with the complexities of the EU system of governance create the conditions for a regulatory nightmare.

The presence of mixed political competences and complexities in EU climate change policy challenge the implementation of a comprehensive set of climate change laws and policies in the EU. Yet, within this intricate framework, the EU has managed to build a rich and evolving set of policies that offer invaluable lessons in success and failure.

Some EU climate change policies have succeeded, including emission reductions from the energy, industry, agricultural, and waste management sectors. Others have failed, resulting in market failures during the first round of the Emissions Trading Scheme, deficient improvements in the market share of renewable energy in electricity production, unrealized use of carbon sinks through land use change and forestry activities, and absolute emission increases from the transport sector.[186] Against this mixed backdrop, what becomes evident is that the willingness of the EU to actively engage with climate change at multiple levels of governance and in multiple sectors of society differentiates it from many—if not most—members of the global community.

---

[185] See ibid; see also Norman J Vig and Michael G Faure (eds), *Green Giants?: Environmental Policies of the United States and the European Union* (MIT Press, Cambridge MA, 2004). But see Jonathan Wiener, 'Convergence, Divergence, and Complexity in US and European Risk Regulation' in Norman J Vig and Michael G Faure (eds), *Green Giants?: Environmental Policies of the United States and the European Union* (MIT Press, Cambridge MA, 2004).

[186] The EU transport sector represents the fastest growing sector for greenhouse gas emissions. EU-15 greenhouse gas emissions from the transport sector are expected to increase up to 31% above 1990 levels. 'Greenhouse gas emission trends and projections in Europe 2005' (n 161) 38 (explaining that most of the growth in emissions from the transport sector is associated with growth in volume that offsets any improvements in energy conservation or efficiency).

For example, while the US has avoided adopting any legislation to trade in or cap greenhouse gas emissions, the EU has established the EU ETS to provide a decreasing cap and trade program for carbon dioxide. Similarly, the EU has implemented renewable energy and biofuels directives that, while limited, provide important steps towards a more comprehensive, legally binding, and renewable energy framework. The US, in contrast, has avoided measures that mandate comparable changes in the energy sector.

Finally, compared to the US, the EU has consistently been an active participant in the international climate regime. As a party to both the UNFCCC and the Kyoto Protocol, the EU has made important, legally binding international commitments. Therefore, the EU is serving as an international leader both in terms of making substantive commitments and in pursuing national, regional, and climate change dialogue.

The EU was a powerful proponent of the Kyoto Protocol; it is now a vocal advocate for negotiating an aggressive, legally binding post-Kyoto agreement.[187] The EU has adopted a global leadership position on climate change and is seeking to set the standard for climate change law and policy both at home and in the international context. Despite resistance from the US and despite modest progress from other parties to the Kyoto Protocol, the EU is proceeding with an increasingly ambitious climate change agenda.

While the EU cannot claim absolute success in its efforts to address climate change, it can point to an increasingly robust system of climate change laws and policies that make it a symbolic and substantive role model for progressive climate change policy in the world.

---

[187] Sebastian Oberthur and Hermann E Ott, *The Kyoto Protocol: International Climate Policy for the 21st Century* (Springer-Verlag, New York, 1999) 14; Joyeeta Gupta and Michael Grubb (eds), *Climate Change and European Leadership: A Sustainable Role for Europe?* (Kluwer Academic Publishers, Dordrecht Netherlands, 2000).

6
Member State Laws and Policies

Introduction

The member states of the EU are diverse in history, character, demography, and politics. Beyond sharing the common categorization of 'European' and a shared goal of advancing regional economic, political, and social stability through a democratically accountable governance system, the member states of the EU reflect social and cultural variation that would seem to belie attempts to create a common rule of law.[1]

Member states vary in terms of linguistic, social, political, and economic history, and legal culture. For this reason, the challenges inherent in negotiating regional European policy must be distinguished from the real but comparatively modest challenges associated with navigating the US federalist system. From questions of sovereignty and devolution to questions of common currency and modes of communication, regional cooperation in Europe raises an entirely separate set of issues than those of negotiating inter-state politics within the United States.

Within this context, this chapter explores how three member states—Germany, the United Kingdom, and Poland—reflect the challenges, progress, and setbacks that define the contours of EU climate policy. This chapter begins by exploring the roles of Germany and the UK as members of the EU-15 and nominal regional leaders in efforts to advance progressive climate policies. The analysis focuses first and foremost on Germany as the EU's most vocal and active proponent of aggressive climate policy. It begins by examining how Germany's history and system of governance have influenced the role it has played in regional and international climate politics. It then more briefly assesses the climate policy of the UK, a second EU member state that has sought both to create a comprehensive set of climate laws and policies and to position itself as a global climate leader. Focusing solely on German and British efforts, however, creates a distorted picture of EU member states in terms of political attitude, legal responsiveness, and social and economic flexibility. In order to create a more inclusive picture of the challenges

---

[1] The foundation of the EU system rests upon shared values: liberty, democracy, respect for human rights and fundamental freedoms, and the rule of law. Eg, Commission (EC), 'On the Charter of Fundamental Rights of the European Union' (Communication) (COM) 2000 /559 final, 13 July 2000.

inherent in creating common climate policies for the full EU-27, this chapter also examines Poland's role in—and response to—efforts to create a common European rule of law with regards to climate change. As a large, populous, former Communist nation currently in the midst of economic transition, the challenges Poland faces in crafting national climate policies vary from those faced by the United Kingdom and Germany and are more widely representative of challenges facing many of the newer EU member states.

Chapter 5 reviewed the intricacies of member state–EU relations in the context of climate change obligations. It is worth reiterating several key points here:

- First, member states are bound by international, supranational, and national obligations to reduce greenhouse gas emissions.
- Second, as sovereign entities, member states are autonomous in determining how they will achieve specified emissions reductions and maintain the right to establish national rules that exceed EU emissions reduction goals, as compared to EU obligations.
- Third, member states' EU obligations vary significantly according to notions of historic responsibility and present capacity.
- Fourth, EU climate policy is a vast sphere coordinated through the ECCP and member states. Member states are tasked with implementing a wide variety of decisions, directives, and regulations ranging from mandatory, legally enforceable emission reduction goals and emissions trading plans to ambitious, policy-based programs for reviewing carbon capture and storage capacities.
- Fifth, all EU climate policy is based on underlying notions of shared authority, with the nature of power-sharing varying according to the specific measure at issue.[2]

The point here is that EU climate policy imposes a vast array of obligations on member states. Some of these obligations take the form of traditional command-and-control style regulation, but most are more facilitative and focus on creating frameworks for overcoming existing barriers to change.

Keeping these common qualities in mind, this chapter examines the specific and representative roles that Germany, the UK, and Poland play in the formulation and fulfilment of EU climate policy. To reiterate, Germany and the UK were chosen because they are two of the most prominent member states in terms of the past, present, and future of EU-15 climate policy. Both States have taken on heavy emissions reduction obligations; both States hold themselves out as regional and international climate leaders; both States are regionally and internationally significant economic and political forces, especially as concerns international relations with the US and its ability to influence international climate negotiations.

Poland was chosen because it is one of the largest, most populated, and most politically significant of the non-EU-15 member states. As a Communist state

---

[2] Eg, Discussion of harmonization *infra* Ch 5.

until 1989, it is also a new democracy still struggling through transitional social and economic growing pains. Poland's struggles resonate with many of the non-EU-15 member states, most of which face similar social and economic transitions that overshadow environmental protection and challenge the eagerness and ability of the State to prioritize measures to address climate change.

Individually and collectively, Germany, the UK, and Poland influence not only the content of EU climate policy, but also the ability of the EU as a whole to fulfill supranational goals and international commitments.

Germany

The Federal Republic of Germany (Germany), is the EU's most populated State and one of its most economically and politically powerful States. With an estimated population of 82.3 million people, Germany out-populates the other EU States nearest in size—France and the UK—by 20+ million.[3] While it does not have the highest GDP per capita, Germany does have the highest GDP in terms of purchasing power, accounting for almost 20% of the EU's total purchasing power.[4] Germany's sheer size and economic prowess make it a powerful and influential figure in regional and international politics. Given that Germany currently accounts for approximately 4% of global greenhouse gas emissions, it is a similarly influential player in regional and international climate politics.[5]

Germany has been particularly influential in the climate change debate, consistently pushing for strong and binding measures since the late 1980s, despite widespread international opposition.[6] Germany has retained an enduring leadership role in climate policy in the face of external pressures, economic downturns, and a recent history of political upheaval.

German Governance and Environmental Politics

Germany is a democratic parliamentary State, with 16 constituent federal states (Länder) operating under a central Federal Constitution (Grundgesetz (GG)).[7]

---

[3] US Central Intelligence Agency, 'The World Factbook: Germany' (General information on Germany, 27 November 2009) <https://www.cia.gov/library/publications/the-world-factbook/geos/gm.html> accessed 31 December 2009.

[4] Europa Eurostat, 'Gross Domestic Product at Market Prices' (2008) <http://epp.eurostat.ec.europa.eu/tgm/refreshTableAction.do?tab=table&plugin=0&init=1&pcode=tec00001&language=en> accessed 4 January 2010.

[5] Germany's greenhouse gas emissions have declined since the 1990s, due in part to reunification and in part to early adoption of targeted climate change policies.

[6] Germany, *Report Under the Kyoto Protocol to the United Nations Framework Convention on Climate Change* (2006) 3 (2006 UNFCCC Report).

[7] The Federal States have constitutions of their own, many of which expand rights and obligations in regard to environmental protection.

Similar to the US, Germany's government consists of legislative, executive, and judicial branches. Unlike the US, however, Germany does not vest central authority in its Executive President. Rather, Germany is lead by its Federal Chancellor who is elected by the Bundestag—the primary legislative body of the State—and who heads the Federal Cabinet (Bundesrat). Germany's President, in contrast, serves primarily as a political figurehead.

The Federal Constitution provides the foundation for Germany's robust system of environmental law.[8] It establishes fundamental rights, but few detailed rules or provisions, focusing primarily on delineating competencies as between different governmental institutions. As amended in 1994, however, Article 20a GG established environmental protection as an underlying state principle, thereby obligating all branches of the government to uphold the goal of environmental protection.[9] The provision, however, did not establish environmental protection as a fundamental right or create a private right of action for citizens.[10]

Germany's ability to advocate aggressive domestic and international responses to climate change is closely tied to its constitutionally rooted system of environmental governance, its current political and economic stability, and its democratic and strongly consensus-based style of decision making.[11] The German system provides for coalition governments, which prevents bi-partisan politics from dominating governmental debate and enables peripheral and single-issue political parties greater opportunities to participate in the political process.[12] Thus, for example, as environmentalism became a growing concern among the German electorate in the 1970s, the previously marginal Green Party saw a corresponding increase in political participation and, ultimately, political power at the federal level.[13]

Germany's present political stability and history of progressive environmental politics belies challenges associated with post-World War II political

---

[8] The Federal Constitution was adopted in 1991 following reunification. Michael Rodi, 'Public Environmental Law in Germany' in J Rene and others (eds), *Public Environmental Law in the European Union and US: A Comparative Analysis* (Kluwer Law International, The Hague, 2002).
[9] Ibid.   [10] Ibid 206.
[11] 'The German political system is noted for its consensual nature. To achieve consensus, dialogue between the various parties involved in policy making is required. This feature of the political system, together with the imperatives of the international management of [global climate change], suggests that discourse theory...may be helpful in explaining how and why Germany, and arguably other [EU] states responds to [global climate change] the way they do'. Lyn Jaggard, 'The Reflexivity of Ideas in Climate Change Policy: German, European and International Politics' in Paul G Harris (ed), *Europe and Global Climate Change: Politics, Foreign Policy and Regional Cooperation* (Edward Elgar Publishing, Northampton MA, 2007) 323.
[12] Lyn Jaggard, *Climate Change Politics in Europe: Germany and the International Relations of the Environment* (Tauris Academic Studies, New York, 2007) 20 (describing the development of Germany's electoral system post World War II). Germany is currently led by a coalition government, dominated by the Christian Democratic Union/Christian Social Union Party (CDU/CSU) and the Social Democratic Party of Germany (SPD).
[13] Ibid 25 (describing the origins of the Green Party in concerns over nuclear energy and the environment in the 1970s).

division and subsequent reunification. Reunification created a myriad of social, economic, and environmental challenges for Germany in the early and mid-1990s. The dismal state of East Germany's natural environment, however, reinforced rather than discouraged public support for environmental protection measures.[14] In addition, reunification of economically depressed East Germany with the more industrialized and prosperous West Germany enabled Germany to press for aggressive greenhouse gas reduction goals in the international arena.

Rule of Law in Germany

Germany's system of administrative and environmental law must be viewed in the context of its system of civil law. German environmental law is comprised of numerous, highly detailed, issue-specific laws and regulatory measures.[15] Unlike the US system of environmental law, which relies on a vast and sprawling adversarial regulatory system and widespread judicial review of administrative agencies' interpretation of environmental statutes, German environmental law is autonomous and administrative agencies, rather than the courts, provide the authoritative voice in legislative interpretation. In contrast to the US system, the German system is largely non-adversarial in nature, relying heavily on detailed enabling statutes and higher levels of trust and power vested in bureaucratic decision-makers.

Further distinguishing the German legal system are distinctions in the basic relationships between the federal government and the federal states. German Föderalismus (akin to, but not equivalent to the US notion of federalism) divides power between federal and state governments horizontally rather than vertically, as is the case in the US.[16] In the US, the federal government has authority over lawmaking and law implementation (in conjunction with the states, as delineated by the Constitution). In contrast, in Germany, the federal government is primarily responsible for lawmaking, with implementation powers reserved for states and local governments, and with a greater degree of institutionalized cooperation between the federal, state, and local governments on all levels of lawmaking, implementation, and enforcement.[17]

In further contrast to the US system, Germany's key executive figures—the Chancellor and the Federal Cabinet—are elected (and dismissible) by the legislature. Thus, unlike the US, where power is carefully separated—yet often wrangled over between the legislative and executive branches—Germany's centralization of power over the executive in the legislative branch reduces inter-governmental

---

[14] Ibid.
[15] There have been various attempts to consolidate Germany's disparate laws into one overarching Environmental Code, but these attempts have failed in large part due to underlying issues about competences as shared by the federal and state governments. Rodi (n 8) 208.
[16] Ibid 212.   [17] Ibid.

conflict. It also minimizes the likelihood that the two branches of government will produce distinctly different political messages, as is often the case between the legislative and executive branches in the US.

Finally, and of particular relevance to the evolution of climate law, German politics function on the basis of cooperation and consensus among governmental entities and interest groups representing a wide range of social and economic interests. Germany's emphasis on collaborative decision-making heavily influences its environmental and climate change policy, and stands in direct contrast to the partisan nature of US climate politics.[18]

Foundations of German Environmental Law

German environmental law is rooted in cultural conceptions of nature—especially forests—as permanent features of the landscape.[19] In the 1980s, widespread forest loss caused by acid rain—a phenomenon known as *Waldsterben*—elicited far-reaching public concern in Germany.[20] The effects of *Waldsterben* in conjunction with pervasive environmental deterioration prompted a lasting evolution in German environmental lawmaking.

Beginning in the early 1980s, prompted by *Waldsterben* and other forms of domestic environmental deterioration as well as by the ongoing development of European environmental law, Germany began an active period of environmental lawmaking.[21] Starting with the Federal Emission Act in 1983, Germany began implementing stringent regulatory regimes, including regimes controlling air pollution from stationary sources, vehicle emissions, and recycling and waste disposal.[22] During the 1980s, Germany simultaneously began pushing for more aggressive regional action on air pollution, signaling its new and continuing role as a European leader on environmental matters.[23]

In 1986, Germany's environmental movement received an unexpected boost following the disaster at the nuclear facility in Chernobyl. The Chernobyl accident widened the framework of environmental policymaking to include intense debate on the safety and viability of nuclear energy. Within two months of the Chernobyl disaster, the German government created the Federal Ministry for the Environment, Nature Conservation and Nuclear Safety (BMU), which is

---

[18] Michael Hatch, 'The Europeanization of German Climate Change Policy' (Conference paper for the European Union Studies Association Biennial Conference in Montreal 2007) <http://aei.pitt.edu/7897/> accessed 4 January 2010.
[19] Jaggard, 'The Reflexivity of Ideas in Climate Change Policy' (n 11). [20] Ibid.
[21] Ibid 328–29. During this era, the Green Party gained its first seats in the Bundestag. The Green Party continues to be active in German politics; it currently holds 68 seats in the Bundestag. Election Resources on the Internet, 'Elections to the German Bundestag' (27 September 2009) <http://electionresources.org/de/bundestag.php?election=2009> accessed 4 January 2010.
[22] Jaggard, 'The Reflexivity of Ideas in Climate Change Policy' (n 11) 329. [23] Ibid.

now the Ministry responsible for handling not only questions of nuclear safety but also climate change.[24]

The BMU is now one of the key federal institutions responsible for domestic environmental policymaking. In Germany, the federal government is responsible for defining the overarching parameters of environmental policy, including defining substantive provisions, procedure, and specific emission and environmental quality standards. In this way, the federal government serves a role similar to the US Congress—it creates the primary legislation that establishes overarching principles and goals with varying degrees of precision. The federal government is assisted in these tasks by the BMU and a number of sub-agencies, including the Federal Environment Agency, which provides technical and scientific support to the BMU, and the *Bundesverband der Deutschen Industrie*, or Federation of German Industries (BDI), which provides a pathway for industry input into the political process.[25]

Once the legislative phase is complete, however, Germany's federal states have primary responsibility for executing the environmental laws. This stands in contrast to the US system, where implementation responsibilities are shared between administrative agencies and state governments. However, while the German federal states are the only entities competent to enforce environmental laws, they are aided in their executive role by regional administrators (eg, Regierungspräsident) and by the Factory Inspectorates (Gewerbeaufsichtsämter).[26] The state environmental ministries also receive technical and scientific assistance from the federal environmental protection agency (Umweltbundesamt) and other federal institutions.[27]

Introduction to German Climate Change Law

As German environmental law evolved during the 1980s, it developed around the key principles of protection, precaution, sustainable development,

---

[24] The Ministry of Transport previously handled global climate change issues, but its early failures—particularly in regards to participation in IPCC deliberations—led to the transfer of responsibility to the BMU in 1988. Ibid 329–30.

[25] See Jaggard, *Climate Change Politics in Europe* (n 12) 26 and 29; Federal Environment Agency, 'Get To Know Us' (General information on its mandate and organization) <http://www.umweltbundesamt.de/uba-info-e/index.htm> accessed 3 May 2009.

[26] Rodi (n 8) 201.

[27] The Federal Environment Agency is responsible for providing scientific advice to the federal government, coordinating research and development initiatives, facilitating implementation of environmental laws, and keeping the public informed on environmental matters. Federal Environment Agency, 'Get to Know Us' (General information on its mandate and organization) <http://www.umweltbundesamt.de/uba-info-e/index.htm> accessed 4 January 2010. The Agency also houses the the German Emissions Trading Authority (DEHSt), which is the national authority responsible for implementing emissions trading and project-based mechanisms of the Kyoto Protocol, including JI and CDM projects. German Emissions Trading Authority, 'Welcome to the German Emissions Trading Authority (DEHSt) at the Federal Environment Agency' (Homepage) <http://www.dehst.de/EN/Home/homepage__node.html?__nnn=true> accessed 4 January 2010.

cooperation, and polluter pays.[28] Reliance on notions of precaution and protection catalyzed Germany's early entry into the climate debate and prompted it to become one of the most aggressive advocates for a precautionary approach to climate change at the international level. Strong domestic support for action on climate change also buoyed Germany's efforts to push for greenhouse gas emission reductions during the early days of the global climate change debate. Its entry into climate politics began early, predating the negotiation of the UNFCCC.

Perceived domestic vulnerabilities and early reliance on notions of precaution prompted Germany to begin debating legal responses to climate change as early as the late 1980s.[29] Much of the impetus for early action arose from concerns over the security of domestic energy supplies. Although Germany is one of the EU's largest energy producers—being the second largest coal producer, the second largest nuclear producer, and the third largest producer of natural gas—it is also one of the EU's biggest energy consumers. As a result, Germany relies heavily on imported sources of energy to sustain high domestic per capita energy consumption levels.[30] Growing energy demand coupled with burgeoning concerns over nuclear energy and fears over the future economic costs associated with climate change prompted Germany to instigate early climate change policymaking efforts at both the domestic and international levels.[31]

As early as 1996, for example, Germany was calling for global CO2 emissions to be reduced by 20% from 1990 levels by 2005 and by 15–20% by 2010. The German proposal represented one of the most aggressive stances leading into negotiations for the Kyoto Protocol. Only a handful of other proposals contained provisions exceeding Germany's suggestion. Despite divergent positions among the various member states, aggressive campaigning on the part of the German government eventually persuaded the EU to support Germany's proposal for a 'no regrets' approach to the stabilization of CO2 concentrations in the atmosphere.[32] This early victory on the part of the German government played a significant role in shaping the EU's aggressive leadership stance on climate change.

---

[28] The principle of protection is often negatively phrased as 'avert[ing] danger to health, life and the environment caused by catastrophes (explosion, fire, flood) or by harmful substances which pollute the soil, the air or the waters'. Rodi (n 8) 210.

[29] Marc Zebisch and others, *Climate Change in Germany: Vulnerability and Adaptation Strategies of Climate-Sensitive Sectors* 7 (Research Report for the Federal Environment Agency 2005) <http://www.umweltdaten.de/publikationen/fpdf-k/k2974.pdf> accessed 4 January 2010.

[30] European Commission—Energy, 'Germany: Energy Mix Fact Sheet' (January 2007) <http://ec.europa.eu/energy/energy_policy/doc/factsheets/mix/mix_de_en.pdf> accessed 10 October 2008.

[31] Germany, for example, was one of the first States to implement domestic climate policies. It was also a key player in initiating UNFCCC negotiations and then later hosted the first meeting of the UNFCCC COP and pushed for the negotiation of a protocol to the UNFCCC. Paul G Harris, 'The European Union and Environmental Change: Sharing the Burdens of Global Warming' (2006) 17 *Colorado J Intl Envtl L & Policy* 309, 331.

[32] Ibid 332.

Domestic Evolution of Climate Policy

As discussed, the German climate change debate dates back to the Chernobyl controversy and the ensuing political dispute over the continuing operation of Germany's nuclear energy facilities. As nuclear energy lost favor, political debate centered on proposals to construct new coal-fired power plants to meet Germany's energy needs in the case that nuclear power sources were abandoned. With the Green Party and the Social Democratic Party (SPD) arguing for immediate or gradual phase-out of nuclear energy, the Christian Democratic Union (CDU)—the governing party at the time—and its Bavarian sister party, the Christian Socialist Union (CSU) drew upon climate change as a powerful environmental protection-based argument for keeping nuclear energy capacity and minimizing construction of coal-fired plants.[33]

During this pivotal era, the nuclear energy and the climate change debates were shrouded by scientific uncertainty. In an effort to focus and clarify the debates, the German government launched a special Inquiry Commission (Enquete) to help members of parliament 'bridge the gap between the knowledge necessary for important parliamentary decisions and the prevailing lack of it'.[34] As part of the Enquete, in December 1987, a special Enquete Commission on Preventive Measures to Protect the Atmosphere was created to respond to climate-related issues.[35] The Commission—made up of leading scientific and political figures chosen not only on the basis of their individual competences but also their links with key interest groups—reflected German reliance on collaborative decision-making.[36]

The Commission held hearings and solicited the opinions of scientists, politicians, industry, and civil society on the question of climate change, ultimately finding that there was 'an extraordinary need for action'.[37] As early as 1988—preceding the negotiation of the UNFCCC—the Commission released a unanimous report concluding:

there is growing evidence suggesting that the expected changes in the Earth's atmosphere and in the climate...will have serious consequences for human living conditions and for the biosphere as a whole...consequences which preventive measures can prevent to a limited extent only...[t]here is such massive and unequivocal scientific evidence on...the man-made greenhouse effect, the resulting climate change and its repercussions...that

---

[33] Jaggard, 'The Reflexivity of Ideas in Climate Change Policy' (n 11). Following the Chernobyl accident, the Green Party demanded an immediate shutdown of nuclear facilities, while the SPD argued for an incremental phase-out. Ibid.
[34] Andreas Vierecke, 'Complex Problem Solving in the Co-operation Between Science and Politics? The workings of the Enquête Commissions of the German Bundestag in the Fields of Technological Development and Environmental Policy' in Elmar Stuhler and Dorien DeTombe (eds) *Complex Problem Solving: Cognitive Psychological Issues and Environment Policy Applications* (Hampp, Mering, Germany, 1999) 110–18. [35] Ibid.
[36] Konrad von Moltke, 'Three Reports on German Environmental Policy' (September 1991) 33(7) *Environment* 1. [37] Ibid 5–6.

Member State Laws and Policies 201

there can be *no doubt* that preventative action *must* be taken immediately, irrespective of any need for further research.[38]

Subsequent reports followed, calling for 'drastic measures' to control the anticipated effects of climate change.[39] The widely respected Commission reports contributed to a growing sense of urgency amongst both politicians and the public on the need to respond quickly to the threats posed by climate change.[40]

Responding to mounting concerns, in January 1990, the Chancellor's Office tasked the Ministry for Environment with creating the framework for a national climate change policy. In June 1990, the Ministry issued a draft policy proposing ambitious reductions in CO2 emissions and laying out a roadmap for achieving them. The draft policies produced by the Ministry defined the contours of the climate change policy that still exists in Germany almost two decades later.

The Ministry proposal led to the June 1990 creation of the Interministerial Working Group (IMA), 'CO2 Reduction' to analyze the best combination of measures for combating climate change.[41] Subsequently, in 1992, the federal government established an independent, scientific advisory body, the German Advisory Council on Climate Change (WGBU), to undertake research on critical climate related issues and to provide the government with data and recommendations on alternative strategies.[42]

Emerging Domestic Strategy

The year 1990 signaled the beginning of concerted efforts in Germany to develop a climate change strategy. Even prior to adoption of the UNFCCC and despite fears of economic harm, Germany began advocating dramatic emissions reductions. In June 1990, the German cabinet adopted plans to cut CO2 emissions

---

[38] German Bundestag (ed), *Protecting the Earth's Atmosphere: An International Challenge* (German Bundestag, Bonn, 1989) 439–41.

[39] Enquete Commission, *Protecting the Earth: A Status Report with Recommendations for a New Energy Policy* (Third Report on preventative measures to protect the earth's atmosphere) (German Bundestag, Bonn, Germany, 1991).

[40] Miranda Schreurs, *Environmental Politics in Japan, Germany, and the United States* (CUP, Cambridge, 2002) 155.

[41] Timothy O'Riordan and Jill Jager (eds), *Politics of Climate Change: A European Perspective* (Routledge, London, 1996). The IMA is tasked with 'drawing up guidelines for climate protection policy activities, identifying the existing need for action, indicating potentials for reducing greenhouse gases, and submitting to the Federal Cabinet comprehensive packages of measures for reducing greenhouse gas emissions in Germany'. Ibid.

[42] WBGU, 'Mission and Concepts' (General information on the German Advisory Council on Climate Change) <http://www.wbgu.de/wbgu_auftrag_en.html> accessed 4 January 2010. The key responsibilities of the WBGU are to 'analyse global environment and development problems and report on these, review and evaluate national and international research in the field of global change, provide early warning of new issue areas, identify gaps in research and to initiate new research, monitor and assess national and international policies for the achievement of sustainable development, elaborate recommendations for action and research and raise public awareness and heighten the media profile of global change issues'. Ibid.

by 25% of 1987 levels by 2005.[43] This early commitment to reducing carbon emissions reflected strong domestic adherence to notions of precaution and protection. Domestic commitment to the precautionary principle similarly led Germany to propose including the precautionary principle within the UNFCCC, a proposal that was ultimately accepted.[44]

Since the early 1990s, Germany has incrementally developed its domestic climate change policy regime by:

- reviewing technical, legal, and economic framework conditions;
- identifying technical potential for reduction;
- identifying obstacles;
- analysing measures to overcome obstacles;
- defining measures to overcome existing obstacles and developing an effective and comprehensive program;
- progressively implementing agreed measures in coordination with the parties concerned (Länder, local authorities, industry, major groups);
- reviewing the effectiveness of the measures.[45]

Germany's efforts target activities and behaviours across multiple sectors—from individual households to large-scale energy producers. In particular, the strategy focuses on improving energy conservation and energy efficiency while increasing the production capacity of alternative energy sources. As discussed in more detail below, German climate policy includes a far-reaching and ambitious collection of measures targeted at transforming the energy sector, including energy taxes and renewable energy requirements.[46]

As discussed, the federal and state governments share power and responsibility for addressing climate change and achieving the desired energy transition. Within the federal government, climate issues are handled primarily through the BMU, currently headed by Federal Cabinet member Norbert Röttgen.[47] At the state level, the federal government works with the states and with local municipalities to coordinate implementation measures and to establish joint national/regional committees, eg, the Conference of Environment Ministers.

The comparatively long and prominent nature of the climate debate in Germany has resulted in climate issues ranking high on both political and public

---

[43] M Hatch (n 18) 10. Also, Hatch noted that 'a subsequent cabinet decision designed to accommodate BMWi's views extended the reduction target to 25–30 percent by the year 2005 in light of German unification'. Ibid.

[44] Jaggard, *Climate Change Politics in Europe* (n 12) 41.

[45] Shuzo Nishioka (Institute for Global Environment Strategies, Keio University), *Detailed Description of Best Practices: Germany Report* (Report given at the G8 Forum on Domestic Best Practices Addressing Climate Change 2000) <http://www.env.go.jp/earth/g8_2000/forum/g8bp/detail/germany/report.html> accessed 4 January 2010.

[46] Harris, 'The European Union and Environmental Change' (n 31) 340.

[47] Federal Ministry for the Environment, Nature Conservation and Nuclear Safety, 'Home' (Homepage) <http://www.bmu.de/english/aktuell/4152.php> accessed 4 January 2010.

agendas, with one 2006 study revealing that '93 percent of surveyed Germans considered climate change an important issue, second only to unemployment on the list of top concerns'.[48] Early and sustained consideration of climate change in German politics and civil society contributed to Germany's early and continuing leadership role in international negotiations.[49]

Components of German Climate Strategy

Even prior to the passage of the EU Burden Sharing Agreement, which obligates Germany to reduce its domestic greenhouse gas emissions to 21 percent below 1990 levels by 2008–2012, as a party to the UNFCCC and the Kyoto Protocol, Germany had adopted a series of measures to meet its legal obligations to reduce greenhouse gas emissions.[50]

NCPP 2000

The Kyoto Protocol was opened for signature on 16 March 1998. Germany signed the Protocol shortly thereafter on 29 April 1998 and ratified it on 31 May 2002. Between signing and ratifying the Protocol, Germany implemented a wide range of climate-focused measures. By the end of 1999, Germany had enacted new legislation and policy programs, to include:

- the Ecological Tax Reform, providing for incremental energy tax increases to incentivize efficiency, conservation, and research and development;
- the Renewable Energies Act, promoting increased use and reliance on renewable energy sources;
- the Market Introduction Programme for Renewable Energies, promoting the use of renewables, generally, and solar panels, specifically;
- the 100,000 Roofs Solar Power Programme, providing financial and research support for photovoltaic systems; and
- measures promoting low-sulfur or no-sulfur automotive fuels.[51]

During this period, Germany also enhanced public information campaigns targeted at raising awareness of climate change issues and providing advice for households on ways to reduce domestic emissions.[52]

As a result of these early measures, between 1990 and 1999, Germany cut CO2 emissions by 15.3%, and aggregate emissions of five other GHGs by 18.5%.

---

[48] Allianz Knowledge, 'Germany Climate Change Profile Part 2: Fact Sheet' (Facts and figures about Germany and climate change, 17 April 2009) <http://knowledge.allianz.com/en/globalissues/climate_profiles/climate_germany/climate_profile_germany_facts.html> accessed 4 January 2010.
[49] Jeannine Cavender and Jill Jäger, 'The History of Germany's Response to Climate Change' (1993) 5 *Intl Envtl Affairs* 3, 6–9 (describing how the domestic German environmental policy was the basis for expansion into international policy).
[50] Thomas Kappe and Alexandra Liebing (eds), *Taking Action Against Global Warming: An Overview of German Climate Policy* (Report by the BMU) (German Ministry for the Environment, Nature Conservation and Nuclear Safety, Berlin, 2007) 4 (Taking Action).
[51] 2006 UNFCCC Report (n 6) 4–5.   [52] Ibid.

Despite these figures, emission reductions in Germany were inconsistent and asymmetrical, with emissions declining in the industrial and energy sectors but jumping markedly in the household and transport sectors.[53] The suite of measures adopted in the 1990s moved Germany towards a more energy efficient society, but fell short of federal targets.

Recognizing that these early measures would not allow Germany to meet its domestic target of reducing domestic greenhouse gas emissions 25% by 2005 (1990 baseline), on 18 October 2000, at the recommendation of the BMU, the federal government adopted the first National Climate Protection Programme (NCPP).[54] The NCPP is the central pillar and umbrella for Germany's climate policy and sets out the federal government's primary climate targets, to include:

- reducing emissions of carbon dioxide by 25% by 2005, compared with 1990 levels;

- reducing emissions of the six greenhouse gases cited in the Kyoto Protocol by 21% between 2008 and 2012, within the context of EU burden-sharing;

- doubling the proportion of renewable energy sources by 2010, compared with current levels, and a further substantial increase in the proportion of renewable energy sources after 2010;

- expanding combined heat and power generation by means of set quotas, aimed at cutting CO2 emissions by an additional 10 million tonnes by 2005, and by 23 million tonnes by 2010;

- significantly increasing energy productivity over the next few years, including reduction targets for individual sectors.[55]

As adopted in 2000, the NCPP embodied cross-sectional measures, as well as 64 measures targeting seven key economic sectors: Households, Transport, Industry, Energy Industry, Renewable Energy, Waste Management, and Agriculture.[56] Key provisions included:

- adopting quota arrangements to expand combined heat and power generation;

- passing the Energy Saving Ordinance, which aims to reduce demand and ensure equal competition among energy sources by allocating 'true' prices to energy, and providing for mandatory upgrades of existing buildings and updated efficiency requirements for new construction;

- creating a subsidy program to reduce CO2 in existing buildings;

---

[53] Between 1990–1998, CO2 declined 31% in the industrial sector and 16.1% in the energy sector, while household emissions grew 6% and transport emissions grew 11.1%. BMU, 'Germany's National Climate Protection Programme: Summary' (2000) (National Climate Protection Programme 2000).   [54] Ibid.
[55] Ibid.   [56] 2006 UNFCCC Report (n 6) 5.

- calling for concrete commitments from German industry, including 19 leading industry organizations and trade associations, to reduce CO2 and GHG emissions by specific amounts by 2012;
- implementing a package of measures designed to reduce emissions from the transport sector;[57]
- establishing a new working party on 'emission inventories' under the umbrella of the existing interministerial working group on CO2 reduction;
- adoption by the federal government of its own independent target of reducing CO2 emissions within the federal sphere of influence by 30% by 2010, and by 25% by 2005 (1990 baseline);
- implementing additional measures to tackle non-CO2 GHG emissions.[58]

Between 2000–2005, the federal government worked to implement the measures laid out in the NCPP.

NCPP 2005

The National Climate Protection Programme of 18 October 2000 was updated on 13 July 2005.[59] As revised, the National Climate Protection Programme 2005 (NCPP 2005) contains a comprehensive catalogue of measures to ensure that the goal of reducing greenhouse gas emissions by 21%, as compared to 1990 levels, between 2008–2012 will be achieved.[60] Germany successfully reduced its greenhouse gas emissions by 18.5% by 2003, as compared with 1990 levels, and thus entered the NCPP 2005 period already having achieved significant emissions reductions.[61] The NCPP 2005 introduced new and expanded measures to help Germany meet its regional and international commitments as well as its more ambitious domestic targets.

Under the umbrella of the 2005 Climate Protection Programme, the need for action within sectors that are not covered by the EU ETS—such as private households, transport, industry, trade and the service sector, agriculture, forestry, and the waste sector—continues to be assessed. Corresponding with the outcome of the evaluation of the Climate Protection Programme 2000, the NCPP 2005

---

[57] The package included measures to: provide additional funds for investments in the rail infrastructure; introduce a distance-based motorway toll for HGVs from 2003; promote fuel-efficient cars within the context of the motor vehicle tax; negotiate agreements with the car manufacturers on contributions to a reduction of consumption; launch a broad-based information and public education campaign on good driving practice; introduce an emission-differentiated landing fee for airports; campaign for the introduction of an EU-wide emission-based air traffic levy; implement integrated transport planning and climate-compatible community planning schemes; implement a transport-efficient energy strategy; and increase use of telematics and fleet management systems. National Climate Protection Programme 2000 (n 53).   [58] Ibid 3–5.
[59] BMU, 'The National Climate Protection Programme 2005: Summary' (2005) <http://www.bmu.de/files/english/climate/downloads/application/pdf/klimaschutzprogramm_2005_en.pdf> accessed 4 January 2010.   [60] Ibid.
[61] Ibid.

formulates a series of measures focused on reducing emissions from the transport sector and from private households.

The Climate Protection Programme 2005 emphasizes the principle of sustainable development as elaborated in Agenda 21, and is premised on the notion that decentralized, local actions are essential to long-term efforts to address global climate change. The 2005 measures also developed Germany's EU ETS program, incorporating into domestic law the EU Emissions Trading Directive, the EU Linking Directive, and existing German measures.[62]

Pursuant to its strategy under the NCPP 2005, Germany now operates an extensive emissions trading operation that is run by the KfW Banking Group, as charged by the Federal Environment Ministry.[63] Having included trading as part of its climate strategy since 2000, when it first established the Working Group on 'Emissions Trading as a Means to Combat the Impacts of Greenhouse Gases' (AGE), Germany now functions as a key player in the EU ETS. Demonstrating its central role to the EU ETS, the European Energy Exchange (EEX), which facilitates region-wide compliance with NAPs approved under the EU ETS is housed in Germany; Germany's Dresdner Bank is one of the leading traders of emission certificates; and, the German government has long been one of the greatest advocates of permit auctioning.[64]

Renewable Energy and Energy Efficiency

Beyond the two National Climate Protection Programmes, German energy policy, more broadly conceived, plays a critical part in German climate policy. Germany's energy program forms the core of its regional and global leadership in the climate arena. Furthermore, the nation's ability to promote a strong energy program based on high levels of renewable energy and great improvements in energy efficiency underpin its global leadership on climate change. Despite having made great strides in improving its energy policy, however, Germany continues to struggle to reduce its dependency on fossil fuels, especially coal.

Germany is a world leader in renewable energy technology, with approximately one-third of all water-power installations in the world being produced in Heidesheim by the German firm Voith, and with every other wind turbine and every third solar cell also being German-made.[65] In September 2004, for example, the largest solar power plant in the world, at that time, was opened in Germany. The plant can provide electricity for approximately 1,800 European

[62] Ibid.
[63] See BMU, 'KfW Banking Group to Handle Sale of CO2 Emission Allowances' (Press Release, 18 December 2007) <http://www.bmu.de/english/press_releases/archive/16th_legislative_period/pm/40645.php> accessed 3 April 2010.
[64] Chicago Climate Exchange, 'Launch Date Set for First EU ETS Futures Contracts' (Press Release, 21 March 2005) <http://www.climatechange.com.au/2005/03/29/launch-date-set-for-first-eu-ets-futures-contracts/> accessed 3 April 2010; Allianz Knowledge, 'Germany Climate Change Profile Part 2: Fact Sheet' (n 48).
[65] Allianz Knowledge, 'Germany Climate Change Profile Part 2: Fact Sheet' (n 48).

homes, reducing annual carbon dioxide emissions by approximately 3,700 tons.[66] Similarly, the German wind industry is thriving and accounts for the largest share of renewable energy in German gross electricity consumption.[67] In the Northern German state of Schleswig-Holstein, wind energy supplies approximately 30% of the energy needs and the local wind industry has 'generated over 300 million euros of state revenue through its contributions to the national power grid, and created 5,000 local jobs over the last few years'.[68]

Germany's green technological prowess is due in great part to governmental support for energy initiatives. Germany's federal government has consistently sought to increase the total share of energy produced by renewable sources and to increase energy efficiency.[69] In line with its domestic policy, Germany has been one of the most vocal European proponents for increasing mandatory shares of domestic and regional electricity supplies from renewable sources to at least 25% by 2020.[70] The EU 20/20/20 objectives, for example, exist in large part as a result of German campaigning.

Beyond its national and regional efforts, Germany has also advanced renewable energy technologies at the global level. The German federal government, for example, was the driving force behind the recently created International Renewable Energies Agency (IRENA).[71] IRENA was developed to 'provide practical advice and support for both industrialized and developing countries who wish to expand their renewable energy sectors and move toward improved regulatory frameworks', with the ultimate aim of 'promoting a rapid transition towards the widespread and sustainable use of renewable energy on a global scale'.[72] The Founding Conference for IRENA took place on 26 January 2009, in Bonn, Germany and the Agency is now operational, with Germany continuing to play an active governance role.

---

[66] Eric Shaffner, 'Repudiation and Regret: Is the United States Sitting Out the Kyoto Protocol to its Economic Detriment?' (Comment) (2007) 37 *Envtl L* 441, 443.
[67] BMU, *Renewable Energy Sources Act (EEG) Progress Report 2007* (Report submitted to the German Bundestag 2007) <http://www.bmu.de/files/pdfs/allgemein/application/pdf/erfahrungsbericht_eeg_2007_zf_en.pdf> accessed 4 January 2010 at 6 (EEG 2007 Progress Report).
[68] Ibid.
[69] 2006 UNFCCC Report (n 6) 12; EurActiv, 'German Climate Plan Gets Mixed Reviews' (News Release, 27 August 2007) <http://www.euractiv.com/en/climate-change/german-climate-plan-gets-mixed-reviews/article-166113> accessed 4 January 2010.
[70] 'German Climate Plan Gets Mixed Reviews' (n 69).
[71] See BMU, 'One More Step Towards Setting Up an International Renewable Energies Agency' (Joint Press Release of BMU and Federal Ministry for Economic Cooperation and Development, 1 July 2008) <http://www.bmu.de/english/current_press_releases/pm/41931.php> accessed 1 June 2010. It should be noted that the Bundestag voted in favor of this program (IRENA) on 19 June 2008 by an overwhelming majority. Ibid.
[72] Initiative for an International Renewable Energy Agency, 'International Renewable Energy Agency (IRENA): Initial Work Programme: Second Draft' (August 2008) 4 (IRENA Draft) (noting that while '[t]here are several organizations and networks—the IEA, UNDP, UNEP, UNIDO, REN21, REEEP and others—working at the local, national and international levels to expand the use of renewable energy', there is 'no international organisation whose work is mainly focused on renewable energy and who, in addition, comprises industrialised as well as developing countries').

At the domestic level, the core of Germany's renewable energy strategy is the Renewable Energy Sources Act (EEG).[73] Initially enacted in 2000 and revised in 2009, the EEG 'guarantees the feed-in of electricity from renewables at a fair and fixed fee'.[74] In this way, the EEG promotes the use of renewable energy by requiring power grids to purchase energy from renewable sources at a fair cost—thus guaranteeing a market and a competitive market price for renewable energy producers.

The EEG mandates that the BMU and other federal departments submit reports of progress to the Bundestag every four years. The first of these reports was submitted in December 2007, and revealed that:

Since the [EEG] entered into force in 2000, the share of renewables in primary energy consumption has more than doubled, from 2.6% in 2000 to around 5.8% in 2006; the same applies to the share of renewables in total final energy consumption, from 3.8% (2000) to around 8.0% (2006). The share of renewable energies in total gross electricity consumption has almost doubled, from 6.3% in 2000 to around 11.6% in 2006. A figure above 13% is forecast for 2007, which means that the expansion target set in the [EEG] for 2010 will be exceeded as early as 2007.[75]

Just two years later, in June 2009, a government report demonstrated further improvements, revealing that renewable sources now accounted for: 9.5 % of total final energy consumption; 15.1 % of gross electricity consumption; 7.4 % of final energy consumption for heat; 5.9 % of fuel consumption; and, 7.0 % of primary energy consumption.[76]

The EEG, complemented by initiatives to improve domestic energy intensity, has spurred the renewable energy industry in Germany, thereby resulting in job growth and significant increases in megawatt capacity of renewable energies, particularly wind energy.[77]

As a result of Germany's success with the Renewable Energy Sources Act, instruments guaranteeing renewable energy feed-ins at fixed fees are now used worldwide, including in the Southern United States.

Within its evolving energy policy, Germany does not categorize nuclear energy as an alternative source of energy, nor does it include nuclear power as a component of its climate governance system.[78] In contrast to certain of its European neighbors, Germany does not view nuclear energy as a viable alternative to fossil fuels. Rather, it has committed to phasing out all nuclear power plants by 2023, despite the fact that nuclear facilities continue to provide 26% of all gross

---

[73] Taking Action (n 50) 10.
[74] See Renewable Energy Sources Act of 25 October 2008 (Federal Law Gazette I p. 2074); EEG 2007 Progress Report (n 67) 5.
[75] EEG 2007 Progress Report (n 67) 6.
[76] BMU, 'Renewable Energy Sources in Figures' (11 June 1990) <http://www.erneuerbare-energien.de/files/english/renewable_energy/downloads/application/pdf/broschuere_ee_zahlen_en_bf.pdf> accessed 31 December 2009.
[77] Allianz Knowledge, 'Germany Climate Change Profile Part 2: Fact Sheet' (n 48).
[78] Ibid.

electricity produced in Germany, while renewable energy sources only account for approximately 14% of gross electricity production.

While it has had great success in mobilizing renewable technologies and current levels of renewable energy consumption brings it near to meeting the EU 2020 renewable energy requirement, Germany's decision to phase out nuclear energy based on health and safety concerns creates the conditions for a looming energy and environmental crisis. To make up for energy shortages due to the loss of nuclear capacity, energy suppliers plan to build 26 new coal-fired plants.[79] Coal-fired power accounts for approximately 45% of Germany's electrical output. This figure will increase to around 57% if the plans for new coal-fired plants go forward.[80] Expected increases in GHG emissions from coal-fired plants are incompatible with the increased reduction requirements that the EU 2020 plan places on Germany, namely, a 14% reduction below 2005 levels. The growth in coal-fired power plants is also incompatible with Germany's own environmental and climate strategies. Tensions between the desire to eliminate nuclear energy and the desire to curb GHG emissions date back to the earliest days of the nuclear-climate debates in Germany and reflect ongoing, competing political priorities. The resolution of this issue is critical to Germany's ability to meet its regional and domestic renewable energy and emissions reduction obligations in the future.

The Integrated Energy and Climate Programme 2007

In 2007, Germany sought to expand the energy component of its climate strategy and to improve upon existing climate instruments through the Integrated Energy and Climate Programme 2007. The Integrated Energy and Climate Programme included a series of 14 proposals for comprehensively updating Germany's energy sector and climate program, including the following measures:

- **Combined Heat and Power (BMWi):** In the electricity sector, the government intends to press ahead with the expansion of cogeneration. In order to use fuel efficiently, the share of high-efficiency CHP plants in electricity production will be doubled by 2020 from the current level of around 12% to around 25%. The amendment to the Combined Heat and Power Act, which promotes the construction of new CHP plants and heat grids, is an important component in achieving this goal. This measure is complemented by voluntary commitments on the part of industry. (IECP Action 1)

- **Amendment to the Renewable Energy Sources Act (EEG):** The government's goal is to increase the share of renewables in the electricity sector from the current level of at least 13% to 25–30% in 2020, steadily raising this percentage in the years beyond. The amendment to the Renewable Energy Sources Act (EEG), which among other things contains new provisions on regulating tariffs for offshore wind farms, serves this goal. (IECP Action 2)

- **Renewable Energies Heat Act (BMU):** Renewable energies in the heat sector offer huge potential for climate protection and savings in fossil fuels. The share

[79] Ibid.   [80] Ibid.

of renewable energies in heat provision will therefore be increased to 14% by 2020. Obligations to use renewable energies in new buildings as far as commercially feasible will be laid down in the Heat Act. (IECP Action 14)

- **Amendment to the Gas Grid Access Ordinance/Biogas Feed (BMWi):** Existing legislation will be fine-tuned and, where necessary, improved to facilitate the feed-in of biogas into the natural gas grid. This includes firming up the existing legal provisions to eliminate any remaining hurdles to biogas feed. This will make it considerably easier and commercially attractive to feed biogas into the natural gas network. (IECP Action 9)

- **Actions for Grid Expansion (BMWi):** The Cabinet agreed on the essentials for a grid expansion package designed to improve the integration of renewables into the grid. To this end, in August 2009, the government adopted the Energy Grid Expansion Act, which facilitates the construction of new grids using new technologies and offers incentives for the development of new storage technologies for renewables. (IECP Action 2)

- **Amendment to the Energy Industry Act (EnWG) on Liberalising Metering (BMWi):** Amending the Energy Industry Act will open electricity metering fully to competition with the aim of encouraging the use of smart devices. Liberalizing electricity metering will facilitate innovative metering methods and load-related, time-variable tariffs. This can help consumers reduce their energy costs and contribute to boosting the efficient utilization of the power station fleet. An ordinance specifying the requirements came into effect in January 2009. (IECP Action 4)

- **Report and Draft Amendment to the Energy Saving Ordinance (EnEV):** In order to increase energy efficiency in buildings, energy standards will be tightened by an average 30% from 2009. As a second step (planned for 2012), these efficiency standards will be tightened by an additional 30%. The Cabinet has adopted corresponding key elements.

- **Clean Power Plants:** By amending the 37th Ordinance on the Implementation of the Federal Immission Control Act (BImSchV), ambitious standards will be laid down for nitrogen oxide emissions from new power plants. This will make sure new plants are not only more efficient, but also cleaner than old ones.

- **Guidelines on the Procurement of Energy-efficient Products and Services:** With the adoption of guidelines on environmentally friendly and energy-efficient procurement, the German government is setting a good example for others to follow. Energy-efficient appliances and services will be promoted through priority procurement.

- **Amendment to the Biofuel Quota Act:** As a contribution to achieving the German government's energy and climate policy goals, the share of biofuels will be increased and from 2015 will be geared more towards reducing greenhouse gas emissions. The amendment to the Biofuel Quota Act will lead to a rise in the biofuels' share to around 20% by volume (17% by energy content) by the year 2020.

- **Sustainability Ordinance:** The Sustainability Ordinance will ensure that when producing biomass for biofuels, minimum requirements for sustainable management of agricultural land and for the conservation of natural habitats are complied with. Furthermore, the entire production, processing, and supply chain must show a certain potential for reducing greenhouse gases.

- **Fuel Quality Ordinance:** The amended Fuel Quality Ordinance will increase the blending limit of bioethanol in petrol fuels from 5 to 10% volume. For biodiesel in diesel fuels, this blending limit will increase from 5 to 7% volume.

- **Hydrogenation Ordinance:** By approving biogenic oils that are hydrogenated together with mineral oil based oils in a refinery process, compliance with the increased blending quotas will be made considerably easier in future.

- **Reform of vehicle tax to a pollutant and CO2 basis:** The vehicle tax was amended in May 2008. For new vehicles, this tax is now calculated on the basis of a vehicle's emissions rather than engine capacity.

- **Chemicals Climate Protection Ordinance:** This Ordinance will reduce emissions of fluorinated greenhouse gases from mobile and stationary cooling installations through provisions on leakproofness and labeling of the installations, and on recovery and return of the refrigerants used.[81]

In presenting this comprehensive plan of action, the German government sought to stimulate both domestic and EU climate policy, and to demonstrate that fighting climate change can be efficient.[82] In doing so, the German government stressed that the key to success lay in improving energy efficiency and increasing and diversifying low-CO2 forms of energy. At the same time that the German government presented its domestic Integrated Energy and Climate Programme 2007, it also offered:

to cut emissions by 2020 to a level 40% below that of 1990. This offer assumes that the European Union will, over the same period, reduce its emissions by 30% compared with 1990 and that other countries will adopt similarly ambitious targets. The package of acts, ordinances and reports that has been adopted here amounts to a major advance towards achieving these goals and sends a clear message to the international climate negotiations in Bali.[83]

Underscoring that the government's energy policy continues to be guided by the triple objective of supply security, economic efficiency, and environmental

---

[81] BMU, *Report on Implementation of the Key Elements of an Integrated Energy and Climate Programme Adopted in the Closed Meeting of the Cabinet on 23/24 August 2007 in Meseberg* (5 December 2007) <http://www.bmu.de/english/climate/downloads/doc/41258.php> accessed 4 January 2010.   [82] Ibid 2.
[83] Ibid.

acceptability, Germany sought to alleviate domestic concerns while simultaneously placing itself in a position to influence regional and global climate negotiations.

June 2008 Climate Law Package

Shortly following the launch of the ambitious 2007 Programme, on 19 June 2008, the BMU presented a comprehensive Climate Protection Initiative consisting of a package of four laws.[84] Then Federal Environmental Minister Sigmar Gabriel introduced the package by stressing the economic opportunities associated with climate protection:

> The Climate Protection Initiative makes an important contribution to reaching our climate protection goal. With it, we are tapping major potential for CO2 savings on a large scale, in schools, small and medium-sized enterprises and private households. We want to demonstrate that climate protection pays off for industry, municipalities and consumers. Those investing today will be rewarded with lower energy costs.[85]

The 2008 climate law package was designed with the objectives of reducing carbon emissions by 40% below 1990 levels by 2020 and encouraging the uptake of effective adaptation measures. As of 2008, Germany had achieved a 20% reduction and expected the four new laws to facilitate at least a further 10% drop.[86]

The package of bills includes provisions aimed at:

- Solar energy: limits on cross-subsidies for solar energy/photovoltaics due to a shifting emphasis on wind power;
- Wind energy: measures intended to double renewable electricity content to 30% by 2020, focusing especially on boosting wind energy capacity;

---

[84] BMU, 'Federal Environment Minister Gabriel Launches 400 Million Euro Climate Protection Programme' (19 June 2008) <http://www.bmu-klimaschutzinitiative.de/en/press?d=100> accessed 3 April 2010.
[85] Ibid.
[86] To achieve the additional emissions reductions, a study commissioned by the federal government suggests that the additional 270 million tone reduction could be achieved by 2020 through measures in eight key areas:

1. An 11% reduction in electricity consumption through a massive increase in energy efficiency;
2. Renewing the power plant park, with more efficient power plants;
3. Increasing the share of renewables in electricity generation to more than 27%;
4. Doubling the efficient usage of combined heat and power (CHP) to 25%;
5. Reducing energy consumption through renovation of buildings, efficient heating systems and in production processes;
6. Increasing the share of renewable energies in the heating sector to 14%;
7. Increasing efficiency in the transport sector and increasing the share of biofuels to 17%;
8. Cutting emissions of other greenhouse gases such as methane.

BMU, 'The Federal Government's Climate Policy in the Wake of the European Council: Climate Agenda 2020' (Speech by Federal Minister Sigmar Gabriel, 26 April 2007) <http://www.bmu.de/english/speeches/doc/39349.php> accessed 4 January 2010 (Federal Minister Sigmar Gabriel's Speech).

- Heating: provisions designed to increase the number of combined heat and power generation plants, allowing competition among electricity and gas meter reader companies, and thereby encouraging tele-heating, where power stations pipe surplus hot water into homes;
- Insulation: measures to encourage improved insulation and energy efficiency in public and private buildings.

The combined effect of the laws will be to improve energy efficiency and increase the uptake of green energy for electricity production. The improvements will not be cost free, however. German consumers have been warned to expect increases in energy prices over the coming years. The new package of laws received lukewarm reception from a domestic constituency whose ardor for climate policy has waned as recession set in and energy prices continued to steadily rise.

Despite economic recession, weakening domestic enthusiasm, and increasing signs of political division, the German government led by the BMU has thus far continued to press forward with ambitious climate objectives.

Progress and Challenges

The combined effect of Germany's past and present climate measures reveals a record of great successes mixed with equally great future challenges. According to figures released by the Federal Environment Agency, greenhouse gas emissions in Germany in 2007 were 22.4% below emissions in the Kyoto base year (1990/1995), while greenhouse gas emissions in 2008 were at their lowest level since 1990. As of 29 March 2009, Germany declared that it had successfully achieved its commitment under the Kyoto Protocol and the EU-15 bubble of reducing emissions 21% by 2008–2012 as against 1990 levels.[87] Further, Germany is driving efforts to improve the EU ETS through increased auctioning of allowances and it is striving to become a world leader in international adaptation aid.[88]

This progress is commendable, yet not enough. First, while successful emission reductions are partially due to the climate protection measures Germany has adopted, they are also partially attributable to low temperatures, economic recession, and political reunification.[89] For example, approximately 15% of Germany's

---

[87] BMU, 'Climate protection: Greenhouse gas emissions in 2008 at their lowest since 1990 Germany reaches its Kyoto target' (Joint Press Release, 29 March 2009) <http://www.bmu.de/english/press_releases/archive/16th_legislative_period/pm/43723.php> accessed 31 December 2009.

[88] BMU, 'The International Climate Initiative of the Federal Republic of Germany' (November 2009) <http://www.bmu.de/files/english/pdf/application/pdf/brochure_iki_en_bf.pdf> accessed 31 December 2009.

[89] Cass R Sunstein, 'Of Montreal and Kyoto: A Tale of Two Protocols' (2007) 31 *Harvard Envtl L Rev* 1, 27 (noting that '[f]or the United Kingdom, the story is not very different. The target

'overall 21-percent-below-1990-level-reduction' was achieved through improving infrastructure and reducing pollution in what was formerly East Germany.[90] Second, Germany continues to fall short of its own more ambitious domestic targets. These shortcomings are due to a combination of factors, including, eg, failure on the part of German industry to honour its pledges on the expansion of combined heat and power, failures in the functioning of the EU ETS, as well as structural failings in the German energy sector.

Recognizing the limits of existing climate measures, Germany continues to review and update its climate policy regime. Yet, for all of its efforts to progressively evolve its system of climate laws and policies, Germany continues to struggle with plans for its energy future with regards to eliminating existing coal subsidies and deriving a low-carbon solution for future losses in nuclear capacity.

Germany's climate policy regime, successes and failures combined, represents one of the most hopeful and forward-looking large-scale climate governance systems worldwide. Further, Germany continues to maintain a strong leadership role in global negotiations. Fully accepting anthropogenic forcing as a primary cause of on-going global warming, Germany estimates that '[m]ankind only has 10 to 15 years to prevent the worst impacts and to respond adequately to this huge challenge'.[91] As such, Germany views itself as having a moral duty to be 'a pioneer in climate protection' and has steadily fought for increasingly progressive climate policy at the domestic, Community, and international levels.[92] Taking as its motto that 'the best economic policy is a smart environmental policy', Germany continues to advocate improved climate policy through domestic and global economic recession.[93]

The ability of the EU to meet its Kyoto obligations turns on the ability of key member states, such as Germany and the UK, to not only meet but also exceed their domestic emissions reduction commitments.[94] Germany was one of the first European nations to adopt far-reaching climate measures and it has continued to be at the vanguard of climate policy. As early as 2000, Germany was on track to reach its 21% domestic commitment within the EU burden-sharing agreement. And, while much of Germany's success is correctly attributed to political reunification, this by no means devalues the very real efforts Germany has taken to create an effective domestic climate governance system and to promote improved

---

reduction of 8% was less severe than it seemed, because in 1997, the United Kingdom was already at a level 5% below that of 1990').

[90] Harris, 'The European Union and Environmental Change' (n 31) 340.
[91] Taking Action (n 50) 5.
[92] Ibid 8. See also BMU, 'International Climate Policy' <http://www.bmu.de/english/climate/international_climate_policy/doc/41824.php> accessed 17 September 2008. But see Spiegel Online, 'German Climate Protection Package at Risk' (24 May 2008) <http://www.spiegel.de/international/germany/0,1518,555187,00.html> accessed 3 April 2010 (arguing that only 28% is achievable). [93] Taking Action (n 50) 9.
[94] Harris, 'The European Union and Environmental Change' (n 31) 339.

climate governance worldwide. As then Federal Environmental Minister Sigmar Gabriel noted:

> implementing Europe's climate protection targets entails no less than the radical restructuring of industrial society. If we want to provide goods and services for a world whose population is expected to grow from 6.5 billion to more than nine billion, and we want to halve greenhouse gas emissions at the same time, this will require a quantum leap in the development of industrial society.[95]

Germany has not yet made this quantum leap, but it has begun to lay the foundations for incrementally restructuring industrial society.

The United Kingdom

Like Germany, the UK's role in climate change policy is symbolically and substantively important in both the European and international contexts. As a stable and prosperous world economy, a long-standing democracy, and a key player in the international economic and security arenas, the UK's response to climate change is central to shaping global climate negotiations.

In contrast to Germany and many of the 'new' European States, the UK has a long history of political and governmental stability. While the UK suffered great social and economic losses during the two World Wars, its underlying system of government remained constant. Unlike Germany and Poland, for example, the UK did not undergo significant shifts in political or geographical structure, allowing it to sustain its long-standing parliamentary democracy even during times of turmoil. The UK's social, economic, and political stability, combined with its colonial history and its enduring influence worldwide, make it a dominate presence in international politics despite its small size and modest population.

Examining the UK's national climate change policy is fundamental to understanding the larger EU climate framework. As the member state with the highest historical per capita greenhouse gas emissions and one of the higher national emission reduction obligations under the EU bubble, the UK occupies an important substantive and symbolic role in the EU climate change regime. The UK's decision to take an active leadership role in efforts to develop climate change law and policy at the regional and global level further elevates the importance of its domestic climate change regime.

Like Germany, the UK has played a prominent role in regional and international climate debates. Its pathway to climate leadership, however, experienced a slower and more turbulent start. Although the UK now holds itself out as a global climate change leader, this was not always the case. In the very early days of international climate negotiations, the UK was reluctant to recognize and respond to

---

[95] Federal Minister Sigmar Gabriel's Speech (n 86).

climate change with legally binding commitments, choosing to align itself more closely with the US than with its fellow EU member state, Germany, in advocating a conservative approach towards climate change. Over time, however, the UK has distanced itself from the US's historically reticent approach to climate change, choosing instead, to advocate more aggressive domestic, regional, and international action on climate change.

UK Climate Policy in Context

The UK is one of the founding members of the EU and is a member of the EU-15 'bubble' under the Kyoto Protocol. Under the EU burden-sharing agreement, the UK agreed to reduce its greenhouse gas emissions by 12.5% below 1990 levels by 2008–2012.[96] Beyond the binding contours of the burden-sharing agreement, the UK adopted an early domestic goal of cutting emissions to 20% below 1990 levels by 2010.[97] This domestic goal has evolved over time. In 2003, the UK government stated its aim of 'achiev[ing] a carbon dioxide emissions reduction of some 60 per cent by 2050, with real progress by 2020'.[98] Five years later, in 2008, the UK government once again revised its domestic objectives when, at the recommendation of the government-appointed Climate Change Committee, the new climate change secretary, Ed Miliband, announced that Britain would aim to cut greenhouse gas emissions by 80% as against 1990 levels by the year 2050.[99] Based on its Fifth Assessment Report to the UNFCCC, the UK is expected to cut emissions by around 23% below 1990 levels in 2010 and at least 34% per cent below 1990 levels for 2020.[100]

The strengthening of the emission reduction objectives reflects a larger evolution in the UK's system of climate policy since the early 1990s. The UK published its first climate change program in January 1994.[101] This initial program was subsequently reviewed and updated by the 2000 and 2006 Climate Change Programmes. The contours of the UK's current climate strategy are based on an early review of policy options led by Lord Marshall in 1998. The 1998 review, 'Economic Instruments

---

[96] Department of the Environment, Transport and the Regions, *Climate Change: The UK Programme Summary* (DETR, London, 2000) <http://www.ieadsm.org/Files/Exco%20File%20Library/Country%20Publications/cclsumm.pdf> accessed 3 April 2010 (DETR, *UK Climate Change Programme 2000*). [97] Ibid 4.
[98] BMU, *Demonstrable Progress Report: 2006 Report Under the UFCCC* (Brochure, June 2006) <http://www.bmu.de/english/climate/downloads/doc/38474.php> accessed 4 January 2010 (Demonstrable Progress).
[99] UK Department of Energy & Climate Change, 'UK on track to double Kyoto target' (Press Release, 5 June 2009) <http://decc.gov.uk/en/content/cms/news/pn058/pn058.aspx> accessed 30 October 2009; UK Department of Energy & Climate Change, 'UK at forefront of a low carbon economic revolution' (Press Release, 15 July 2009) <http://www.decc.gov.uk/en/content/cms/news/pn081/pn081.aspx> accessed 30 October 2009.
[100] Department of Energy & Climate Change, '5NC: The UK's 5th National Communication under the United Nations Framework Convention on Climate Change' (London, 2009) (5NC UNFCCC). [101] DETR, *UK Climate Change Programme 2000* (n 96).

and the Business Use of Energy', recommended incorporating emissions trading as a central component of a long-term strategy to address climate change.[102] In his report, Lord Marshall acknowledged the complexities inherent in structuring climate policy and emphasized the importance of formulating a strategy that integrated 'all sectors of the economy—business, domestic and transport' in domestic emissions reduction efforts.[103] To this end, his review advocated a 'mixed approach' to climate policy that integrates economic instruments, regulation, voluntary and negotiated agreements, and other measures in an effort to maximize the combined power of mandatory and incentive-based mechanisms. Following the release of Lord Marshall's report in November 1998, the UK government passed a series of initiatives to address climate change, including an energy tax (ie, the Climate Change levy), energy efficiency incentives, and an emissions trading scheme.[104] The portfolio-style approach to climate policy advocated by Lord Marshall in 1998 emerged as a dominant theme in UK climate policy over the next decade.

Prompted by Lord Marshall's Review, in 2000, the UK published its first Climate Change Programme (UKCCP) in November 2000. The 2000 UKCCP represented the UK's first concerted effort to develop a comprehensive domestic climate strategy. It set out a range of cross-sectoral policies and measures designed to help the UK meet its emissions reduction targets under the Kyoto Protocol and its own more aggressive domestic targets.[105] The 2000 UKCCP was reviewed and supplanted just over five years later by the 2006 UKCCP, which provides the parameters for the UK's current strategy for mitigating and adapting to climate change.[106]

The UK's 2006 climate change program included provisions for:

- a climate change levy package;

- agreements with energy intensive sectors to improve businesses' use of energy and to stimulate investment and cut costs;

- a domestic emissions trading scheme to compliment the EU ETS;

- regulations to stimulate new, more efficient sources of power generation;

- the creation of a Carbon Trust;

- obligations to cut emissions from the transport sector;

- a Great Britain-wide wholesale electricity market;

- agreements to promote new energy efficiency in the domestic sector.[107]

[102] Lord Marshall, 'Economic Instruments and the Business Use of Energy' (November 1998) <http://archive.treasury.gov.uk/pub/html/prebudgetNov98/marshall.pdf> accessed 1 june 2010.
[103] Ibid 1.
[104] Matthew Rees and Rainer Evers, 'Proposals for Emissions Trading in the United Kingdom' (2000) 9 *Rev Eur Community & Intl Envtl L* 232.
[105] DETR, *UK Climate Change Programme 2000* (n 96).
[106] Secretary of State for the Environment, Food and Rural Affairs, *Climate Change: The UK Programme 2006* (Report presented to Parliament by the Command of Her Majesty) (The Stationery Office, London, 2006) SE/2006/43.
[107] See DETR, *UK Climate Change Programme 2000* (n 96).

In developing its climate change strategy, the UK has focused on involving multiple stakeholders, including government, industry, and the domestic sector. In an effort to improve the political palatability of climate measures, the UK program emphasizes the benefits climate change policies will bring, including the following:

- improved energy efficiency and lower costs for businesses and householders;
- more employment opportunities through the development of new, environmental technologies;
- a better transport system;
- better local air quality;
- less fuel poverty; and
- improved international competitiveness for the UK.[108]

By focusing on the positive impacts of reducing greenhouse gas emissions, the UK seeks to provide incentives for active private sector participation in efforts to address climate change, as reflected in the policies and measures set out in the current UK climate change program. Policies and measures in the UK Climate Change Strategy fall into seven basic categories targeting business, power generation, transport, domestic, building, agricultural, and public sectors.[109]

The first category of policies and measures aims to facilitate business' improved use of energy, as well as to stimulate investments and reduce the costs of energy-efficient operations.[110] In order to achieve these goals, the strategy adopts six measures that constitute the backbone of the UK strategy.

- First, the government has created a climate change levy package. The package consists of improvement targets for energy intensive sectors via climate change agreements as well as supplementary assistance for energy efficiency measures in the business sector.[111]
- Second, the government created a fixed term domestic emissions trading scheme that created an early model for the EU ETS.[112] The government launched the program in 2003–2004 by investing £30 million in the scheme and by establishing financial incentives for businesses to assume binding emission reduction obligations.[113]
- Third, the UK set up a Carbon Trust.[114] The objective of the Carbon Trust is to encourage businesses to utilize cost effective, low-carbon technologies.[115]

[108] Ibid. [109] Ibid 6–7. [110] See generally ibid.
[111] Benjamin J Richardson and Kiri L Chanwai, 'The UK's Climate Change Levy: Is It Working?' (2003) 15 *J Envtl L* 39.
[112] DETR, *UK Climate Change Programme 2000* (n 96) 6. [113] Ibid. [114] Ibid.
[115] The Carbon Trust was set up by Government in 2001 as an independent company; its mission statement is 'to accelerate the move to a low carbon economy by working with organisations

- Fourth, as part of the climate change strategy, the government will exempt high quality CHP and renewable sources of electricity from the climate change levy.[116]
- Fifth, the UK government has established a sophisticated system of energy labels, standards, and product-related measures intended to facilitate 'market transformation' in the energy efficiency of lighting, appliances and other heavily traded goods.[117]
- Sixth, and finally, the strategy includes establishing a system for integrated pollution prevention and control.[118]

The second arm of the strategy seeks to encourage the development of new and increasingly efficient sources of power generation.[119] To accomplish this goal, the UK has created a scheme known as the Renewables Obligation Certificates Program (ROCs).[120] This program requires electricity suppliers to increase the percentage of electricity provided by renewable sources to 10% by 2010.[121] The UK has also set a target of more than doubling its Combined Heat and Power (CHP) capacity by 2010.[122] CHP is a fuel-efficient energy technology that is believed to be able to increase the efficiency of fuel use to 75%+ versus the 40% efficiency achieved from conventional electricity generation.[123] Thus, the UK strategy is combining fiscal incentives, grant support, a regulatory framework, and government leadership and partnerships with industry to promote the growth of CHP.

The third goal of the UK strategy is to cut emissions from the transport sector.[124] To do this, the UK took part in an EU-level agreement with car manufacturers to improve the average fuel efficiency on new cars by a minimum of 25% by 2008–2009.[125] This effort, while largely ineffective, was bolstered by accompanying modifications to vehicle excise duties and company car taxation schemes.[126] In addition, the UK strategy includes a '10 Year Plan' for

---

to reduce carbon emissions and develop commercial low carbon technologies'. Carbon Trust, *2006/07 Annual Report* (London, 2007) (Carbon Trust 06/07 Report). According to the 2006 to 2007 report of the Carbon Trust, during those years, the program was able to identify carbon savings of at least 4.6 million tonnes of CO2. Ibid 1 and 6. The Carbon Trust has a dual accountability: '... [t]o the country as a whole, to help meet the nation's carbon reduction commitments, [and] [t]o private and public sector organisations, to help them meet the challenges of climate change and seize the opportunities that it creates'. Ibid 3. The Carbon Trust estimates that '25% of the energy used by businesses could be saved', making this 'a clear business opportunity'. Ibid 9.

[116] DETR, *UK Climate Change Programme 2000* (n 96) 6.
[117] Ibid.   [118] Ibid.   [119] Ibid.
[120] See UK Utilities Act 2000 c 27. On 15 July 2009, the government published 'The Renewable Energy Strategy' (RES), which includes announcements on expanding and extending the program to enable it to deliver 30% renewable electricity or more by 2020. UK Department of Energy & Climate Change, 'Renewables Obligation' (About the Government's main mechanism for supporting renewable electricity) <http://www.decc.gov.uk/en/content/cms/what_we_do/uk_supply/energy_mix/renewable/policy/renew_obs/renew_obs.aspx> accessed 31 December 2009.
[121] Ibid.
[122] See Department for Environment, Food and Rural Affairs (DEFRA), 'The Government's Strategy for Combined Heat and Power to 2010' (2004) <http://www.chpa.co.uk/news/reports_pubs/government_reports/chp-strategy.pdf> accessed 17 April 2010.   [123] See ibid.
[124] DETR, *UK Climate Change Programme 2000* (n 96) 6.   [125] Ibid 10–11.
[126] Ibid.

the transport sector.[127] This plan includes investing £180 billion to improve the energy efficiency of the transport sector, with the objective of reducing both congestion and pollution.[128]

In addition to cutting emissions from the transport sector, the fourth objective of the UK strategy is to promote increased energy efficiency in the domestic sector, thereby reducing domestic emissions and enabling households to save money.[129] In order to involve the domestic sector in the climate change strategy, the UK has initiated a new 'Energy Efficiency Commitment' program.[130] This program promotes cooperation between gas and electricity suppliers and domestic customers, with the goal of helping elderly and low-income customers conserve energy and reduce their energy bills.[131]

The domestic sector strategy also encompasses the 'New Home Energy Efficiency Scheme' in England, as well as comparable programs in Wales and Northern Ireland, and the 'Warm Deal Initiative' in Scotland.[132] These programs help fuel-poor households in the private sector access affordable energy. To do so, the program provides grants for insulation and improvements that will increase the energy efficiency of private homes.[133] Thus, the schemes aim both to alleviate fuel poverty and to increase energy efficiency. Additionally, the 'Affordable Warmth Programme' will coordinate the installation of efficient gas central heating systems and insulation in a million homes.[134] Finally, the UK domestic scheme promotes the growth and improvement of community heating systems and increasingly efficient lighting, heating, and other domestic appliances.[135]

The fifth objective of the UK strategy is to improve the energy efficiency requirements of the Building Regulations.[136] The Building Regulations are issued by the Office of the Deputy Prime Minister and are intended to 'ensure the health and safety of people in and around buildings by providing functional requirements for building design and construction'.[137] In addition, the regulations promote energy efficiency in buildings.[138]

Sixth, the UK strategy targets the agricultural sector.[139] One of the key goals of the UK strategy is to reduce emissions from agriculture by improving countryside management, reducing fertilizer use, preserving and managing forests, and improving energy efficiency across the board using sustainable agricultural practices. This is expected to have the dual effect of reducing carbon emissions

---

[127] Ibid.   [128] Ibid 7.   [129] Ibid.
[130] Ibid. The Energy Efficiency Commitment program is the successor to Energy Efficiency Standards of Performance (EESOP). Ibid.   [131] Ibid.
[132] Ibid 11.   [133] Ibid 8.   [134] Ibid.   [135] Ibid 8.   [136] Ibid 7.
[137] Department of Planning, Building and the Environment, 'Building Regulations' (Information about building regulations, building control, and other departmental responsibilities) <http://www.communities.gov.uk/planningandbuilding/buildingregulations/> accessed 4 January 2010.   [138] Ibid.
[139] DETR, *UK Climate Change Programme 2000* (n 96) 7.

from the agricultural sector while simultaneously improving the viability of the agricultural sector.[140]

The seventh and final objective of the UK strategy is to ensure that the public sector assumes a leading role in managing climate change.[141] To promote this objective, the UK government has set new targets for improving the energy efficiency of public buildings, local authorities, schools, and hospitals, and is developing green travel plans for public officials.[142] It has also initiated the 'Act on CO2' program, which focuses on increasing the general public's involvement in reducing carbon emissions.[143] Specifically, the initiative focuses on helping individuals simultaneously save money, conserve energy, and reduce CO2 emissions.

While one of the primary objectives of the strategy is to meet the UK's Kyoto Protocol obligations, the policies and measures the UK government is putting into place focus on addressing climate change beyond 2012.

Evolution of the UK Climate Strategy

The Climate Change Act 2008

Recognizing the limits of the UKCCP, including the absence of a reliable regulatory framework, in November 2007, a new climate bill was introduced in Parliament. The UK heralded the bill, enacted as the Climate Change Act, as a national and international milestone, characterizing it as 'the first time anywhere in the world' that a State had introduced 'a long term legally binding framework to tackle the dangers of climate change'.[144] Upon receiving Royal Assent on 26 November 2008, Parliament began the long process of overhauling the existing framework for climate change decision-making in the UK.[145] The backbone of the Act is a greenhouse gas emission reduction target for 2050, as well as complementary provisions doing the following:

- establishing a carbon budgeting system,
- creating a Committee on Climate Change,

[140] Ibid; See also, National Farmers Union, 'Agriculture & Climate Change' (2005) <http://www.cfeonline.org.uk/x9725.xml> accessed 17 April 2010.
[141] DETR, *UK Climate Change Programme 2000* (n 96) 7.   [142] Ibid.
[143] The 'ACT ON CO2' program is a cross-government initiative, currently involving the Department for Environment, Food and Rural Affairs (DEFRA), the Department for Transport (DfT) and the Department for Communities and Local Government (DCLG). ACT ON CO2, 'About ACT ON CO2' (Informational Webpage) <http://actonco2.direct.gov.uk/actonco2/home/about-us.html> accessed 4 January 2010. This program includes a 'package of long term assistance to households to help them tackle rising energy prices...through improved energy efficiency and other measures'. Ibid.
[144] DEFRA, 'Implementing the Climate Change Act of 2008' (18 June 2009) <http://www.defra.gov.uk/environment/climatechange/uk/legislation/> accessed 4 January 2010.
[145] Climate Change Act 2008 c 27. The impact assessment for the Act was released in March 2009. See Department of Energy & Climate Change, 'Climate Change Act 2008 Impact Assessment' (March 2009) <http://www.decc.gov.uk/en/content/cms/legislation/en/content/cms/legislation/cc_act_08/cc_act_08.aspx> accessed 31 December 2009.

- enumerating authority for establishing new emissions trading schemes,
- delineating powers for crafting climate adaptation policies,
- conferring authority for new waste minimization, recycling, and collection schemes,
- providing powers for charging fees for the use of single use carrier bags, and
- amending existing legal provisions related to renewable transport fuel obligations and previously set carbon emissions reduction targets.[146]

In key part, the Act sets legally binding targets for greenhouse gas reductions, specifying that through domestic actions and actions abroad, the UK must reduce emissions by at least 80% by 2050, including reducing CO2 emissions by at least 34% by 2020, against a 1990 baseline.[147] To help achieve these reductions, the Act establishes an incremental carbon budgeting system, which requires emission levels to be set and capped—by law—over five year periods, beginning in the period 2008–2012.[148] The Committee on Climate Change (CCC), an independent body, is tasked with providing expert advice to the government on setting appropriate carbon budgets, including the numerous scientific, economic, and legal questions that budget setting will raise.[149]

Expanding to the intersection of domestic and international law, the Act requires the government to include international aviation and shipping emissions in the Act, or provide reasons for its failure to do so, by 2012. Regulating international aviation and shipping emissions is controversial both domestically and internationally. Within the international context, complexities arise over the compatibility of any proposed regime with international trade law, for example. The UK's regulatory efforts in these areas will be closely watched and critiqued at domestic, regional, and international levels.

At the jurisdictional margins, the Act also directs the government, in consultation with the CCC, to establish the appropriate balance between achieving emissions reductions through domestic, European, and international ventures. Again, the balance here has implications well beyond the UK and will provide an observable blueprint for assessing particular types of emission reduction techniques in terms of actual reductions, equity, leakage, and additionality.

The Act also streamlines the process for creating new domestic emissions trading schemes and provides new powers to address waste management as an integral part of climate change mitigation efforts.

The Act's mitigation components are complemented by provisions for adaptation. The adaptation measures are largely information based, mandating that the government report every five years on the threats climate change poses to the UK and how these threats will be addressed, giving the government powers to require

---

[146] Climate Change Act 2008.   [147] Ibid.   [148] Ibid.
[149] Committee on Climate Change, 'Climate Change Act' <http://www.theccc.org.uk/about-the-ccc/climate-change-act> accessed 4 January 2010.

public bodies and utilities to undertake individual risk assessments, and creating an Adaptation Sub-Committee of the CCC.

Finally, the Act requires new rules regarding private reporting on greenhouse gas emissions and public reporting on the overall efficiency and sustainability of the government estate.

To date, the Act has prompted discernible responses including the establishment of the CCC on 1 December 2008, initial communications between the CCC and the government on the first round of carbon budgeting, new guidelines for businesses detailing how they should measure, report and reduce greenhouse gas emissions, new requirements for the government to issue guidance to companies on how to report their greenhouse gas emissions, and the setting of spring and summer deadlines for the establishment of the first three carbon budgets (2008–2022).

The Act establishes an underlying emission reduction goal, an institutional framework, authority, and—most importantly—a mandate for creating a more detailed regulatory regime. The success of the Act is wholly dependent on effective implementation and enforcement measures. The Act creates a clear framework for subsequent development of implementing laws and regulations; it is less clear whether it also provides mechanisms to compel compliance or ensure adequate enforcement.

Energy Bill 2007–2008

Just prior to passage of the Climate Change Act, on 26 November 2008, the UK published a new Energy Bill that implements the legislative aspects of the 2007 Energy White Paper. The Bill, which received Royal Assent on 8 April 2010 focuses on the dual goals enumerated in the White Paper—improving energy security and combating climate change by reducing CO2 emissions from the energy sector.[150]

Expanding on these primary goals, the bill attempts to balance four key subsidiary goals: 'put[ting] [the UK] on a path to cutting CO2 emissions by...60% by about 2050, with real progress by 2020'; 'maintain[ing] the reliability of energy supplies'; 'promoting competitive markets in the UK and beyond'; and 'ensur[ing] that every home is adequately and affordably heated'.[151] To promote energy security and affordability, the Energy Bill provides for exploration for and storage of natural gas, successful decommissioning of energy facilities, safe storage of nuclear waste, better management of petroleum licenses and oil and gas pipelines, and improved metering. The bill also enhances security measures and enforcement provisions for license violations.

---

[150] Energy Act 2010. See also Department of Trade and Industry, 'Meeting the Energy Challenge: A White Paper on Energy' (Cm 7124, 2007) (Energy White Paper 2007). Subsequently, on 19 November 2009, the government introduced a new Energy Bill that seeks to develop on the 2008 Bill protecting the poorest and most vulnerable consumers and by incentivizing carbon capture and storage. Energy HC Bill (2009–10) [7].   [151] Energy White Paper 2007 (n 150).

Simultaneous with provisions regarding conventional forms of energy, the bill complements the GHG emission reduction obligations established in the Climate Change Act by strengthening programs for carbon capture and storage, by bestowing power to impose renewable obligation orders, and by mandating that energy providers supply a specified portion of their energy from designated renewable sources.[152] The bill further mandates the continuing provision of annual energy reports on progress towards sustainable energy aims.[153]

Governmental Restructuring

In addition to adopting new primary and supplemental climate change and energy legislation, the UK government has restructured its departmental offices in response to climate change concerns. First, in September 2006, the government set up the Office for Climate Change (OCC), with the mandate of coordinating governmental efforts to develop climate change policies and strategies.[154] The primary objective driving the creation of the OCC was to improve intradepartmental communication and to coordinate and maximize climate change programs across governmental units. The OCC ultimately served as the first step to establishing a new overarching departmental level unit.

Created by then Prime Minister Gordon Brown on 3 October 2008, the Department of Energy & Climate Change (DECC) is designed to 'bring together much of Defra's existing climate change responsibilities with the energy component from [the *UK* Department for Business, Enterprise & Regulatory Reform], to focus on solving the challenges of climate change and energy supply'.[155] The DECC is in its infancy and its mandate and future role in climate change policy is still evolving and remains unclear and untested at this early point.[156] By combining energy and climate responsibilities in one department, the government is sending a signal about the primary links between short and long-term energy and environmental security.

Progress and Challenges

By 2004, UK greenhouse gas emissions were down 14.6% below base year levels. A year later, in 2005, the UK government stated that greenhouse gas levels had been further reduced to 15.3% below base levels, meaning that emissions

---

[152] Energy HC Bill (2009–10) [7].   [153] Ibid.

[154] Office of Climate Change, 'Welcome to the Office of Climate Change' (8 October 2008) <www.occ.gov.uk> accessed 4 January 2010.

[155] DEFRA, 'Gordon Brown Announces Changes to Defra' (3 October 2008) <http://www.defra.gov.uk/news/latest/2008/defra-1003.htm> accessed 4 January 2010.

[156] In an effort to lead UK efforts to reduce greenhouse gas emissions, in fall 2009, the DECC pledged to cut its own carbon emissions by 10% in 2010. Department of Energy & Climate Change, 'DECC Signs Up to 10:10 Campaign' (Press Release, 5 November 2009) <http://www.decc.gov.uk/en/content/cms/news/pn128/pn128.aspx> accessed 31 December 2009.

decreased by 0.5% between 2004 and 2005 alone.[157] The UK government further declared that, 'taking into account the net impact of allowances and credits surrendered through the EU emissions trading scheme, emissions of the basket of six gases controlled under the Kyoto Protocol might be about 23% below base year levels in 2010'.[158] In its 2006 Kyoto progress report, the UK attributed early and continuing reductions to 'restructuring of the energy supply industry; energy efficiency and energy intensity improvements; pollution control measures in the industrial sector and other policies that reduced emissions of non-carbon dioxide greenhouse gases'.[159]

Two years later, in 2008, the UK released a new progress report finding that by relying on a combination of regulations, partnerships, incentives, and innovation, it was on track to reduce greenhouse gas emissions to 23% below 1990 levels in 2010 and in excess of 34% per cent below 1990 levels by 2020.[160] Although these estimates put the UK on target to meeting its Kyoto Protocol obligations the report also revealed that absent additional measures and heavy reliance on the Kyoto Protocol's flexibility mechanisms, the UK will fall short of meeting its more ambitious domestic goals as currently outlined in the Climate Change Act.

Beyond the EU-15: The Case of Poland

Germany and the UK are regarded as keystone nations in EU efforts to achieve Kyoto obligations. The inability of these states to meet, if not exceed, individual emission reduction commitments foretells trouble for region-wide efforts to comply with Kyoto obligations, just as emission reduction successes on the part of these two states could lift the EU to overall compliance despite failures elsewhere. Yet, focusing exclusively on German and British climate governance systems ignores the complexities of EU member state relations and only reveals a very narrow segment of EU climate politics. While it is beyond the scope of this book to review the complex world of EU member states' climate politics, it is worth briefly examining the case of Poland as a case study in contrast to Germany and the UK.

The Republic of Poland (Poland) is one of Europe's largest and most populous nations. Bordering Germany on the west, Russia on the north, Lithuania, Belarus, and Ukraine in the east, and the Czech Republic and Slovakia on the south, it occupies a geographically and politically important corner of Europe.

---

[157] DEFRA, News Release, 31 January 2007, 'Greenhouse Gas Statistics Show UK on Track to Double Kyoto Target' <http://docs.middevon.gov.uk/pap/showimage.asp?j=07/02262/MFUL&index=302299> accessed 17 April 2010.
[158] DEFRA, *UK Climate Change Programme: Annual Report to Parliament, July 2008* (DEFRA, London 2008) 26.   [159] Demonstrable Progress (n 98) 8.
[160] 5NC UNFCCC (n 100) foreword at 2; 'UK on track to double Kyoto target'.

Following the political, social, and economic devastation of World War II, Poland was subject to Communist rule until 1989. Between 1989 and the present, Poland has 'transformed itself into a stable democracy with a multiparty political system... and made one of the most successful transitions to a free market economy'.[161] Poland is a constitutional republic governed by a presidential-parliamentary democracy, under which the President heads the state. A bicameral parliament—consisting of the Sejm (Chamber of Deputies) and the Senate (Chamber of Senators) sitting together in National Assembly— holds legislative powers. Finally, a Council of Ministers carries out executive functions.[162]

Poland first applied for membership in the European Union in 1994 and formally joined the EU at the end of 2002. While Polish law and policy have undergone significant Europeanization over the past decade, substantive changes are still required to enable Polish law and policy to exist and evolve compatibly within the EU system. Economic concerns dominate Polish decision making regarding prioritization of where, when, and how it chooses to integrate European legal and political culture into Polish governance systems. Yet, Poland has invested great effort in transforming its political and economic systems in order to more fully integrate with Western political and economic structures.[163]

Although Polish integration efforts still proceed, they are hampered by problems of over-centralization, competing competencies, corruption, regulatory capture, lack of public participation and transparency, cultural resistance to change, an anemic civil society, and low levels of social capital.[164] While these challenges are common to overarching EU integration efforts, they pose particular challenges to integrating Polish environmental law and policy with EU systems.

Upon emerging from Communist rule, Polish environmental law and policy was virtually non-existent and policies promoting heavy industry and unmitigated natural resource consumption were the norm.[165] Poland, thus, was tasked with building a system of environmental law and policy from the ground up. This involved the dual tasks of first, overcoming the entrenched view of the natural environment as merely a subset of economic policy and second, creating the legal and political foundations for integrating environmental protection into State policy. Poland accepted this challenge in

---

[161] European Commission, 'EU Enlargement and Multi level Governance in European Regional and Environment Policies: Patterns of Institutional Learning, Adaptation and Europeanization Among Cohesion Countries (Greece, Ireland and Portugal) and Lessons for New Members (Hungary and Poland)' (Final Report on Research of Socio-economic Sciences and Humanities 2007) 183 (EC Environmental Study).
[162] Ibid.   [163] Ibid.   [164] Ibid 239–41.
[165] Anita Bokwa, 'Climactic Issues in Polish Foreign Policy' in Paul G Harris (ed), *Europe and Global Climate Change: Politics, Foreign Policy and Regional Cooperation* (Edward Elgar Publishing, Northampton MA, 2007) 113.

1991 with the publication of its first strategic planning document, the *State Ecological Policy*.[166] Since that time, Poland has sought to harmonize its environmental governance system with that of the EU, although the legislative process did not begin in earnest until 2001, in anticipation of accession into the EU.

While Poland has made great strides in surface-level, legal harmonization, it has struggled to effectively implement the vast array of requirements inuring in EU environmental law, largely due to very high implementation costs. Challenges plague Polish integration efforts; problems of overlapping competencies, regulatory capture, weak civil society, and opaque decision-making structures impede Polish environmental law and policymaking efforts.[167] And, not surprisingly, environmental policy occupies a much lower position on the political agenda, compared to competing economic and industrial development policies. Accordingly, while legal and institutional infrastructure now exists for environmental protection in Poland, this infrastructure remains underutilized and under-funded.[168]

The picture is not all negative, however. The quality of Poland's environment has improved and has begun to recover from the legacy of Communist rule. Policymaking efforts coupled with economic recession in the early 1990s have resulted in sharp decreases in air pollution, water consumption, and waste generation.[169]

Polish Climate Policy

Poland faces great challenges, but also great opportunities, in responding to climate change. As a former Communist state and a relative newcomer to EU legal culture and economic and political ideologies, Poland's political focus remains on fundamental structural integration, rather than on integration in specific and costly arenas, such as climate policy. Poland is also the EU's largest coal producer and, thus, has a vested interest in ensuring that there are healthy markets for coal consumption.

---

[166] The State Ecological Policy was a progressive document that identified a series of fundamental principles that would underpin Poland's evolving system of environmental policy, including the sustainable development principle; the rule of law in ecological policy; the polluter pays principle; the elimination of pollution at the source principle; the social partnership rule; collaborative problem solving; and the principle of staging of environmental measures. EC Environmental Study (n 161) 185; See also Robert Szczepankowski, 'Public Policy Activity in Polish Public Administration—General Overview' (Paper presented on the 13th NISPAcee Annual Conference, 19–21 May 2005) <http://unpan1.un.org/intradoc/groups/public/documents/nispacee/unpan021846.pdf> accessed 4 January 2010 at 4.

[167] EC Environmental Study (n 161) 239–40.   [168] See ibid 184.

[169] Poland's existing environmental programs are funded by fees and fines imposed for exceeding emissions limits; these funds are also used to incentivize green activities, including air protection, climate protection, and environmental education. Ibid 185.

In contrast, Poland also has the potential to generate significant amounts of wind and solar energy.[170] Further, like other former Soviet Bloc countries, Poland's carbon intensity levels declined precipitously following the end of Communist rule due to the accompanying onset of economic recession as Poland underwent political and economic restructuring (eg, privatization and market liberalization). This emission decline created 'hot air'—ie, opportunities for economic growth despite GHG emission limitations since emission reduction levels were set using the baseline year 1988, when carbon intensity was still high in Poland. For example, Poland's GHG emissions decreased by 18% between 1990 and 2005 (a 32% decline since 1988), meaning that meeting GHG emission reduction commitments is not as economically onerous for Poland as they might initially appear to be.

Despite a sharp decline in carbon intensity between 1990–2005, Poland is now the sixth largest emitter in the EU-27, accounting for approximately 8% of total EU-27 GHG emissions and reciving approximately 10% of all emission allowances allocated under the EU ETS.[171] Moreover, the trend in Poland towards emission decreases has begun to reverse. Between 2004 and 2005, Poland experienced the largest emissions increase in absolute terms among the EU-12 (ie, the 12 member states not participating in the Kyoto bubble), experiencing a rise of 0.6% or 2.3 million tonnes of CO2 equivalent.[172]

Climate Commitments and Policies

Poland is an Annex I party to the UNFCCC and an Annex B party to the Kyoto Protocol. Under the UNFCCC, Poland committed to ensuring that its 2000 levels did not exceed its 1988 levels.[173] As an Annex B party to Kyoto, Poland has legally binding emission reduction obligations. Under Kyoto, Poland has committed to reducing GHG emissions by 6% as against 1988 levels by the first Kyoto compliance period.[174] In agreeing to a 6% emission reduction in the 1997 Kyoto negotiations, however, Poland warned that 'the commitment to stabilise greenhouse gas emissions by 2000 was not easy for a country in the process of

---

[170] Republic of Poland, 'Fourth National Communication Under the United Nations Framework Convention on Climate Change' (Warsaw, 2006) 17 (Poland's 4th National Communication).

[171] Commission (EC), 'Progress Towards Achieving the Kyoto Objectives' (Communication) COM (2007) 757 final, 27 November 2007.

[172] 'This was mainly due to a 1% increase in fugitive CH4 emissions from energy and rises in CH4 and N2O emissions from the agriculture sector of 5% and 4.5% respectively.' Ibid 8.

[173] United Nations Framework Convention on Climate Change (adopted 9 May 1992, entered into force 21 March 1994) 31 ILM 849.

[174] Under Kyoto, Poland is required to reduce its greenhouse gas emissions by 6% based upon 1988 as a base year level for CO2, methane, and NO2 and 1995 as a base year level for industrial fluorinated gases, HFCs, PFCs, and sulfur hexafluoride. Kyoto Protocol to the United Nations Framework Convention on Climate Change (adopted 11 December 1997, entered into force 16 February 2005) 37 ILM 32.

economic transformation', and that '[n]o significant reduction of coal consumption in the economy can be expected until the year 2010'.[175]

During negotiations for the Kyoto Protocol, Poland repeatedly stressed the particular challenges economies in transition face in attempting to reduce GHG emissions during a period of great social, political, and economic turmoil. Poland also backed efforts to extend Kyoto negotiations to include new commitments for developing countries in subsequent compliance periods. Despite expressing significant hesitations during negotiations, Poland signed and subsequently ratified the Kyoto Protocol. Poland's ratification of the Protocol was significant given that it is responsible for approximately 3% of global GHG emissions and is one of the top 25 major emitters.[176]

Following Poland's ratification of the Kyoto Protocol, between the years 1998 and 2004, Polish GHG emissions decreased as much as 31.7% below the base year.[177] Poland argues reductions are attributable to domestic implementation of 'a package of policies and measures primarily leading to the improvement of energy efficiency and restructuring of fuel consumption'.[178] As previously discussed, however, this claim is contested, with most commentators suggesting that the end of Communist rule coupled with recession accounts for the vast majority of emission reductions.

Futher, while Poland has heralded past emission reductions as a sign of successful climate policy, it also cautions that due to continuing economic modernization, it is likely to see future increases in greenhouse gas emissions. This is in large part due to continuing reliance on heavy fossil fuel infrastructure, eg, hard coal and lignite.[179] As of 2008, for example, it was estimated that Poland produces 96% of its energy from high-polluting coal plants.[180] Despite heavy reliance on coal-fired power plants, Poland claims that the consumption of coal has been decreasing by approximately 3% per year between 1995 and 2003, with total primary energy use decreasing by an average of 0.8% annually, and with oil and natural gas increasing by 3.5 and 2.3% respectively.[181] Poland further maintains that it will see a 5.1% increase in gross domestic product by 2010, and more during the years following, and that continuing economic modernization has 'always been targeted at energy-saving and environmentally friendly measures'.[182]

---

[175] Euro-East, 'EU/East Europe: Balancing the Needs for Growth Against Climate Change' (News Release, 18 March 1997), quoting then Polish Environment Minister Stanislaw Zelichowski.
[176] Kevin Baumert and Jonathan Pershing, 'Climate Data: Insights and Observations' (December 2004) <http://www.pewclimate.org/docUploads/Climate%20Data%20new.pdf> accessed 31 December 2009; Tamara L Harswick, 'Developments in Climate Change' (2002) *Colorado J Intl Envtl L & Policy* 25, 31.
[177] Poland's 4th National Communication (n 170) 7.   [178] Ibid.
[179] Poland's 4th National Communication (n 170) 10.
[180] Philippa Runner, 'Poland Keen to Block EU's CO2 Auction Scheme' (EU Observer News Release, 12 August 2008) <http://euobserver.com/9/26586> accessed 4 January 2010.
[181] Poland's 4th National Communication (n 170) 23.   [182] Ibid 11.

Current Activities

Poland hosted[183] the December 2008 UNFCCC Conference of the Parties (COP), the last full COP before the December 2009 Copenhagen Climate Change Conference. Poland's hosting duties were greeted with scepticism, given Poland's alignment with Italy in threatening to utilize its veto power on the EU's proposed new package of climate measures, as discussed *infra* in Chapter 7. Citing recession and the costly business of combating climate change, Italy's Prime Minister Silvio Berlusconi said that Italy's and Poland's 'businesses are in absolutely no position at the moment to absorb the costs of the regulations that have been proposed'.[184] He also said 'We do not believe that the time has come to play Don Quixote, to go forth alone when the big CO2 emitters like the United States and China are absolutely opposed to joining our drive... In this time of crisis, we need a bit of flexibility.'[185] Poland's Europe Minister, Mikolaj Dowgielewicz, supported Berlusconi's concerns, noting that: '[w]e certainly don't see the conditions for early agreement if we don't find a better burden-sharing inside the package'.[186] Poland and Italy were joined in their opposition to the proposed package by Bulgaria, Estonia, Latvia, Hungary, Lithuania, Romania, and Slovakia.[187]

As the largest coal producer in the EU, Poland opposes EU climate policies that constrain its energy choices as well as those that negatively impact the price of coal. Among the key concerns to Poland, as well as to other member states—including Germany (the original proponent of the plan), Italy, Austria, and most of the Central and Eastern European countries—were planned modifications to the EU ETS as part of the post-2012 strategy. In particular, the most contentious question involved plans to require the full auctioning of allowances in Phase III of the EU ETS for energy generators, with a gradual shift to full auctioning for other covered industries. Germany, Italy, and Austria expressed concerns over the impact on business, especially in the manufacturing sector, and argued that full auctioning would drive industry out of the EU. The Central and Eastern European states futher criticized the proposal as disproportionately impacting their economies due to their heavy reliance on coal and the continuing legacy of Communism, as evidenced by vestiges of heavy industry. In criticizing plans to auction upwards of 90% of allowances, Poland expressed fears that the modified EU ETS would cripple growth in its emerging energy sector and cause domestic energy prices to spike, with negative implications for consumers and for economic development.[188]

---

[183] 'Poland Keen to Block EU's CO2 Auction Scheme' (n 180).
[184] Leigh Phillips, 'Italy, Poland Threaten to Veto EU Climate Package' (EU Observer News Release, 16 October 2008) <http://euobserver.com/9/26945> accessed 4 January 2010.
[185] 'Climate Change/European Council: Climate Change a Collateral Victim of Financial Crisis' *European Report* (17 October 2008).
[186] 'Italy, Poland Threaten to Veto EU Climate Package' (n 184).
[187] Ibid.
[188] 'Poland Keen to Block EU's CO2 Auction Scheme' (n 180).

Poland's opposition to the 2008 package of measures was ultimately resolved through President Sarkozy—occupying the French Presidency of the Council at the time—negotiating a compromise whereby, among other changes, the term 'balance in the fundamental parameters' was deleted from the guidelines for internal discussions in exchange for keeping the December 2008 deadline. As amended, the proposal created a set of differentiated emission targets with reference to existing energy mixes, topography, and GDP per capita, as well as including financial compensation mechanisms for the less well-off member states.[189] The modified proposal afforded more opportunities for flexibility in shaping climate policy and meeting commitments, as well as for transferring final decision-making authority for the proposal from the Energy and Environment Councils to the European Council, thus requiring unanimity.[190]

Poland's opposition to increasingly stringent climate policy predates the December 2008 measures. Poland has consistently aligned itself with other economies in transition in appealing for flexibility and financial assistance in European climate policy. For example, as further discussed in Chapter 7, Poland has opposed new binding targets for renewable energy based on fears that the cost of complying would be disproportionately prohibitive and unrealistic for certain countries. Similar to the compromise negotiated by President Sarkozy in December 2008, German Chancellor Merkel was able to negotiate Polish support for the 20/20/20 proposal in 2007 by leaving questions of implementation up to the Commission, allowing more flexibility for determining how to distribute energy obligations equitably among the individual member states.

Beyond these specific examples, since joining the EU in 2002, Poland has consistently taken the position that EU energy and climate policy must 'reconcile environmental objectives and the need to ensure sustainable economic growth' and that the EU must 'avoid adopting measures that do not take account of differences in member states' economic potential'.[191] Poland, thus, does not oppose climate policy per se; it opposes climate policies that it perceives to be at the expense of economic development. Poland faces real challenges in recovering from the economic and environmental upheaval experienced during and immediately following the period of Communist rule. While Poland has begun to structure the framework for a functioning system of environmental law, it should come as no surprise that environmental protection—including climate change—continues to be overshadowed by questions of economic development when there are perceived conflicts between the two policy arenas.

Poland represents a microcosm of the challenges confronting EU and global efforts to improve climate policy. Poland seeks to integrate economically and

---

[189] Simon Tilford, 'The EU's Climate Agenda Hangs in the Balance' (Center for European Reform Bulletin December 2008/January 2009) <http://www.cer.org.uk/articles/63_tilford.html> accessed 4 January 2010.
[190] 'Climate Change/European Council: Climate Change a Collateral Victim of Financial Crisis' (n 185). [191] Ibid.

politically with Western culture; and in so doing, it prioritizes economic development, improved standards of living, and political stability. Climate policy, by modifying patterns of energy production and consumption, requires physical and cultural changes that appear costly and politically unpalatable.

Gaining Polish support for progressively stringent climate policies—as in other EU economies in transition and developing countries—rests on a twofold strategy of convincing politicians and their constituents, first of the social and economic costs of *failing* to address climate change and second, that short-term expenditures create opportunities for long-term benefits.

In addition, within the EU—just as between developed and developing countries—to achieve requisite levels of political support, climate policies must not only differentiate among nations according to notions of common but differentiated responsibilities, but also include mechanisms for technology and financial transfer. During each round of European climate negotiations, Poland's ultimate willingness to support EU measures has revolved around differentiation of targets and financial assistance. It is regularly joined in these demands by fellow Eastern European States.

The addition of 12 new member states since the negotiation of the Kyoto Protocol, coupled with the pressures of recession, places increasing constraints on the EU's ability to negotiate aggressive climate policies. Yet, as of spring 2010, the EU continued to proceed with efforts to progressively evolve its climate strategy. Progress comes at great effort and with great compromise, yet progress continues. The ability of the EU to maintain momentum despite political and economic constraints is significant and attests to the existence of governance structures that support extended negotiation, flexibility, and consensus building. The complex and often convoluted nature of EU governance limits transparency and defies efficiency. It also affords ample opportunities for communication, cooperation, and negotiation at every step in the decision-making process—traits that have proved essential to developing climate policy at the local and global levels.

Conclusion

The EU, much like the US, is much greater than the sum of its parts. Just as examining the climate policies of California or Texas reveals only a small and often distorted picture of US federal policy, so exploring the climate policies of Germany, the UK, and Poland tells an incomplete tale of the larger workings of EU policy. Yet, these snapshots are essential to understanding how and why federal and regional policies emerge.

Too often, EU and US laws and policies are analyzed without reference to the contexts within which they emerge and evolve. The global view of the US as a climate laggard ignores great legal and technological innovation taking place below the surface level. Similarly, the depiction of the EU as a climate leader glosses over internal inconsistencies and discord, and overshadows the rich political, legal, and cultural history that the EU negotiates to create region-wide policy.

The case studies of Germany, the UK, and Poland reveal much about how factors exogenous to climate policy have shaped climate policy preferences and policy successes, how variations in political processes impact domestic policymaking in the climate milieu, and how domestic policy choices affect EU-wide climate policy development. The contrasting social, political, and cultural contexts within which German, UK, and Polish climate policies arise and then converge in EU strategies show that meaningful change is possible.

PART IV

US AND EU CLIMATE CHANGE LAWS AND POLICIES COMPARED

7
US and EU Laws and Policies Compared

Our strong friendship is essential to peace and prosperity around the globe. No temporary debate, to passing disagreement among nations, no power on earth, will ever divide us.

President George W. Bush

The relationship between the United States and Europe is the world's strongest, most comprehensive, and strategically important partnership. The United States, and a united Europe—this is really the indispensable partnership.

President of the European Commission, José Manuel Barroso

Introduction

The disparities—both perceived and real—between United States and European Union responses to climate change play a central role in defining the political parameters of on-going negotiations over a post-Kyoto agreement. With the EU playing the role of the vocal, aggressive, and experienced elder climate leader and the US playing the part of the cautiously eager younger sibling ready to engage but wanting to do so slowly and conditionally, the level of unpredictability in international legal negotiations remains high.

While the EU and the US are only two actors at a sprawling negotiating table increasingly dominated by the rapidly developing economies, they are two of the players whose early expressions of interest and commitment helped shape the framework of the debate as well as the willingness of other key players—notably Australia and Japan among Annex I countries and China, India, and Brazil among non-Annex I countries—to come to the table. For this reason, how and why the EU and the US respond to climate change matters both substantively, ie, in terms of their actual ability to lower GHG emissions, and symbolically, ie, in terms of the message they send to the international community. And, while the EU and the US are often portrayed as sitting at opposite ends of the negotiating table, this depiction is neither precise nor helpful.

This chapter attempts to more thoroughly analyze convergences and divergences in US and EU climate strategies and to begin examining some of the key factors that influence the two regimes' legal and political approaches to climate change.

Leading up to 2009, the EU and the US appeared at logger-heads over the appropriate international response to climate change. In the waning hours of President George W. Bush's administration, the political discord became so heated that it prompted two commentators to note that 'excluding the US intervention in Iraq in 2003, climate policy is the most prominent example of a transatlantic rift... since WWII'.[1] The juxtaposition of climate policy with the Second World War suggests the scale of the debate and the importance of more critically analysing transatlantic approaches to climate change. Because, while the EU has undoubtedly assumed a leadership role in international climate negotiations, both Europe's leadership and the US's omissions require more careful scrutiny.[2]

Europe's leadership has filled a critical void. Had the EU failed to advocate aggressively for ratification of the Kyoto Protocol, it is unlikely that either Russia or Australia would have ratified the agreement, thus ensuring its legal failure. Similarly, if the EU had neglected to adopt a comprehensive supranational climate policy regime or neglected to press for increasingly progressive GHG emission reduction goals both regionally and internationally, the starting point for post-2012 climate negotiations would be a decade behind. In these varied ways, Europe has undoubtedly played a crucial leadership role. This leadership, however, should not mask very real struggles to structure and implement effective GHG emissions reduction strategies across a range of member states with varying economic, social, and political interests and capacities and limited authority at the international level.

The US's omissions at the international level have helped create the void within which Europe has acted. The US's failure to lead international negotiations, however, should not be confused with the US's absence from negotiations or with domestic inaction. The US has continued to participate in international negotiations. Until 2009, however, US negotiators advocated a cautious approach to climate change and continued to emphasize scientific uncertainties and the negative economic consequences associated with aggressive abatement actions. Further, until 2009, the US openly opposed entering into any international agreement that would create legally binding emissions reduction obligations, especially in the absence of legally binding commitments for the rapidly developing economies.

This is where popular depictions of US climate policy often stop, but the reality of US climate policy is considerably more complex. Just as Europe's international

---

[1] A Ochs and D Sprinz, 'Europa riding the hegemon? Transatlantic climate policy relations' in DB Bobrow and W Keller (eds), *Hegemony Constraint: Evasion, Modification, and Resistance to American Foreign Policy* (University of Pittsburgh Press, Pittsburgh, 2008) 144.

[2] Matthew Paterson, 'Post-Hegemonic Climate Politics?' (2009) 11 *British J Politics and Intl Relations* 140; M Schreurs and Y Tiberghien, 'Multi-level reinforcement: Explaining European Union leadership in climate change mitigation' (2007) 7[4] *Global Envtl Politics* 19.

leadership often masks mixed success internally, US omissions at the international and federal levels frequently overshadow noteworthy efforts to address climate change within the US. Inaction by the federal legislative and executive branches, in fact, has created a legal void that has allowed many states, cities, and regional entities to adopt diverse laws and policies aimed at addressing climate change and prompting federal engagement.

The realities of EU and US climate change law and policy are vastly more nuanced than depicted in common narratives. Yet, the differences are real and are telling. Political change will inevitably bring about shifts in climate politics, but the legacy of US and EU actions and omissions over the past 12 years will continue to shape the ability of the global community to address climate change for years to come.

The remainder of the chapter provides a detailed analysis of how US and EU approaches to climate law and policy have converged and diverged over time and analyzes the implications of these convergences and divergences for past and future efforts to craft a legal framework for addressing climate change at the domestic and international levels.

EU and US Climate Change Policy Pathways Compared

Early Negotiations

Beginning with negotiations for the UNFCCC in the early 1990's, the EU has been a steady driving force behind the shaping and implementation of the international climate change regime.[3] Despite internal inconsistencies in member state stances on climate change, as early as 1990, the EU was engaged in formulating regional GHG stabilization goals and was pushing for early action and legally binding commitments to halt climate change.

Following negotiation of the UNFCCC, the EU began formulating regional initiatives to address climate change and advance international goals. Further confirming Europe's attempt to lead climate negotiations, Germany volunteered to host the first meeting of the UNFCCC COP in Berlin. In its role as global leader on climate policy, the EU has focused on pressing for binding international commitments, maintaining the 'environmental integrity' of the Kyoto Protocol by promoting domestic policies and measures, calling for new rules on supplementarity and the use of carbon capture and storage, and by adopting aggressive

---

[3] See Nuno S Lacasta, Suraje Dessai, and Eva Powroslo, 'Rio's Decade: Reassessing the 1992 Earth Summit: Reassessing the 1992 Climate Change Agreement: Consensus Among Many Voices: Articulating the European Union's Position on Climate Change' (2002) 32 *Golden Gate U L Rev* 351, 352; see generally Joyeeta Gupta and Michael Grubb (eds), *Climate Change and European Leadership: A Sustainable Role for Europe?* (Kluwer Academic Publishers, Dordrecht Netherlands, 2000).

regional emission reduction commitments.[4] Europe's leadership has varied across time and issue, but it has been the most prominent and consistent proponent for proactive domestic and international action among industrialized countries.

In contrast, the US played a prominent role in early climate change discussions, including the negotiations for the UNFCCC and the Kyoto Protocol, but gradually became one of the most vocal opponents to the Kyoto Protocol and its possible successor. For example, while Vice-President Al Gore was in Japan personally participating in negotiations for the Kyoto Protocol, the American Senate was passing a unanimous resolution opposing participation any international agreement that would impose legal emission reduction obligations and, thus, economic burdens on the US, especially in the absence of mandatory emission reduction obligations for the rapidly developing economies of China, India, and Brazil.[5]

With the election of President George W. Bush to office in 2001, US resistance towards international climate negotiations intensified. As President Obama's Kyoto negotiator, Todd Stern, commented in April 2009, the Bush Administration was 'not fundamentally looking for an international agreement'.[6] Signing onto an international climate agreement with any type of mandatory framework was virtually inconceivable between the years 2001–2009.

Opposition to the Kyoto Protocol, however, was not based solely on the legally binding framework. During and following the negotiations for the Kyoto Protocol a significant percentage of politicians and the public continued to challenge the scientific certainty of climate change. In fact, continuing to this day, many US politicians consistently emphasized continuing scientific uncertainties associated with theories of human-induced climate change at home and abroad as justification for inaction.[7] Further, while the EU and its member states have approved and/or ratified the Kyoto Protocol, the US has steadily opposed ratification of the Kyoto Protocol and has publicly repudiated the Protocol despite the fact that the US—through then Vice-President Al Gore—played a leading role in drafting the Protocol to meet US specifications.[8]

The nuances of historical climate negotiations are many, but as of early 2010, the EU continued to occupy the role of international climate leader. And, while

---

[4] Sebastian Oberthur and Claire Roche Kelly, 'EU Leadership in International Climate Policy: Achievements and Challenges' (2008) 43[3] *The Intl Spectator* 35.

[5] The Byrd-Hagel Resolution, '[e]xpressing the sense of the Senate regarding the conditions for the United States becoming a signatory to any international agreement on greenhouse gas emissions under the United Nations Framework Convention on Climate Change', unanimously passed in the Senate 95-0. S Res 98, 105th Cong (1997) (enacted).

[6] Jeff Mason, 'U.S. to Take Reins in Global Climate Talks' *Reuters* (Washington, 24 April 2009) <http://www.reuters.com/article/idUSTRE53N12720090424> accessed 2 January 2010.

[7] See The White House, 'Global Climate Change Policy Book' (2002) <http://georgewbush-whitehouse.archives.gov/news/releases/2002/02/climatechange.html> accessed 3 April 2010 (Global Climate Change Policy Book).

[8] See UNFCCC, 'Kyoto Protocol Status of Ratification' <http://unfccc.int/files/essential_background/kyoto_protocol/application/pdf/kpstats.pdf> accessed 15 April 2006.

the US has now actively re-engaged with global climate negotiations, the fact that the US remains a non-signatory to the Kyoto Protocol and continues to lack a comprehensive domestic legal framework for climate change hamstrings US efforts to hold itself out as a global climate leader.

Europe's position as the leader and the US's position as the laggard in the first 20 years of global climate change governance are final verdicts fixed in most scholarly and popular assessments. This discussion does not dispute this overarching conclusion, but it does seek to condition this assessment in order to more effectively move forward drawing upon past successes and failures.

Behind the Scenes

While the EU—as a regional bloc—has kept international climate negotiations moving forward, individual member states have played different roles in shaping regional leadership. The addition of 12 new member states since 2004 has further complicated the ability of the EU to speak with a unified voice. Even among the original EU-15 member states, however, significant differences have and continue to exist. For example, while the UK is often regarded as a global leader in driving and implementing aggressive climate policies, its early leadership was less consistent. The UK was a vocal force at negotiations for the UNFCCC but has been criticized for misrepresenting itself as more progressive than it really was. During the early 1990s, the UK adopted a goal of stabilizing CO2 emissions by the year 2005, in direct conflict with a more ambitious EU commitment to stabilizing emissions by the year 2000. The conflict almost led the UK to block an entire package of EU climate initiatives. Even after the European Environment and Energy Councils agreed that the EU would stabilize CO2 emissions at 1990 levels by the year 2000, the UK was able to negotiate allowances enabling Britain to keep its original, more lenient target date of 2005.[9]

Further, critics have challenged the UK and Germany's moral right to hold themselves out as climate change figureheads. As the two largest emitters of GHG in the EU, UK and German actions dominate European policy and progress. Both the UK and Germany—moderate differences aside—have unquestionably played key roles in encouraging Europe's climate leadership. Both countries, however, underwent economic changes in the 1980s and early 1990s that led to dramatic GHG emissions declines.[10]

In the UK, the breaking of the coal unions and the liberalization of the energy markets led to the infamous UK 'dash for gas'—a widespread switch from coal to North Sea natural gas. Extensive fuel switches from oil and coal

---

[9] Paul G Harris (ed), *Europe and Global Climate Change: Politics, Foreign Policy and Regional Cooperation* (Edward Elgar Publishing, Northampton MA, 2007) 74.
[10] Commission (EC), 'Progress Towards Achieving the Kyoto Objectives' (Communication) COM (2007) 757 final, 27 November 2007.

to gas resulted in steep emission declines in the 1990s. In Germany, political reunification and the concomitant collapse of East German heavy industry led to similarly precipitous emissions declines.[11] These energy shifts eased economic and political pressures associated with emissions reduction commitments and allowed the UK and Germany to argue for accelerated emissions reductions from a position of economic power. This is in contrast to the remaining 13 members of the EU-15, who experienced emissions increases of roughly 12% between 1990 and 2005.

Political and economic changes created space for the UK and Germany to maneuver.[12] This added flexibility often serves as a basis for criticisms of the UK's and Germany's climate leadership claims. As discussed in Chapter 6, while these two factors accounted for a significant percentage of actual emissions reductions, regional and domestic initiatives have also played an essential role in emissions reductions.

Further, in the case of the UK, the process of fuel switching also demonstrates that meaningful technological change is economically and politically feasible. And, in Germany, reunification often overshadows pre-existing German commitments to eliminating inefficient processes, developing new energy efficient technologies, and promoting sustainable resources.[13] Further, while the UK and Germany may have lobbied from positions of power, this in no way should deflect from the fact that they still faced political and economic obstacles in proposing aggressive regional and international climate policies and that both States have proposed and complied with very real and significant emissions reduction obligations. In the absence of their leadership and support, it is unlikely that the EU could have played the role it has in international negotiations.

The UK and Germany have played a central role in directing EU climate policy, but very little can be achieved without region-wide consensus. Looking beyond the UK and Germany, EU policy has grown increasingly complex due to the accession of 12 new member states between the years of 2004–2007, including multiple former Eastern Bloc nations. During the formative period for international climate negotiations, the EU represented only the EU-15, which while still culturally and economically diverse did not pose the range of challenges that the accession of the new member states poses in terms of formulating a unified approach to economic and environmental policymaking. Because the EU's climate policy was in place at the time the new member states acceded, the new States had to agree to a range of initiatives and strategies that they would very

---

[11] Warwick J McKibbin and Peter J Wilcoxen, *Climate Change Policy After Kyoto: Blueprint for a Realistic Approach* (Brookings Institution Press, Washington DC, 2002) 43–44.

[12] Hugh Ward, 'Game Theory and the Politics of Global Warming: the State of Play and Beyond' (2006) 44 *Political Studies* 850, 861.

[13] See Adam M Dinnell and Adam J Russ, 'The Legal Hurdles to Developing Wind Power as an Alternative Energy Source in the United States: Creative and Comparative Solutions' (2007) 27 *Northwestern J Intl L & Business* 535, 569.

likely have opposed if they had been member states during the early years of regional and international climate negotiations.

This is not to ignore the challenges the EU-15 faced in creating a unified approach to climate change. Internal divisions could have easily overwhelmed supranational efforts, but the EU headed off coordination problems by creating a flexible system that seeks to coordinate Commission and member state positions regionally before the onset of international negotiations. A European Council working group serves as the hub for formulating regional climate policy positions. As climate negotiations have become increasingly complex, the EU has responded by 'gradually diversif[ying] the system of expert groups supporting the Council working group and has delegated more authority to them to develop negotiating positions' and by creating a tiered system of negotiators with authority over different components of the negotiations.[14]

Despite early successes in developing a unified regional policy, the EU did not play a prominent role in shaping the form and substance of the UNFCCC and the Kyoto Protocol. The US, in fact, was the primary architect of both the UNFCCC and the Kyoto Protocol.[15] The EU's primary influence on climate negotiations only truly began following the Bush Administration's secession from the Kyoto Protocol in 2001. Thus, while the EU has been widely heralded as an international climate leader and the US as an international climate laggard, the US, in fact, played a more dominant role in constructing the existing international legal regime.

The US's role as institutional engineer is due in large part to the presence of former Vice-President Al Gore at negotiations for the Kyoto Protocol. Despite facing staunch domestic resistance to the ratification of a legally binding protocol to the UNFCCC, Vice-President Gore was one of the primary architects of the Kyoto Protocol. Aggressive US negotiating was responsible for the inclusion of the flexibility mechanisms, which have become the backbone of climate initiatives both globally and locally, including within the EU.

Vice-President Gore negotiated from a position of weakness. He knew that the US was unlikely to ratify the Protocol he helped negotiate; he knew that even among the Clinton Administration his position was weakly supported. Despite this knowledge, he worked until the end to structure the Protocol along lines favored by his negotiating team. His ability to influence climate negotiations was profound; the existing international framework is, in large part, a by-product of a US—but, more specifically, Vice-President Al Gore's—vision.

In this way, the US was indeed a profound early climate leader in terms of its actual ability to shape climate negotiations. Vice-President Gore's work in Kyoto helped create the framework of the existing international climate regime but it

---

[14] Oberthur and Kelly (n 4) 38.
[15] S Andresen and S Agrawala, 'Leaders, pushers and laggards in the making of the climate regime' (2002) 12 *Global Envtl Change* 41.

also signaled a turning point in America's climate politics. His aggressive stance was widely rejected as counter to American interests. The Protocol that Gore helped negotiate was so anathema to American interests that it was never even presented to the US senate. Further, following his ascendency to the American Presidency, President George W. Bush rescinded America's signature to the Protocol and abandoned Protocol ratification and implementation efforts domestically and globally.

By mid-2001, the US role had changed from one of reluctant architect to one of open antagonist. Having shaped the parameters of the debate, the US vacated any proactive role in international negotiations, opening the door for the EU to evolve from a symbolic but ineffective leader into an accepted and relied-upon agenda setter.

The UNFCCC and the Kyoto Protocol exist as a result of over a decade of multilateral negotiations. These institutions reflect evolving principles of international environmental law and establish the baselines for international efforts to combat global climate change. The creation of these institutions, however, came at a high cost. The paths of the US and the EU only truly diverged following the creation of the basic international legal framework, leaving a framework riddled with compromises yet failing to generate consensus. This ensuing transatlantic divergence reflects both the complexities inherent in climate policymaking and fundamental differences in American and European perceptions of the problem.

As will be discussed in Chapter 8, societal perceptions of science, risk, economic costs and benefits, social welfare and political capital, drove the US and the EU down divergent paths following the adoption of the Kyoto Protocol. The following section examines specific divergences and convergences in US and EU climate strategies between the adoption of the Kyoto Protocol and the final phases of negotiations for a post-Kyoto agreement.

From Kyoto to Copenhagen: Transatlantic Divergences

Once the EU—and each of the EU-15 member states—ratified the Kyoto Protocol and the US decided to abstain from ratification, the overarching context for EU and US climate policy diverged.

EU member states are operating in an entirely different legal framework from the US. The EU is bound by international emissions reduction obligations, while individual member states are subject to international, supranational, and national emissions reduction obligations. The US, on the other hand, is not bound by international or national emissions reduction obligations.

This fundamental distinction is not to be underestimated. As a result of this fundamental difference, the legal cultures of climate change in the US and the EU have emerged and evolved along diverging paths, with myriad implications for the development of social norms surrounding climate change.

At a very basic level, the Kyoto Protocol establishes a framework of overarching obligations and institutional mechanisms for EU efforts to reduce greenhouse gas emissions. Within this larger framework, the EU creates a system of common coordinated policies and measures (CCPMs) that provide a structure—including both binding and non-binding measures—for individual member states to implement policies and measures. Within the CCPMs, the ECCP creates as the overarching policy umbrella, establishing *minimum* GHG emission reduction requirements for member states, serving as a vehicle for coordinating Community-wide action, and overseeing the administration of the EU ETS.

Under the ECCP member states establish national climate change strategies. Member states are autonomous in determining how they will conform to EU standards. They must enact laws and policies to meet minimum EU emission reduction standards and to comply with EU initiatives across a range of sectors, eg, renewable energy, biofuels, transport, building. Beyond this, however, in most cases member states may establish additional national rules that exceed baseline EU obligations. Member states thus operate within a three-tiered system of climate obligations. This triadic framework requires specific legal action but it also encourages the development of widespread social awareness of climate issues, including how climate change implicates policy choices across the domestic, transport, energy, and industrial sectors.

The US's climate policy, in contrast, exists outside a discrete legal framework. With no obligatory emission reduction obligations at either the international or the federal level, the legal culture surrounding climate change has largely evolved through the federal judiciary, a hodgepodge of federal administrative decisions and a variety of localized state laws. The absence of a coherent legal framework has spawned creative law-making but has resulted in a) almost two decades passing without the development of a federal emissions reduction framework, with the result that GHG emissions have continued to increase, and b) almost two decades of conflicting, bi-partisan debate over the realities of climate change and the appropriateness of any legal response to climate change, thus creating a legal culture characterized by lawlessness and uncertainty and a social culture of disaffection and frustration around the issue of climate policy.

The multi-tiered versus un-tiered distinction between transatlantic legal infrastructures, creates the foundation for legal and social deviation on climate change in the EU and the US. Within this framework, several key divergences emerge.

Measures for Progress and Commitment to GHG Emission Reductions

To begin at a significant point of departure, the EU has legally committed to achieving *absolute* emissions reductions. The US has made no legal commitment to reduce absolute greenhouse gas emissions. In fact, until 2009 the federal government abandoned even any rhetorical commitment to *absolute* emissions reductions speaking only in terms of a GHG *intensity* standard.

The intensity metric is indicative of the US's approach to climate policy during the 2000s in many ways: it emphasizes the dominance of economic development over environmental concerns; it reflects reluctance towards committing to an internationally-based, objective standard; it signals a willingness to respond to climate change in a peripheral, non-intrusive manner. While the US has now abandoned the intensity metric and begun speaking in terms of absolute emission reduction under President Obama, the language of intensity characterized the US approach to climate policy during the formative years of the international climate regime, often derailing efforts to further international dialogue on a post-Kyoto emissions reduction framework.

Beyond the language of emissions reductions, the EU and the US diverge in their willingness to commit to emissions reduction obligations at the domestic/supranational and international levels. Currently, the EU-15 is jointly obligated to reduce emissions 8% below 1990 levels between the years of 2008–2012. Within this bloc, each member state has an independent emission reduction obligation, ranging from 28% reductions to 27% increases as against 1990 levels between the years of 2008–2012.

The US, on the other hand, has not established a specific GHG emissions reduction objective, using either an absolute or intensity standard. During the Bush Administration, the Executive Branch stated that US GHG 'intensity' was projected to decrease by 18%, translating to an overall GHG increase of 14% by 2012.[16] The Obama Administration has established a goal of reducing US GHG emissions 80% by 2050; this is, as yet, just a goal. Although it continued to inch closer to implementing domestic legislation, as of early summer 2010, the US continued to abstain from making any concrete, measurable, or enforceable emissions reduction commitments at the international or domestic level.

Policy Mix

Both the EU and the US advocate using a combination of traditional and non-traditional policy approaches to address climate change.

EU climate policies have embodied a mixture of initiatives, including:

- mandatory command and control style laws (eg, Directive 96/57/EC, determining the energy efficiency of domestic appliances and Directive 2009/28/EC establishing mandatory targets for renewable energy);
- market-based flexible programs (eg, EU ETS and Linking Directive);

---

[16] US State Department, *The Fourth United States Climate Action Report to the UN Framework Convention on Climate Change* (27 July 2009) <http://www.state.gov/g/oes/rls/rpts/car4/index.htm> accessed 3 April 2010.

- hybrid approaches (eg, Decision 1513/2002/EC promoting research and development on climate change);
- voluntary agreements (eg, voluntary agreements with car manufacturers[17]).

While it continues to include a diverse array of measures, over time, the EU's package of measures has become increasingly mandatory, moving away from voluntary agreements towards command and control style regulation of, for example, renewable energy production and greenhouse gas emissions from automobiles.

The US has advocated using a similar mixture of policies to tackle climate change. While the US strategy continues to rely heavily on voluntary and incentive based programs, due to heavy reliance on EPA regulation under the CAA, the portfolio of measures in the US has similarly shifted towards traditional command and control style regulation. The US strategy includes:

- command and control style laws (eg, new CAFE standards and appliance and lighting energy efficiency standards established in the Energy Independence and Security Act of 2007, and proposed regulation of stationary source greenhouse gas emissions under the CAA);
- promotion of market-based programs (eg, myriad of legislative cap-and-trade proposals, including the proposal approved by the House in ACES and the proposal put forward by the Senate in the APA);
- hybrid approaches (eg, incentives for the development of plug in hybrids in the 2007 Energy Act);
- voluntary agreements (eg, US voluntary carbon offset market and the Climate Leaders and the Climate VISION Programs).

In both the EU and the US, policy choices and policy proposals reflect widespread recognition that the nature of climate change requires programs that target emissions reductions in a variety of ways across a multitude of sectors, eg, ranging from education and incentive-based programs, encouraging lifestyle changes in the domestic sector, to regulatory programs controlling automobile fuel economy standards and manufacturing standards, to governmental support programs for research and development, to voluntary regulation to encourage greenhouse gas emissions from not-yet-regulated industrial and manufacturing sectors.

While the EU and the US converge in embracing a mixture of policies to reduce greenhouse gas emissions, key divergences exist in their choices of component parts. At a fundamental level, the EU strategy differs in that it has consistently revolved around (1) clearly defined objectives, and (2) emissions reduction commitments enshrined by law. That is, the EU climate change strategy has a defined pathway and a legally enforceable backbone. Flexible and voluntary measures

---

[17] These voluntary agreements were valid between 1998/1999–2009, but are only of historical significance now, due to the passage of Regulation 443/209/EC establishing fleet-average CO2 emission targets for passenger cars.

serve as tools to meet defined goals and legally-binding obligations. In contrast, the US strategy is the amalgamation of a collection of expressed policy choices, cobbled together with judicial decisions, voluntary programs, and indirectly germane legislative choices. Throughout the 1990s and 2000s, the US's policy mix lacked politically agreed-upon objectives and a legal or regulatory backbone and, thus enforceability.

While the US strategy lacks coherent goals and enforceability, efforts to mould a long-term strategy benefit from two unique characteristics of the US political system—adversarial legalism and the US system of administrative law. Inactivity in the US Congress belies a flurry of activity within the judiciary and the system of administrative law.

Litigation is central to the evolution of US law and policy and nowhere is this as evident as in the field of climate change. In the absence of overarching federal legislation, individual citizens, NGOs, states, and industry have all turned to the courts to answer discrete climate-related questions over roles and responsibilities under common law, federal statutory law, state law, and public international law. Since US judges have historically been more willing than their European counterparts to review and overturn legislative and administrative decisions, the courts have become an active playground for interest groups hoping to advance specific policy goals.[18] The range of cases challenging governmental and private actions and omissions on climate change demonstrates this trend.

One of the leading experts on US climate change law, Michael Gerrard, has helped assemble a climate change litigation chart detailing the plethora of past and on-going cases. What this chart and other studies of US climate change litigation reveal is that, while the US may lack the climate-specific legal framework that the EU possesses, pre-existing legal infrastructure provides numerous avenues for both prompting and preventing—or at least re-directing—governmental and private actions on climate change.

The US system of adversarial legalism does not offer a direct path forward in structuring a cohesive US climate change system, but it does provide opportunities for defining the reach of existing laws and for prompting governmental responses on questions that might otherwise stall on the political agenda. The *Massachusetts v EPA* case demonstrates this phenomenon aptly. The decision in this case reshaped US climate policy by, in effect, directing the EPA to act on climate change. The Court's decision reached well beyond the EPA, however. By forcing the EPA to reassess its responsibility to regulate greenhouse gases across a range of sectors, the Court also prompted the legislature to act. The possibility of the EPA creating a national greenhouse gas regulatory regime under the CAA is seen as one of the primary factors driving Congressional efforts to

---

[18] Robert Kagan, 'Globalization and legal change: The "Americanization" of European Law?' (2007) 1 *Regulation & Governance* 99.

enact overarching climate change legislation. It is, in essence, a turf war that was pressed forward by judicial decree.

This instance of adversarial legalism redefining the climate agenda simultaneously reveals the other distinctive aspect of the US political system—the power of the sprawling system of administrative law. It is no exaggeration to say that much US legislation and, in particular, environmental legislation is only as effective as the administrative agency charged with implementing and enforcing that legislation.

The creation of the EPA in 1970 underpinned the rapid growth of US environmental law in the 1970s. Absent the creation of the EPA—and the establishment or empowerment of parallel federal agencies—the federal government would lack the capacity to regulate the vast range of public and private activities that impact environmental quality.

The basic structure of US environmental law was set in the 1970s through the enactment of a series of medium-based environmental laws. The everyday process of environmental regulation, however, is conducted by administrative agencies, such as the EPA, the DOI, and the USFWS.[19] These agencies exercise great influence in interpreting the meaning and scope of federal environmental law. The extent to which administrative agencies shape federal law was aptly described by Judge Jackson when he noted that '[t]he rise of administrative bodies probably has been the most significant legal trend of the last century.... They have become a veritable fourth branch of the Constitution, which has deranged our three-branch legal theories.'[20] Judge Jackson made this comment in 1952, even before the birth of federal environmental law and the attendant surge in administrative law that accompanied it.

Since the 1970s, Congress has enacted very few pieces of major environmental legislation and has adopted only a handful of truly substantive amendments to existing environmental laws. Yet, since the 1970s, the face of US environmental law has continuously evolved to address new challenges, to modify approaches to existing challenges, and to accommodate fluxes in popular and political interest in environmental protection. This evolution is due in significant part to regulations issued and decisions made by administrative agencies in conjunction with judicial review of these administrative decisions.

The effects of high-level regulation and litigation are mixed. The highly detailed regulatory regime allows administrative agencies to develop highly-technical, medium, and location specific environmental rules. This facilitates

---

[19] Federal agencies with environmental responsibilities include, eg, the Environmental Protection Agency (EPA), the Department of Interior (DOI), the Department of Transportation (DOT), the Council on Environmental Quality (CEQ), the Army Corps of Engineers (Corps), the Department of Energy (DOE), the Food & Drug Administration (FDA), the Occupational Safety & Health Administration (OSHA), the Nuclear Regulatory Commission (NRC), and Fish & Wildlife Service (FWS).
[20] *Federal Trade Commission v Ruberoid Co.*, 343 US 470, 487 (1952) (Jackson J, dissenting).

science-driven, location-based strategies for reducing environmental degradation. Detailed regulatory regimes offer more room for considering a host of factors, to include: science, economics, and equity concerns. De-centralized regulation, however, also leads to instances of regulatory capture, inconsistent implementation and enforcement nationwide, and over-burdened agencies. In addition, regulatory choices are politically laden. The US President appoints the heads of many administrative agencies, eg, the Administrator of the EPA, while Congress controls agency appropriations. Agency priorities and capacities, thus, vary greatly across time and political administration. This political fluctuation creates regulatory uncertainty and undermines intra-agency efforts to implement long-term strategies. Despite political obstacles, administrative agencies are powerful political entities that exercise great influence over the shape of environmental law.

Similarly, heavy reliance on litigation offers benefits and drawbacks. Litigation allows the courts to, eg, monitor administrative agencies to ensure they are properly interpreting, implementing, and enforcing environmental law; review the reach of common law; resolve the compatibility of state and federal environmental laws; and determine the Constitutional legitimacy of Congressional actions. Adversarial legalism provides incentives for agencies to fulfil their legislative mandates and to do so in a transparent and even-handed manner. Adversarial legalism also permits both governmental agents and public citizens to challenge regulatory decisions and private actions and to play a part in monitoring the implementation and enforcement of federal environmental law. Adversarial legalism, however, comes at great economic, administrative, and social costs. Litigation costs frequently swamp agency budgets, dominate agency time, and slow down regulatory implementation. These costs may or may not be warranted depending on the basis of the case at hand and the purpose of the litigation, eg, substantive challenge, delay technique, moral/political opposition. Litigation also increases the adversarial nature of policymaking, forcing interested parties into opposite corners and discouraging cooperation and consultation.

Regardless of the perceived productivity of litigation and the purpose behind the legal challenge, environmental litigation is endemic. The process of US environmental lawmaking involves a continuous interface between legislation, regulation, and litigation.

US environmental law is a product of adversarial legalism and administrative law. As a consequence, American laws and regulations are generally highly detailed, complex, and prescriptive.[21] In contrast, although there has been some movement towards more prescriptive rulemaking, EU rules normally tend to be less detailed, focusing more on establishing overarching goals and obligations and allowing member states flexibility to choose how to meet those obligations. Further, the EU system of environmental law evolves through different channels. EU environmental lawmaking is not characterized by adversarial legalism and

---

[21] Kagan (n 18).

is not filtered through an empowered administrative law system. European lawmaking is more generally characterized by cooperative decision-making, multiple internal checks on executive power, and decentralized State-based implementation and enforcement.

There is no direct EU equivalent to US administrative agencies. The European Commission is in many ways the closest counterpart to the American executive and to executive administrative agencies in that it is bestowed with executive rule-making powers. The Commission's executive and administrative powers, however, differ from those possessed by the American President and US administrative agencies. They are subject to additional substantive and procedural constraints and originate and flow through different channels.

The majority of EU regulations are not enacted as primary legislation by the Council and the European Parliament but, rather, as secondary implementation measures by the Commission—eg, proposals for regulations, directives, and decisions. The Commission possesses primary rule implementation powers in the EU and bears the brunt of European rule-making. Each year, the Council of Ministers and the Parliament delegate executive powers to the Commission to issue hundreds of implementing rules. As discussed in Chapter 5, the Commission is constrained by the principles of subsidiarity and proportionality, meaning that the Commission is only permitted to act when supranational rules are required and, even then, measures must be no more strict than necessary to ensure policy objectives.

The Commission's authority is further constrained by a dauntingly complex comitology—or committee—system.[22] The 200–300 committees making up the system consist of member state representatives charged with overseeing and keeping in check the Commission's rule-making powers. While originally created to control Commission powers, these committees often play the role of policy experts and advisers to the Commission on complex policy questions. The committees, however, do possess political power as they are able to vote to either approve Commission proposals or to refer proposals to the Council to receive further analysis.[23] In the case of environmental decision-making, studies of environmental comitology suggest that these committees focus on devising effective solutions to environmental policy problems rather than on serving as an external control mechanism on Commission implementation powers.[24] Yet, the impact varies from issue to issue with the effect of both constraining the commission's authority and impeding transparency in the decision-making process.

---

[22] This committee system—known as comitology—comprises 200–300 committees that the Commission must consult before implementing rules, or adopting secondary legislation. Jens Blom-Hansen, 'The EU Comitology System: Who Guards the Guardian?' (Paper presented at the Fourth Pan-European Conference on EU Politics, Riga, Latvia, 25–27 September 2008) <http://www.jhubc.it/ecpr-riga/virtualpaperroom/085.pdf> accessed 2 January 2010.   [23] Ibid 2.
[24] Ibid 5.

Despite these constraints, like US administrative agencies, the Commission possesses secondary rule-making powers. In the EU, however, secondary rules serve a very different function; and the Commission, thus, also serves a different function. EU rule-making does not flow from overarching, detailed, and prescriptive primary legislation, as does US administrative rule-making. EU rule-making focuses on establishing agreed-upon regional goals and establishing time frames and general parameters within which member states are free to enact detailed regulatory regimes. In contrast, US administrative agencies create detailed and prescriptive regulatory regimes that dictate modes of compliance and establish monitoring and enforcement mechanisms. Administrative agency rules reach the most intimate actions of everyday life nation-wide. The Commission's powers are neither this far-reaching, this detailed, nor this prescriptive.

Thus, while both US administrative agencies and the EU Commission propose and implement secondary rules, the nature of these rules varies. The US system of primary legislation is far more extensive and sets more detailed parameters for administrative rule-making. Within these detailed parameters, however, US administrative agencies then have extensive scope to implement, monitor, and enforce comprehensive highly detailed rules. US administrative rules control the daily activities of public and private actors nationwide.

EU Commission rules, on the other hand, resemble a hybrid of US primary and secondary rules. Similar to US primary legislation, they establish overarching policy objectives for member states. Unlike US primary legislation, they focus on harmonizing—rather than prescribing—member state rules and often lack the same degree of specificity. In common with US administrative regulation, Commission decisions are treated as secondary rules and the Commission requires delegations of authority from the Council and the Parliament in order to act. This is where the similarity with US administrative rules ends; Commission rules do not create a regulatory framework anywhere near as extensive, detailed, compulsory, or directly enforceable as the system of US administrative law.

Similarly, the EU Environment Agency—often perceived as a close relative to the US EPA—lacks the reach and power of US administrative agencies. Created in 1990, by mandate, the EEA is directed to assist in Community and member state environmental decision-making and coordination of the European environment information and observation network (Eionet).[25] The EEA functions largely as a source of information provision; its daily activities in no way resemble the rule making, implementation, and enforcement responsibilities of the US EPA, or its US counterparts.

The US system of administrative law is distinct and potentially very powerful. In the context of US climate change law and policymaking, US administrative

[25] The EEA was created by regulation in 1990; the regulation came into force in late 1993, following the decision to locate the EEA in Copenhagen; and the EEA began its work in 1994. European Environment Agency, 'Who We Are' <http://www.eea.europa.eu/about-us/who> accessed 2 January 2010.

agencies with authority over environmental issues—eg, the EPA, DOI, FWS—were very slow to enter into the domain of climate change regulation. This began to change in 2007, when judicial challenges to agency inaction resulted in the listing of the polar bear as a threatened species under the ESA due to the dangers climate change poses to the bear's habitat, and the Supreme Court decision in *Massachusetts v EPA* forced the EPA to reassess its regulatory roles and responsibilities in the context of greenhouse gas emissions. Together, these two decisions changed the pathway of US climate change policymaking and laid the foundations for widespread regulation of greenhouse gas emissions, even absent legislative rule-making. These events demonstrate the profound ability of litigation and regulation to alter policymaking agendas.

Source of Leadership

One of the most evident distinguishing characteristics of EU and US climate change policymaking is the source of political leadership. The EU, as a supranational entity, has consistently dominated European climate change policymaking. The UK, Germany, and other member states have played central roles in facilitating EU leadership and in acting as independent leaders on the international stage. Since the 1990s, however, the EU has put in place regional laws and policies, coordinated member state strategies, and played a dominant role in setting the agenda for international climate negotiations.

Member state policies emanate from supranational policies. EU measures create obligations that member states must meet, at a minimum, but may generally exceed or expand upon. Just as US states function as the laboratories for democracy, so do member states serve as laboratories for evaluating and improving upon EU measures. This is especially true given that the underlying principle of subsidiarity allows member states flexibility to adopt different pathways to meeting common goals and, thus, allows supranational policymakers to assess the pros and cons of divergent strategies. Member state approaches to meeting EU climate obligations vary greatly, and projections about different States' ability to meet emission reduction obligations by 2012 suggest equally great variations in effectiveness.[26]

The combined efforts of the 27 member states contribute to the EU's climate change leadership role. Yet, the EU is something much greater than its component parts. In international negotiations, the EU speaks with a unified voice representing 27 independent States; yet, EU climate policy formulation is riddled with inter-State bickering. In formulating regional policies, however, the EU is able to mask

---

[26] By 2009, EU reports indicated that the EU-15 and the EU-27 were on track to meeting Kyoto commitments and that measures put in place to meet emission reduction targets ensured continuing emission reductions and improved abilities to decouple economic growth from emissions growth. Commission (EC), 'Progress Towards Achieving the Kyoto Objectives' (Communication) COM (2009) 630 final, 12 November 2009.

this internal wrangling and present its position with a united front. For example, in the much heralded 2007 European Commission Communication to the Council, the Commission stated that: the 'EU must adopt the necessary domestic measures and take the lead internationally to ensure that global average temperature increases do not exceed pre-industrial levels by more than 2ºC'. The EU's adoption of the 2ºC goal became the foundation for global climate negotiations and for EU efforts to shape a post-Kyoto agreement around a goal of reducing global greenhouse gas emissions by 20–30% by 2020. In furtherance of this goal, in 2007, EU Heads of State committed to reducing GHG emissions to 20% below 1990 levels, to improving energy efficiency 20%, and to increase the percentage of renewable in the regional energy mix to 20% by 2020. Subsequently, in 2009, the EU adopted a comprehensive set of legislation designed to meet these new goals.

The presentation and follow-up to the 2ºC goal encapsulates the strength in EU climate policy and reveals how the EU has been successful in representing itself as a climate leader at the international level. First, the EU adopts an aggressive internal policy position that includes a statement of what the EU believes is necessary at the international level. Second, it follows up with proposals that directly address regional actions while simultaneously linking region-wide policy with initiatives to spur more aggressive actions by other developed countries. Third, the EU then 'advertises' this new position widely. In this way, the EU presents itself as leading by doing, giving it political legitimacy to influence the international agenda and to exert pressure on perceived political laggards.

The situation is very different in the US. A careful analysis of US climate change policymaking reveals a complicated picture of pushes and pulls—of stagnation and resistance to change at the federal level meeting innovation and pressure for progress from sub-federal actors. Dominating the picture of American climate change politics is the federal government's failure to develop widespread consensus on the appropriate national response to climate change. As a result, the federal government does not speak with a unified voice on behalf of its 50 states. As of 2009, the US's national policy remained undefined and national climate priorities remained elusive. Federal policy has consistently lacked a legislative framework and been characterized by fluctuating and opaque expressions of policy preferences emanating from the Executive Office.

Internally, as with EU climate policy, US climate policy has been riddled with inter-state bickering.[27] In contrast to EU policy, however, the results of internal wrangling have been very different. First, the US has not been able to mask domestic policy differences to present a unified front at the international level. Second, internal political strife has been the primary driver of progressive policy development. Despite federal inactivity, state and local policymakers have

---

[27] As recently as March 2010, for example, Governors from 18 states and two US territories were asking Congress to halt EPA's efforts to regulate greenhouse gas emissions under the US CAA while an equal number of States were proceeding with their own regulatory regimes to limit greenhouse gas emissions from automobiles.

increasingly adopted climate change laws and policies that mimic the most progressive policies adopted worldwide.

Local, state, and civil society efforts to transform climate change policymaking in the US have taken many forms and, at times, have met with resistance at the federal level. The US federal government was once the leader in domestic environmental policymaking, but it has fallen behind regional, state, and local climate change policymaking efforts. Federal apathy has prompted sub-federal actors to find increasingly inspired ways to push for more progressive climate change policies, employing state and federal law. From adopting local policies, to employing common law and tort-based litigation, to using existing federal environmental laws, to invoking the jurisdiction of international institutions, US civil society has utilized every possible mechanism to overcome stagnation and resistance at the national level and thereby drive a progressive climate change policy agenda from the bottom up. These new and creative uses of state, federal, and international law directly address root causes of climate change while simultaneously forcing legal transformations in federal climate change policymaking.

In this way sub-federal actors served as the source of political leadership in the US, leading up to 2009. In contrast, in Europe, the EU creates a common policy framework within which member states act as well as adopting unified international negotiating positions. Thus, the EU acts as the primary source of political leadership while simultaneously creating room for member states to act as parallel sources of leadership. In the US, with the federal government unable to agree on a common policy framework and, thus, unable to legitimately represent such a perspective in international negotiations, sub-federal actors act as the primary source of political leadership. They fulfil the leadership role by enacting legal and political structures that demonstrate the political and economic feasibility of initiatives to curb greenhouse gas emissions; they employ existing state and federal laws to challenge acts and omission of the federal government; they enter into voluntary climate agreements with foreign governments, and they seek redress for domestic failures in international legal forums.

In the US, state and local governments are currently acting in an open legal field. They are neither bound by international greenhouse gas emissions reduction commitments nor constrained by federal climate change legislation. Within this open field, sub-federal leadership directly and indirectly encourages federal leadership. Direct challenges to federal acts and omissions force legal change, eg, *Massachusetts v EPA*, while the enactment of state laws and policies indirectly compels legal change, eg, California's adoption of automobile tailpipe greenhouse gas emission limits. As discussed in Chapter 4, the faith the American public once vested in the federal government as an environmental law and policy-leader is now vested with state and local governments and civil society actors as the potential harbingers of change. Sub-federal actors are responding by taking varied steps—both substantive and symbolic—to fill the federal leadership void.

The question that plagues claims of sub-federal leadership in the US is whether the agreements, laws, policies, and judicial challenges are merely 'symbolic' or whether they are substantive enough to bring about real change.[28] Many efforts are, in large part, 'symbolic' in that they create ambitious policy goals but often omit specific obligations or discrete measures for implementation and compliance.[29] Symbolic efforts, however, are increasingly outnumbered by substantive victories, eg, enactment of tailpipe emission limits, adoption of greenhouse gas emission limits from different sectors of the economy, and litigation victories that force federal regulatory shifts. Even the more 'symbolic' acts, however, serve an important leadership purpose.

In environmental law, the line between procedural and substantive measures is fine, and often difficult to discern. Many laws and policies that are largely procedural in nature create substantive rights in practice. In the context of climate change laws and policies, the difference between symbolic and substantive policies perhaps turns on whether the law/policy creates enforceable rights and/or obligations. The creation of substantive and enforceable rights is critical to positive action to address climate change. In the context of US federalism—where states are often constrained by federal pre-emption—nominally 'symbolic' actions take on substantive meaning.

Political leadership on climate change is an amalgamation of effort and achievement. Sub-federal actors have not and cannot develop a comprehensive, robust regulatory structure for climate change. The combination of strategies employed by sub-federal actors is ultimately limited; sub-federal actors cannot create unified national policies nor negotiate on behalf of the US in international legal forums. Sub-federal leadership, however, raises the social and political profile of the climate change debate and facilitates active inter-state communication and cooperation. It also creates momentum and incentives for the federal government to assume a climate change leadership role. The power in sub-federal leadership, thus, lies in creating momentum and foundations for a federal climate change framework.

Diverging sources of political leadership in the EU and the US—like divergences in measures of progress and obligatory frameworks—create uneven

---

[28] See Kristen H Engel and Scott R Saleska, 'Subglobal Regulation of the Global Commons: The Case of Climate Change ' (2005) 32 *Ecology LQ* 183, 215 (suggesting that 'at least for now' much of the current environmental legislation is 'symbolic' and 'more show than substance').

[29] For example, Engel and Saleska noted that:

> With several important exceptions, few state and local climate initiatives currently impose mandatory emissions reductions from existing sources. Instead, many constitute preliminary steps toward eventual regulation, affording state and local politicians notoriety on an issue being largely ignored by the federal government without committing the locality to costly controls on in-state greenhouse gas sources. While the federal government is clearly not out ahead of the states on the issue of climate change, nor are the states quite as far out ahead of the federal government as one might believe as a result of the recent attention garnered by state and local climate change initiatives.
>
> Ibid 222–23.

starting points for addressing climate change. With supranational leadership guiding climate strategies at multiple governance levels, Europe benefits from the existence of mechanisms facilitating communication, cooperation, and consensus building. Thus, despite very real internal conflicts, the EU can set minimal performance requirements, facilitate inter-state knowledge and technology transfer, and accommodate varying social and economic needs all while presenting a united front in international dialogue. In contrast, in the absence of federal leadership, the US is hamstrung domestically and internationally. It lacks legal authority to implement coordinated nationwide initiatives to mitigate climate change and it lacks moral authority to influence the international political agenda. Climate leadership exists in the US; many political, regulatory, and technological responses to climate change originate from US sources. The point of origin of US leadership, however, lies below the surface. Consequently, unlike the EU, the US cannot draw upon this leadership to facilitate internal political cohesion or to direct external policy debates.

Policy–Evolution

Ensuing from divergences in policy frameworks and sources of leadership, one of the principle differences between the EU and US climate policy programs lies in the enactment of programs to enforce, monitor, and evolve GHG emissions mitigation strategies.

Under international and European law, the EU is obligated to monitor GHG emissions, create a regional GHG inventory, and evaluate regional progress toward meeting mandatory emissions reduction commitments.[30] These obligations have been transposed into Community law, with concomitant obligations for member states. The Commission, with the assistance of the EEA, is tasked with providing for 'effective cooperation and coordination in relation to the compilation of the Community greenhouse gas inventory, the evaluation of progress, the preparation of reports, as well as review and compliance procedures enabling the Community to comply with its reporting obligations under the Kyoto Protocol'.[31] Additionally, under the ECCP—the main instrument for developing EU climate policy—the Commission must regularly evaluate and update EU climate policy in response to identified needs and shortcomings.

The Commission is currently implementing the Second ECCP. One of the defining features of the Second ECCP is the focus on building upon lessons learned in the First ECCP and to incrementally expanding the scope of European

---

[30] See Commission Decision (EC) 2005/166 of 10 February 2005 laying down the rules implementing Decision No 280/2004/EC of the European Parliament and of the Council concerning a mechanism for monitoring Community greenhouse gas emissions and for implementing the Kyoto Protocol [2005] OJ L55/57; Council Decision (EC) 280/2004 of 11 February 2004 concerning a mechanism for monitoring Community greenhouse gas emissions and for implementing the Kyoto Protocol [2004] OJ L49/1.   [31] Commission Decision (EC) 2005/166.

climate policies. The Working Groups created under the Second ECCP reflect this emphasis on self-evaluation and progressive improvement. The two dominant themes of the new working groups are to review existing policies, eg, ECCP I review (with five subgroups: transport, energy supply, energy demand, non-CO2 gases, agriculture); EU Emission Trading Scheme review; and to develop policy strategies in underserved and/or underperforming areas, eg, aviation, transportation, carbon capture and storage, adaptation.

This on-going process of assessment and expansion enables the EU and its member states to identify policy successes, failures, and omissions and to create a policy-feedback loop that allows decision-makers to refine short and long-term strategies to minimize costs and maximize effectiveness. Mid-term assessment of the EU ETS, for example, revealed tragic flaws in the emissions allocation system resulting first, in a market crash, but second, in the revamping of the national emissions allocation system to address numerous identified defects.

Monitoring and assessment does not always result in successful policy refinement, as demonstrated by regional recognition that many member states continue to fall short of achieving mandatory emissions reductions. The availability of monitoring mechanisms and clearly defined channels for continually assessing and modifying policy choices, however, provides powerful procedural and substantive tools. By increasing transparency, state and regional governments are forced to account for policy failures and to suggest how they will modify and improve policy strategies. The combined effect of existing monitoring and reporting requirements creates a comprehensive record that, in essence, functions as a clearinghouse for information about climate change policy successes and failures.

The availability of information on early climate policy results should not be underestimated. Mid-way through post-Kyoto negotiations the two most evident trends are (1) growing recognition of the threats posed by human-induced climate change, and (2) widespread failures to implement effective climate change mitigation strategies. Growing international concerns are intensifying efforts to erect comprehensive climate change strategies, and to do so quickly. Nations worldwide will be looking for proven strategies. In this context, the availability of detailed assessments of climate policies at multiple jurisdictional levels and in varying social and political contexts provides an invaluable record, with the failure stories being as relevant and informative as the success stories.

The US stands to be one of the largest beneficiaries. Sharing many cultural and political similarities, the US will be able to look to Europe's climate record as it devises a comprehensive federal climate strategy. This symbiotic relationship is already evident as US Congressional leaders look to the early successes and failures of the EU ETS as they draft legislative proposals on climate change.

The US federal government will also benefit from looking inward to records emerging on state-based climate policies. The US, however, will be starting out afresh in testing what legal and political strategies will work at the federal level. In fact, currently the US even lacks the capacity to monitor greenhouse gas emissions. Because, while the federal government has developed the 'Inventory of US Greenhouse Gas Emissions and Sinks' to comply with its obligations under the UNFCCC, leading up to 2009, the federal strategy placed little emphasis on developing comprehensive GHG monitoring mechanisms, leaving much of the onus on states to monitor GHG emissions. However, in December 2009, EPA Administrator Jackson signed a proposed mandatory GHG reporting rule.[32] Although limited to large sources, the rule would facilitate federal efforts to develop a comprehensive GHG emissions database and to begin more accurately monitoring emission trends.[33] Yet, it is of great consequence to both the United States and the global community that, as late as 2010, the US lacked the capacity to adequately monitor domestic greenhouse gas emissions.

Even given successful implementation of the reporting rule, the US lags behind the EU in terms of monitoring, assessment, and enforcement capacities. The US is still at the starter's block attempting to track and monitor emissions, while the EU has progressed to tracking and monitoring emissions reduction policies, updating legal initiatives, and enforcing policy requirements. The US, as of mid-2010, lacked legal or regulatory greenhouse gas emissions reduction, reporting, or monitoring policies to assess, update or enforce.

The EU and the US occupy very different places in the climate change policymaking cycle. The EU is in mid-Phase II; it has adopted, implemented, and assessed its first round of climate policies and is now modifying, expanding, and enforcing its policy choices. Its successes and failures are recorded and offer lessons in ways forward. The US is initiating Phase I; having abstained from domestic and international efforts to craft comprehensive climate change mitigation strategies for over 15 years, it is now drafting (albeit in pencil) a framework for federal efforts to monitor and control greenhouse gas emissions. The US can look inward to California and other states for lessons on domestically-tested strategies and it can look outward to Europe for examples of tested regional and State-based approaches to addressing climate change. As of yet, it has no lessons of its own to offer the global community.

Commitment to International Negotiations, Past and Present

As a final key point of divergence, underlying differences in how the EU and the US engage with international law and international lawmaking processes impact regional approaches to climate change. American politics have historically

---

[32] US Environmental Protection Agency, 'Final Mandatory Reporting of Greenhouse Gases Rule' <http://www.epa.gov/climatechange/emissions/ghgrulemaking.html> accessed 2 January 2010.

[33] Reporting requirements—in the form of annual reports—would be limited to suppliers of fossil fuels or industrial greenhouse gases, manufacturers of vehicles and engines, and facilities that emit 25,000 metric tons or more per year of GHG emissions submit annual reports to EPA. Ibid.

been much more inward-looking than European politics, by and large. That is, American citizens—and, consequently, American politicians—tend to focus more closely on questions of domestic policy rather than on questions of international policy. US engagement in international relations reflects Americans' tendency to be wary of—even hostile to—committing American resources to resolving questions of international law and policy, even when those international issues affect American interests.[34] America's 'hubris' in sidelining international politics reflects long-standing isolationist tendencies, dating back to the early twentieth century and characterizing American sentiment between the two World Wars. Following the Second World War, the US found itself heavily imbedded in isolationist politics and initially unwilling to join the League of Nations, despite President Warren Wilson's support for the new international institution and legal order.[35] Since that time, American engagement with international affairs has ebbed and waned across time.[36]

During the critical era for international climate negotiations, 1997–2009, American climate politics were characterized by isolationist and unilateral tendencies. Regarding the state of political relations during the administration of George W. Bush, for example, Bush's overall approach to environmental politics saw him dubbed the 'Toxic Texan'.[37] Similarly, his hegemonic approach to international relations has been widely criticized by foreign leaders and commentators, alike.[38] Former French Prime Minister Lionel Jospin characterized the Bush Administration as not so much of an isolationist administration as a unilateralist administration, in part due to President Bush's decision to rescind the US's signature to the Kyoto Protocol and to pursue unilateral climate discussions outside the auspices of the UNFCCC framework.[39]

While unilateralism characterized President George W. Bush's general approach towards international relations, his approach to international climate

---

[34] Joseph S Nye Jr, 'The American National Interest and Global Public Goods' (2002) 78 *Intl Affairs* 233.

[35] The US's hostility towards international law at the beginning of the twentieth century has ebbed and waned over time. M Sterio, 'The Evolution of International Law' (2008) 31 *Boston College Intl & Comparative L Rev* 213, 245.

[36] President Truman, for example, was more successful in overcoming traditional American isolationist tendencies after World War II. Recognizing the growing threat the Soviet Union posed to global legal order and US security, President Truman sought to rebuild European economies and develop international allies—thus, requiring US monies and troops. He overcame traditional domestic isolationist tendencies by 'scaring [the] hell out of the country' and by forming alliances with a powerful Republican congressman who could ensure legislative support for the President's strategies. Eric A Posner and Adrian Vermeule, 'The Credible Executive' (2007) 74 *U Chicago L Rev* 865, 871, quoting Arthur M Schlesinger Jr, *The Imperial Presidency* (Houghton Mifflin, New York, 1973) 128.

[37] Katty Kay, '"Toxic Texan" Has Poor Green Record' *Times* (London, 23 August 2002) 19.

[38] Ian Clark, 'Bringing Hegemony Back In: The United States and International Order' (2009) 85 *Intl Affairs* 23; see also Barry Buzan, 'A leader without followers? The United States in world politics after Bush' (2008) 45 *Intl Politics* 554.

[39] BBC News, 'France turns heat on Bush' (4 April 2001) <http://news.bbc.co.uk/1/hi/world/americas/1260499.stm> accessed 2 January 2010.

change law and politics displayed a combination of isolationist and unilateral tendencies.[40] Although the Bush Administration expressed commitment to 'find[ing] a workable solution to this serious problem that affects all of us in the global community', the Administration chose to work outside the parameters of the existing multilateral framework, thereby undermining international negotiations and attempting to re-shape the debate unilaterally.

The election of President Obama was widely lauded as initiating a new, more engaged era in international politics, particularly with regard to climate negotiations. Even during more engaged, outward-looking Presidential administrations, however, the American public often retains its intense inward-looking focus. This inward focus is reflected in federal-level politics, where low-levels of public interest and awareness coupled with suspicion towards international interference with domestic issues often subordinates questions of global engagement on the political agenda, decreases incentives for politicians to actively engage with international politics, and creates more opportunities for special interest groups to influence questions of international policy behind closed doors.

In the context of international climate policy, it is becoming increasingly difficult for the US to sustain isolationist and/or unilateral policies due to mounting scientific evidence, sub-federal policymaking and international pressure. Further, as a consequence of the on-going 'war on terror' and US reliance on its European allies in this regard, the US is increasingly forced to abandon isolationist and unilateral approaches to foreign policy, eg, climate change policy, in order to maintain European cooperation on matters of national and international security.[41] To date, however, US and EU support for the international law-making process and respect for the international legal order in the climate context have diverged markedly.

The US rejected the Kyoto Protocol as 'fundamentally flawed' and 'not the correct vehicle with which to produce real environmental solutions'. In rejecting the Protocol, the US sought to avoid constraining domestic choices by committing to a legally binding international regime that embodied international environmental norms that the US opposed, eg, the precautionary principle and the principle of common but differentiated responsibilities.[42]

---

[40] Greg Kahn, 'The Fate of the Kyoto Protocol Under the Bush Administration' (2003) 21 *Berkeley J Intl L* 548 (describing variations in Bush's willingness to engage in bilateral and multilateral approaches depending on the issue at hand); RC Longworth, ' "Bush Doctrine" Arises From the Ashes of Sept. 11' *Chicago Tribune* (Chicago, 7 March 2002) 4 (reporting that President Bush's post 9/11/01 foreign policy has become increasingly unilateral and isolationist).

[41] Kathryn F King, 'The Death Penalty, Extradition, and the War Against Terrorism: US Responses to European Opinion about Capital Punishment' (2003) 9 *Buffalo Human Rights L Rev* 161, 162.

[42] Christopher Joyner and others, 'Common But Differentiated Responsibility' (2002) 96 *Am Society Intl L Proceedings* 358, 362; 'Rio Declaration on Environment and Development' UN Conference on Environment and Development (Rio de Janeiro 3–14 June 1992) (12 August 1992) UN Doc A/CONF.151/26 (Vol I).

Differences between the approaches of the US and the EU to precaution create rifts in the scholarly literature over the meaning of precaution and the extent to which the two regions have adopted precautionary approaches to environmental protection over time.[43] What is evident, however, is that in the context of international climate policy, between the negotiation of the Kyoto Protocol and the year 2009, the US not only opposed the specific emissions reduction objectives enunciated by the Protocol, but also questioned the principles underpinning these objectives. Until 2007, for example, the US adopted a 'sound science' approach to climate change that stood in direct opposition to a precautionary approach in demanding sound science on climate change before legal action would be justified. The 'sound science' approach has been interpreted by one prominent environmental law scholar as meaning 'we will not act until the science is conclusive, i.e., a cold day in hell'.[44] Similarly, the US rejected the principle of common but differentiated responsibilities, which is the underlying normative principle that the Protocol relies on to assign legal roles and responsibilities. Until 2009, with the election of President Obama, the US refused to shoulder blame or disproportionate burdens for climate change and rejected the Protocol's manner of assigning emissions reduction obligations based on historical responsibility for GHG emissions coupled with social and economic capacity to respond.

In these various ways, during the core development years for the international climate regime, the US renounced not only the international legal framework for climate change, but also the normative principles upon which it is grounded. In so doing, the US adopted a unilateral approach to international and domestic climate policymaking, whereby it sought to shape policies and proposals around norms and notions external to the legal order established by the UN climate regime.

In direct contrast, the EU has been one of the most vocal advocates and closest adherents to the normative foundations and legal framework established by the UN climate regime. Since the earliest days of international climate negotiations, the EU has actively worked to create, sustain, and further develop the architecture of the UNFCCC and the Kyoto Protocol.[45] The EU's adherence to the UN

---

[43] See, eg, Jonathan Wiener and Michael Rogers, 'Comparing Precaution in the United States and Europe' (2002) 5 *J Risk Research* 317; Jonathan Wiener, 'Whose Precaution after All? A Comment on the Comparison and Evolution of Risk Regulatory Systems' (2003) 13 *Duke J Comparative & Intl L* 207.

[44] Oliver Houck, 'How Industry Hijacked "Sound Science"', *New Orleans Times-Picayune* (30 January 2004); see also President George W. Bush, 'Remarks to the National Oceanic and Atmospheric Administration in Silver Spring' (Speech in Maryland, 14 February 2002) <http://www.guardian.co.uk/environment/2002/feb/14/usnews.globalwarming> accessed 3 April 2010 (stating that decisions about climate change should be made 'on sound science; not what sounds good, but what is real' and suggesting that current research does not 'justify' curtailing greenhouse emissions); Patrick Parenteau, 'Anything Industry Wants: Environmental Policy Under Bush II' (2003) 14 *Duke Envtl L & Policy Forum* 363, 364; Chris Mooney, 'Beware "Sound Science." It's Doublespeak for Trouble' *Washington Post* (Washington DC, 29 February 2004) B2.

[45] Jutta Brunnée, 'Europe, the United States, and Global Climate Regime: All Together Now?' (2008) 24 *J Land Use & Envtl L* 1.

regime is further reflected in European regional climate policy, where the EU has explicitly framed climate policies in terms of the objectives and underlying principles of the UNFCCC, eg, avoiding dangerous climate change premised on notions of precaution and common but differentiated responsibilities.[46] The EU, thus, has not only been an engaged participant in multilateral legal negotiations, it has been a driving force in developing and implementing the normative and legal components at the heart of the UN climate regime.

The US and the EU have played incongruous roles in global efforts to shape legal norms and legal order in the climate context. This critical policy divergence creates rifts that affect every aspect of regional approaches to addressing climate change.

Summary of Divergences

EU and US strategies to address climate change diverge from the point of origin. Since the negotiation of the Kyoto Protocol in 1997, EU and US climate strategies have evolved in reference to different legal norms and in line with different legal objectives. This discussion highlights key macro-level ways that the policies differ, but it is in no way comprehensive. EU and US climate laws and policies are inherently multi-faceted and evolving and the details of particular initiatives reveal many further divergences.

The details of specific initiatives, especially key programs such as the EU ETS and the US CAFE program, are relevant and worth exploring. Such detailed analysis offers significant insight for governmental entities seeking to craft and/or update specific strategies and should be done on a case-by-case basis so that the most up-to-date and specific information can be used. Again, however, the ability to analyze policy measures on a case-by-case basis reveals one of the divergences between existing EU and US approaches—that being significant variations in the availability of legally enforceable, monitored, and reported-on programs to analyze.

At the international level, macro-level divergences shape the global agenda and the relative ability of the EU and the US to negotiate from positions of power and authority. The EU's global leadership, engagement with the UN climate regime, commitment to legally binding emissions reduction obligations, enactment of measurable climate initiatives, and creation of a transparent policy record of successes *and* failures on climate change has had immeasurable effect on global debate. The EU's actions have helped maintain international momentum, provided a policy laboratory, exerted political pressure on laggards, functioned as a developed–developing country mediator, and served as the backbone for the existing UN climate regime and for efforts to negotiate a post-Kyoto framework.

In contrast, the US's decision to abstain from participating in the Kyoto Protocol and to disavow international environmental norms underlying the UN climate

---

[46] Jonathan B Wiener, 'Precaution' in Daniel Bodansky, Jutta Brunnée, and Ellen Hey (eds), *Oxford Handbook of International Environmental Law* (OUP, Oxford, 2007) 597, 599.

regime at the domestic level, combined with its inability to develop national consensus or comprehensive legal structures to address climate change has had a mixed effect on global negotiations. The US's perceived failures have helped drive policy negotiations forward by creating moral outrage. This moral outrage has stopped short of instigating true change however. While US inaction has helped spur Europe's leadership and has outraged and motivated many developing countries, especially least developed countries, to advocate for strong international climate policies, it has also alienated the key developing countries of China, India, and Brazil. Absent active participation by the US, these States have opposed legally binding emissions reduction commitments on moral and economic grounds.

The ability of an international climate regime to achieve the UNFCCC goal of preventing dangerous anthropogenic interference with the climate depends on participation by the big GHG emitters, which necessarily includes the US, China, India, and Brazil. Leading up to 2009, the US's macro-level policies, thus, had the effect of sabotaging necessary global consensus—even if a post-Kyoto agreement can still be negotiated, without the US, China, India, and Brazil fully onboard, the agreement will be purely symbolic. As will be discussed further in Chapters 8 and 9, however, the US's political about-face in 2009 has spurred greater global participation in post-Kyoto negotiations, with mixed results.

The primary policy divergences of the US and the EU are greater than the sum of their parts. The decisions by the US and the EU to pursue different policy courses domestically and internationally has impacted the global communities' ability to achieve emissions reductions, to generate a constructive policy record during the first Kyoto compliance period, and to achieve global consensus on the way forward in a post-Kyoto world.

The divergences between EU and US climate policy are many and have discernible physical and moral impacts. The extent to which EU and US policy strategies differ, however, often masks underlying similarities, which are equally important to understanding how European and American policy choices have and will impact long-term global efforts to address climate change.

From Kyoto to Copenhagen: Transatlantic Convergences

Within the diverging architectures of EU and US climate policy, similar strategies, emphases, and limitations emerge. The similarities between the two systems reveal essential elements and critical challenges to developing a post-Kyoto global strategy.

Systemic Pushes and Pulls

Self-evident in US politics and hiding not far behind the veneer of the EU's united front, internal pushes and pulls characterize policymaking in both regions. As previously discussed, domestic politics have resulted in regional variations in terms of which governmental entities have assumed leadership roles in the US and the EU. The variations in policy leadership actually belie similarities in relation to the labyrinth of internal legal infrastructures, political perspectives, and social systems that the EU and the US federal government must maneuver to develop necessary political consensus.

The intense debates taking place in EU and US domestic politics reflect fundamental differences in member state/state willingness and capacity to adopt multifaceted regulatory strategies to curb climate change. Similarities in internal EU and US climate debates are indicative of larger debates going on within the international community, both in terms of intra- and inter-state politics. In particular, understanding how variations among EU and US States in reference to economic stability, political perspectives, and capacity to diversify energy sources impacts climate policy offers useful lessons for dealing with more extreme versions of these variations at the international level.

Pushes and Pulls: The United States

Within the US, for example, a majority of states have individually and collectively embarked on a widespread and coordinated campaign to develop climate change laws in the absence of federal leadership. While traditional economic theory would argue against the rationality of independent or regional state efforts to regulate the global commons in the absence of national coordination and oversight to prevent problems of competitive disadvantage, the flurry of activity at the state level is undeniable. States from coast to coast have adopted a variety of legal, regulatory, and policy measures to address climate change.[47]

As discussed in Chapter 4, the motivation for widespread state action is not evident and there is considerable disagreement about how political, social, and geographic commonalities influence these efforts. Genuine concern for a global environmental problem that threatens the health and well-being of state citizenry is a likely factor, especially given that many of the most active states have coast lines and coastal and agricultural industries that face climate-related threats. States are also likely to be motivated by a desire to garner political good will from concerned constituencies. Other state actions, such as many of California's measures, are taken with the express intention of helping generate models for the federal government to use in creating a comprehensive climate change framework. Equally, states are inevitably motivated by the desire to gain early economic advantages by creating incentives for research and development and, thus, capturing markets in green technologies and minimizing costs associated with reducing GHG emissions in the long-term.

Genuine concern, political maneuvering, political pressure, and economic advantage are not mutually exclusive drivers. States are likely motivated by all

---

[47] Pew Center on Global Climate Change, 'About U.S. States and Regions' <http://www.pewclimate.org/states-regions/about> accessed 16 April 2010. Greenhouse gas inventories are currently one of the most popular ways with which states seek to engage with the climate change debate. Engel and Saleska (n 28) 216.

of these factors to different degrees. The relative importance of political motivations is not always clear. Unlike federal regulation, which has been extensively researched and written about, state regulation has received far less attention and analysis, particularly in the context of environmental law.[48] Thus, there is very little data suggesting why states adopt particular environmental regulatory strategies. What is clear, however, is that a variety of factors are driving US states to both prompt and resist federal climate change legislation.

As discussed in Chapter 3, states such as California, New York, New Jersey, Massachusetts, and Oregon more closely mimic EU climate policy than American climate policy. For example, as the first state to establish mandatory carbon dioxide emissions caps for power plants, and the lead petitioner in the most prominent judicial challenge to federal environmental failures in recent history, Massachusetts, has not only managed to do what the federal government still has not done—regulate carbon dioxide emissions—but has also played a key role in forcing the federal government to reassess its legal responsibility to regulate GHG emissions.[49] Similarly, California has played a key leadership role in domestic efforts both to address climate change, and to overcome social and political resistance to climate change legislation. California's policy framework for addressing climate change is analogous to the most progressive initiatives worldwide, including:

- legally defined limits on GHG emissions from motor vehicles;
- targets for bioenergy use and production as part of California's Renewable Portfolio Standards;
- a 'Million Solar Roofs' project including provisions to create one million new solar roofs for California by 2018;
- a low carbon fuel standard, and legislation;
- an enforceable cap on greenhouse gas emissions—mandating that the State of California reduce its greenhouse gas emissions to 1990 levels by the year 2020 (a 25% reduction); and
- reduction of emissions to 80% below 1990 levels by 2050.[50]

California's greenhouse gas emissions reduction legislation mirrors the most rigorous Climate Change bills proposed in the 110th and 111th Congress. California's body of climate change legislation is not only comparable to EU

---

[48] Paul Teske, *Regulation in the States* (The Brookings Institution, Washington DC, 2004) 8.
[49] Progressive Policy Institute, 'State and Local Governments and Climate Change' (2003) <http://www.ppionline.org/ppi_ci.cfm?knlgAreaID=116&subsecID=900039&contentID=251285> accessed 3 April 2010.
[50] California Air Resources Board, 'Climate Change Draft Scoping Plan: a framework for change' (Discussion Draft pursuant to the California Global Warming Solutions Act of 2006, June 2008) <http://www.arb.ca.gov/cc/scopingplan/document/draftscopingplan.pdf> accessed 2 January 2010.

initiatives, but until very recently also surpassed the reach of EU law in its creation of a regulatory regime for controlling GHG emissions from the automobile sector.

Even within the confines of the US federalist system, California's actions significantly affect national policy choices and international perceptions. Domestically, California's efforts have paved the way for other states to enact climate change laws and policies and have compelled the federal government to act to create federal-level climate laws and policies.[51] Internationally, California's initiatives coupled with similar actions by fellow US states have demonstrated to the international community that resistance to internationally agreed norms does not extend to the entire US population.

Individual states such as California and Massachusetts have created legislative models and political momentum for addressing climate change. Recognizing inherent limits in individual state action, numerous states also have sought out regional partnerships to maximize the political and economic benefits of collective action.[52]

The progressive efforts of these various states and regional actors have received considerable attention at the domestic and international levels. The actions of California and like-minded states are heralded as signs that there is growing consensus within the US for creating a comprehensive climate change regime. The desire for progress, however, is not universal either within proactive states or across the 50 states. For example, in the case of *Massachusetts v EPA*—the most important US climate change litigation to date—while the state of Massachusetts served as the main petitioner and was joined in its case by the states of California, Connecticut, Illinois, Maine, New Jersey, New Mexico, New York, Oregon, Rhode Island, Vermont, and Washington, the EPA was joined in efforts to oppose mandatory regulation of CO2 by not only the Alliance of Automobile Manufacturers, the National Automobile Dealers Association, the Engine Manufacturers Association, the Truck Manufacturers Association, the CO2 Litigation Group, and the Utility Air Regulatory Group, but also by the states of Michigan, Alaska, Idaho, Kansas, Nebraska, North Dakota, Ohio, South Dakota, Texas, and Utah.

In addition to opposing administrative regulation of CO2, many states—especially, coal and automobile producing states, eg, West Virginia and Michigan, states with active oil and gas extraction industries, eg, Alaska, Wyoming, Oklahoma, Louisiana, and Texas, states with large fossil-fuel power plants, eg, Texas and Ohio, and states with modest capacity to produce renewable energy,

---

[51] Jonathan L Ramseur, *CRS Report for Congress: Climate Change: Action by States To Address Greenhouse Gas Emissions* (18 January 2007) <http://fpc.state.gov/documents/organization/80733.pdf> accessed 1 June 2010. States adopting similar regulations include Oregon, Washington, and eight states in the Northeast. Ibid.
[52] About US States and Regions (n 47).

eg, South Carolina—oppose climate change legislation.[53] Some traditionally fossil fuel-heavy states, such as Texas, continue to oppose federal climate change legislation while simultaneously benefiting from early investment in renewable energy industries.[54] Other states are outwardly hostile towards climate change policymaking and actively resist efforts to adopt binding limits on greenhouse gas emissions or mandatory renewable energy programs.[55]

State leaders and congressional representatives from both political parties have opposed federal efforts to limit GHG emissions, pointing to perceived economic disadvantages associated with energy and automobile regulation, uncertainty surrounding the necessary components of a comprehensive regulatory regime, and anticipated regulatory complexity. For example, in Spring 2009, House Agriculture Chairman Collin Peterson—a Democrat from Minnesota—said that he would not support a pending proposal for federal climate legislation because it vests too much power in administrative agencies, such as the EPA. Mary L. Landrieu, a Democratic Senator from Louisiana has consistently opposed proposals for federal climate change legislation based on fears that climate regulations will be economically inefficient and will impose disproportionate costs on petrochemical companies, including many petrochemical manufacturers located in the state of Louisiana. Similarly, Democratic Senator Robert C. Byrd, from the coal-producing state of West Virginia, has consistently emphasized the continuing importance of coal to the American economy and has expressed hesitancy towards federal efforts to regulate GHG that would come at the expense of economic development, noting that 'to be successful, a national climate change effort must have broad public support, and that it cannot be achieved by the regulatory actions of an agency; and that congress has a much broader mandate that includes protecting jobs, communities, and livelihoods'.[56] Senators Peterson, Landrieu, and Byrd's comments reflect widely held, bi-partisan concerns in Congress about the costs and complexities involved in designing and implementing a nationwide climate policy, with particular regard for states whose economies depend on fossil-fuel industries.

Thus, while there are easily identifiable domestic pushes for federal actions to curb climate change, there also remain significant pulls backwards towards

---

[53] See Barry G Rabe, *Statehouse and Greenhouse: The Emerging Politics of American Climate Change Policy* (Brookings Institution Press, Washington DC, 2004) xii–xiii and 47–49 (describing the apparent indifference of Louisiana and Florida to reducing greenhouse gases). In contrast, however, South Carolina Senator, Lindsey Graham has emerged as one of the Senate leaders in pushing for comprehensive climate change legislation.

[54] Ibid 49–62 (detailing Texas's enactment of legislation mandating increased generation of renewable energy for economic reasons).

[55] Ibid 40–47 (describing the hostility of Michigan and Colorado toward environmental regulation).

[56] US Senator Robert C Byrd, 'Byrd Talks Coal with New Cabinet Officials' (Media Release, 5 February 2009) <http://byrd.senate.gov/mediacenter/view_article.cfm?ID=314> accessed 2 January 2010.

the status quo. An overview of state federal action reveals the domestic pushes and pulls:

- 46 states have GHG inventories *but* 4 states continue to lack greenhouse gas inventories;
- 33 states have adopted climate action plans *but* 17 states have not adopted climate action plans;
- 20 states have public benefit funds *but* 30 states have not yet created public benefit funds;
- 20 states have adopted GHG emission targets *but* 30 states have not yet adopted GHG emission targets;
- 28 states have renewable energy portfolios *but* 22 states have not developed renewable energy portfolios;
- 16 states have adopted GHG emission standards for automobiles *but* 34 states have not yet adopted GHG emission standards for automobiles;
- 18 states have mandatory CO2 reporting programs *but* 32 have not yet developed mandatory reporting requirements;
- 24 states have formed climate change advisory boards *but* 26 states continue to work without climate advisory boards;
- 33 states are participating in one or more regional initiatives *but* 17 states are not participating in any regional climate change initiatives.[57]

States failing to adopt basic climate measures, eg, greenhouse gas inventories, and/or to either adopt GHG emission targets or to participate in regional climate initiatives include: Alaska, North Dakota, South Dakota, Nebraska, Idaho, Wyoming, Arkansas, South Carolina, Oklahoma, Missouri, Arkansas, Louisiana, Mississippi, Alabama, Georgia, North Carolina, West Virginia, Kentucky, and Tennessee.

The number of states and Congressional representatives stalling or actively opposing progressive climate change policies are declining yet continue to exist. With resistance still rife, it remains unknown whether there will be sufficient Congressional support to obtain the majority vote needed to pass federal climate change legislation. On the whole, there is a discernible trend for states to adopt a range of climate change laws and policies that promote renewable energy and to inventory, register, and—increasingly—limit greenhouse gas emissions. However, numerous states and Congressional representatives continue to resist mandatory greenhouse gas emissions reductions, largely on economic grounds.

---

[57] About US States & Regions (n 47).

Thus, over the period 1997–2009, while the policies and ideologies of some of the most prominent US states have often shared more commonalities with one another and with the EU than they have with the federal government, it remains unclear whether the state-led push for change will outweigh more dispersed but deeply entrenched resistance to federal climate change legislation.

Pushes and Pulls: The European Union

European climate politics are similarly typified by conflict and contradiction.[58] Despite providing global leadership in efforts to push forward climate negotiations, internal discord challenges efforts to develop and dispatch coordinated European policy.[59] Differences among member states and between the Council and the Commission have frequently stalled or softened climate policymaking efforts. In the early 1990s, near the beginning of international climate negotiations, for example, the Commission advocated employing a combined energy and carbon tax as the primary mechanism for stabilizing GHG. The tax proposal split member states and failed to garner Council support, temporarily stalling the development of EU climate policy.[60] This early impasse threatened regional efforts to develop a coordinated strategy while also 'dent[ing] EC leadership in international negotiations' leading up to the first meeting of the UNFCCC COP.[61] Similarly, during key negotiations over the Kyoto Protocol flexibility mechanisms in 1997, lack of internal coordination and preparedness resulted in EU negotiators spending valuable time behind closed doors attempting to hammer out regional views while international negotiations proceeded without the EU delegates at the table.[62]

The ability of member states to cooperate and reach consensus on climate change policy is essential not just to the formulation of European policy but also to the EU's continuing ability to serve in a leadership role in international forums. This is especially true in the field of climate change since the EU has limited formal competency in the area of energy policy. On questions of international climate policy, the ability of the EU to negotiate from a position of regional power depends on the ability of its member states to agree on common EU policies and positions. For this reason EU member states—with the help of the Commission—attempt to agree upon regional positions prior to arriving at international negotiations.[63] Settled EU positions are then presented in international

---

[58] Thomas Heller, 'The Path to EU Climate Change Policy' in Jonathan Golub (ed), *Global Competition and EU Environmental Policy* (Routledge, New York, 1998).
[59] Farhana Yamin, 'The Role of the EU in Climate Negotiations', in Joyeeta Gupta and Michael Grubb (eds) *Climate Change and European Leadership: A Sustainable Role for Europe?* (Kluwer Academic Publishers, Dordrecht, Netherlands, 2000).
[60] Ibid 49.   [61] Ibid.   [62] Ibid 61.
[63] Sebastian Oberthür and Hermann Ott, 'The Kyoto Protocol: International Climate Policy for the 21st Century' (Springer, New York, 1999).

forums by the current President of the European Council. EU member states retain the right to present independent positions, but this has become increasingly uncommon as the EU has sought to solidify its position as a united supranational entity capable of both coordinating regional climate policy and leading international climate negotiations.

Supranational and international climate policymaking has been hindered at times—especially in the early days—by member states' inability to find agreed-upon solutions. As one commentator has noted, however, in the majority of cases, 'the EU's internal coordination is a building block for international negotiations and strengthens the domestic process of constituency building and stakeholder participation'.[64]

Internal negotiations over the EU's packages of climate change measures as initially agreed in March 2007 and supplemented in December 2008 and April 2009 exemplify the pushes and pulls of internal EU climate politics. Debate over these measures reveals the inherent difficulty of achieving member state coordination and the unique benefits that accrue when member states are able to overcome internal divisions to reach agreement on questions of climate change.

At the Spring 2007 European Summit, German Chancellor Angela Merkel—then President of the European Council—faced an uphill battle in negotiating agreement among the 27 member state heads of state on formulating the EU's first ever comprehensive agreement on climate and energy policy. Merkel introduced a plan supported by Germany, the UK, Italy, and the Scandinavian countries that proposed new binding targets for renewable energy. Key Eastern European countries, including Poland, Slovakia, the Czech Republic, and Hungary opposed the new renewable targets based on fears that the cost of complying would be disproportionately prohibitive and unrealistic for certain countries. In expressing concern about the blanket 20% proposal, Slovak Prime Minister Robert Fico said: 'We are able to support the word binding under conditions that specifics of each country are taken into consideration.... The 20 percent should be effective for the EU as a whole. We can guarantee a 12 percent share by 2020. Let's not put unrealistic pledges on the table.'[65] The Eastern European countries were joined in their opposition to flat rate strictly defined renewable targets by nuclear energy dependant France. In expressing concern about the proposal, then French President Jacques Chirac emphasized that '[t]he distribution of the effort on renewable energies must take into account the share of low-carbon energies—nuclear and clean coal—in our national energy choices'.[66]

Chancellor Merkel ultimately parleyed an agreement based on the principles of 20/20/20, which would require the EU by 2020 to cut greenhouse gases emissions by at least 20% of 1990 levels (or 30% if other developed countries commit

---

[64] Ibid.
[65] Matthew Newman, 'EU Edges Toward Accord on Renewable Energy Target' (News Release, 9 March 2007) <http://www.bloomberg.com/apps/news?pid=20601085&sid=a7fQbHw0LBfQ&refer=europe> accessed 2 January 2010.
[66] Ibid.

to comparable cuts), to increase use of renewables (wind, solar, biomass, etc) to 20% of total energy production, and to cut energy consumption by 20% of projected 2020 levels by improving energy efficiency. The final agreement established overarching goals but left the details of implementation up to the Commission, including determining what sources of energy will count toward renewable goals and how energy obligations will be distributed among the individual member states. Despite the hard work to come in implementing the agreement, Chancellor Merkel heralded the new climate agreement as a 'new dimension of European cooperation'.[67] In similarly extolling the virtues of all 27 member states reaching consensus, European Commission President Jose Manuel Barroso said: 'We can say to the rest of the world, Europe is taking the lead.... You should join us in fighting climate change.'[68]

Following the negotiation of the 20/20/20 plan in 2007, Europe—along with the rest of the world—faced an economic downturn that amplified existing concerns over the economic costs of implementing the climate change agreement. Leading up to the December 2008 European Summit, the always fragile European consensus over climate change seemed on the brink of collapse under the weight of the on-going global recession.[69] Heading into the summit, member state leaders warned of the costs to Europe of the proposed climate policies. Italian Prime Minister Silvio Belusconi colorfully voiced widespread concerns in stating that: '[n]ow is not the time to be playing the role of Don Quixote, when the big producers of CO2, such as the US or China, are totally against adherence to our targets'.[70] Italy took matters one step further when, in November 2008 just prior to the December European Summit, it threatened to use its Council veto to oppose draft legislation containing the requirement that the EU reduce its emissions 20% from 1990 levels by 2020 if the Council failed to address Italy's concerns over the economic costs of the proposed cuts.[71]

Poland expressed similar concerns. As discussed in chapter 6, as the largest coal producer in the EU, Poland has consistently resisted EU climate policies that would constrain Poland's energy choices or negatively impact the price of coal. Among the key concerns to Poland as well as to other member states—including

---

[67] European Commission CORDIS, 'Research high on agenda at Spring Council' (News Release, 12 March 2007) <http://cordis.europa.eu/search/index.cfm?fuseaction=news.document&N_RCN=27286> accessed 2 January 2010.

[68] Carsten Volkery, 'Europe Takes the Lead in Fighting Climate Change' (Spiegel Online International News Release, 9 March 2007) <http://www.spiegel.de/international/0,1518,470926,00.html> accessed 2 January 2010.

[69] Simon Tilford, 'The EU's climate agenda hangs in the balance' (Center for European Reform Bulletin, December 2008/January 2009) <http://www.cer.org.uk/articles/63_tilford.html> accessed 2 January 2010.

[70] Franck Lirzin, 'Quotes: US-Europe politicians talk climate change' (28 October 2008) <http://www.cafebabel.co.uk/article/26962/europe-us-leaders-stances-environment-climate.html> accessed 3 April 2010.

[71] Eg, ENDS Europe, 'Italy threatens to veto EU climate package' (News Release, 26 November 2008) <http://www.endseurope.com/17172> accessed 2 January 2010.

Germany, the original proponent of the plan, Italy, Austria, and most of the Central and Eastern European countries—were planned modifications to the EU ETS as part of the post 2012 strategy. In particular, the most contentious question involved plans to require the full auctioning of allowances in Phase III of the EU ETS for energy generators, with a gradual shift to full auctioning for other covered industries. Germany, Italy, and Austria expressed concerns over the impact on business, especially in the manufacturing sector and argued that full auctioning would drive industry out of the EU, creating a global race to the bottom. The Central and Eastern European States criticized the proposal as disproportionately impacting their economies due to heavy reliance on coal and the continuing legacy of Communism, as evidenced by vestiges of heavy industry.

In response, the Commission drafted a proposal that acknowledged these economic concerns. In order to accommodate varying circumstances, the Commission's proposal set differentiated emission targets with reference to existing energy mixes, topography, and GDP per capita as well as including financial compensation mechanisms for the less well-off member states.[72]

Beyond facing policy critiques raised by various member states, EU political consensus on climate change faced a new challenge in 2009 with the ascendancy of the Czech Republic to the presidency of the European Council. The Czech Presidency of the Council would be held by then Czech President Václav Klaus—an open climate and environmental skeptic whose recent book, *Blue Planet in Green Shackles—What Is Endangered: Climate or Freedom?*, argued that modern environmentalism poses a threat to basic human freedoms. President Klaus, one of Europe's most vocal opponents to EU and UN climate policies, compared environmentalism to communism, arguing that '[l]ike their (communist) predecessors, [environmentalists] will be certain that they have the right to sacrifice man and his freedom to make their idea reality. . . . In the past, it was in the name of the Marxists or of the proletariat—this time, in the name of the planet.'[73] Klaus—an economist—advocated a free market approach to environmental protection, including climate change. He opposed climate regulations, including mandatory emission caps and went so far as to argue that: '[i]t could be even true that we are now at a stage where mere facts, reason and truths are powerless in the face of the global warming propaganda'. While President Klaus's views on climate change were critiqued by fellow Czech politicians, including Czech Minister of the Environment Martin Bursik who wrote an op-ed to *The Guardian* newspaper stating that the newspaper's leader column discussing Klaus's views had 'made the mistake of believing that the extreme personal opinions of the Czech president, Václav Klaus, are relevant to the official views of the Czech Republic or its current presidency of the EU

---

[72] Tilford (n 69).
[73] DPA, 'Czech President Klaus ready to debate Gore on climate change' (29 May 2008) <http://www.infowars.com/czech-president-klaus-ready-to-debate-gore-on-climate-change/> accessed 2 January 2010.

Council' and cautioning that:

while he holds a very important office, it is nonetheless a non-executive and non-accountable office elected by parliament, not directly by the people. His views are headline-catching because they are designed to be, and the only way he could be more transparent would be to wear the logo of Luxoil (a major sponsor of his book) on his shirt.... There is no evidence of his views having any impact beyond a select group of acolytes and ideologues. He will continue his world tour in the cause of denial and inaction and it will continue to be a sideshow to the scientific, public and political consensus that serious and sustained action is urgently required.[74]

Bursik's column sought to ease widespread concerns that Klaus's would use his Presidency to undermine the substance and presentation of the EU's united front on climate change. Klaus himself, however, continued to advocate climate skepticism well into his term, publicly revealing divisions in European climate politics.[75]

Despite these economic and political challenges, EU leaders overcame internal obstacles to negotiate in December 2008, and then to adopt in April 2009, the final legal texts of an energy and climate change package of legislation intended to achieve the previously agreed 20/20/20 goals. As discussed in Chapter 5, the final package of legislation embodied six key measures:

- A revision of the EU ETS to mandate full auctioning of allowances for the power sector post-2013 and to phase in auctioning for other sectors between 2013–2027;
- An 'effort-sharing' agreement setting out binding emission-reduction targets by member states for sectors outside the ETS, eg, transport and agriculture, in order to reach an overall cut of 10% by 2020;
- A regulatory framework for carbon capture and sequestration to support the new technologies before they are commercially viable;
- A directive for the promotion of energy from renewable sources, establishing individual targets for the proportion of renewables in member states' final energy consumption, to reach 20% EU-wide, stipulating that each member state must ensure a 10% renewable share in domestic transport sectors, and establishing criteria on the sustainable use of biofuels;
- CO2 limits for new automobiles;
- Standards for fuel quality.[76]

---

[74] Martin Bursik, 'Response: The Czech president's climate change denial is irrelevant' *Guardian* (25 March 2009) <http://www.guardian.co.uk/commentisfree/2009/mar/25/climate-change-eu-czech-republic> accessed 2 January 2010.

[75] Patrick Henningsen, 'EU President Klaus Opens International Climate Skeptics Conference' (News Release, 12 March 2009) <http://www.opednews.com/articles/EU-President-Klaus-Opens-I-by-Patrick-Henningsen-090309-863.html> accessed 2 January 2010.

[76] EurActiv, 'EU wraps up climate and energy policy' (7 April 2009) <http://www.euractiv.com/en/climate-change/eu-wraps-climate-energy-policy/article-181068> accessed 2 January 2010.

In agreeing to this legislative package despite economic woes and Klaus's skeptical presence at the helm, the EP Council demonstrated that while the EU faces intrinsic challenges in coordinating policy in the broad fields of energy and climate change, it has laid the groundwork for continuing internal coordination and global leadership.[77]

Around the same time the EU debated how renewable energy obligations would be distributed and when and how the ETS would move to an auctioning system, the US Congress was debating framework legislative proposals and wrangling over provisions in President Obama's proposed budget that would provide new cash flows for the development of energy-efficient and carbon-free technologies.[78] In both regions, internal debates rage over appropriate social and political responses to climate change. In common with one another, the acceptance of social, cultural, and political diversity by the EU and the US encourages vast differences in perspective regarding climate change.

Similarities in political diversity, however, simultaneously reveal key differences. Two of these differences are worth mentioning here. First, the EU has laid the groundwork for cooperation that facilitates current climate policymaking efforts. The US still struggles to find this common ground. The two regions are at very different stages in internal negotiating.

Second, the EU faces unprecedented challenges in creating political consensus among sovereign nations. However, in this task, the EU benefits from underlying notions of subsidiarity. That is, while the EU may face greater challenges in coordinating climate policies among nations as diverse as Poland, France, Malta, and Denmark, it also benefits from being able to create overarching objectives which can then be differentiated and variously implemented by individual member states. The US is diverse, but its political, social, and cultural diversity pales next to that of the EU. The US benefits from a shared political history and a more stable and unified cultural personality. In crafting climate law, however, the US must formulate legislation that applies equally across the 50 states, with less room for accommodating economic and social differences. Unlike European legislative negotiations, US negotiations offer fewer opportunities for addressing particularized needs, making compromise more difficult.

The EU and the US are similarly shaped by their political diversity and intense internal political debates but reveal different strategies and success rates in using domestic politics to their advantage.

---

[77] Ibid.
[78] Christa Marshall, John Fialka, and Lea Radick, 'Renewable industry cheers Obama budget while coal and nuclear jeer' *New York Times* (New York, 8 May 2009) <http://www.nytimes.com/cwire/2009/05/08/08climatewire-renewable-industry-cheers-obama-budget-while-12208.html> accessed 2 January 2010.

Economic Centrality

The US's influence on the UN climate regime is best illustrated by the inclusion of the flexibility mechanisms in the Kyoto Protocol. The flexibility mechanisms—including Joint Implementation, Emissions Trading, and the Clean Development Mechanism—frame global efforts to reduce greenhouse gas emissions in economic terms, placing cost-effectiveness at the centre of GHG mitigation decision-making processes. During Protocol negotiations, the US fought for the inclusion of provisions allowing international trading in carbon emissions allocations as a way to improve the cost-effectiveness of the climate regime.[79] The draft language of the US negotiating text stated that: '[a] Party may authorize any domestic entity...to participate in actions leading to transfer and receipt...of tonnes of carbon equivalent emissions allowed'.[80]

Reactions to the US proposals were mixed, with critics questioning both the morality of conferring a transferrable right to pollute and the practicality of devising such a complex mechanism within the negotiating timeframe for the Protocol. While concerns about the morality and practicality of the proposal were settled relatively early in negotiations, two enduring concerns emerged. First, many participants worried that emissions trading offered a way for developed nations, such as the US, to invest in emissions reduction schemes abroad at the expense of investing in initiatives to reduce domestic emissions.[81] This was seen by some as 'cheating on the basic commitment'.[82] The EU was particularly concerned that the proposal would allow the US to shirk domestic emissions reduction responsibilities.[83]

Leading up to the June 1997 G8 Denver Summit—the G8 summit immediately preceding the COP meeting in Kyoto—the Council met to outline its response to the US proposal. The Council concluded that 'mechanisms such as emissions trading are supplementary to domestic action and common coordinated policies and measures, and that the inclusion of any trading system in the Protocol and the level of the targets to be achieved are interdependent. It therefore calls upon all industrialized countries to indicate the targets they envisage for 2005 and 2010'.[84] In framing its position as such, the EU provided room for incorporating economic flexibility mechanisms into the Protocol while stipulating that such mechanisms should be supplemental to—rather than supplanting—domestic emissions reduction measures. Questions of supplementarity continue to plague international negotiations; current rules are vague at best, and parties continue to disagree about the appropriate balance between domestic emission reductions and external emission reductions achieved through ET, JI,

---

[79] Michael Grubb and Duncan Brack, *The Kyoto Protocol: A Guide and Assessment* (Brookings Institution Press, Washington DC, 1999) 89–114.   [80] Ibid 89.
[81] C Warbrick, D McGoldrick, and P Davies, 'Global Warming and the Kyoto Protocol' (1998) 47 *ICLQ* 446, 458.   [82] Ibid.
[83] A Dessler and E Parson, *The Science and Politics of Global Climate Change: A Guide to the Debate* (CUP, New York, 2006) 15.   [84] Ibid 94.

and CDM. The US, for example, advocates unlimited emissions trading while the EU seeks defined limits on the extent to which the flexibility mechanisms can be used to meet domestic obligations.

Beyond questions of supplementarity, what is especially relevant here is how the US succeeded in embedding economic efficiency at the heart of the Protocol and how the EU first hesitated then went on to champion the efficacy of emissions trading. Since the early days of international negotiations, the US has advocated a decentralized, market-based approach to reducing GHG emissions. This approach, commonly termed 'ecological modernization' emphasizes flexibility and cost-effectiveness and shies away from traditional regulatory approaches to limiting GHG emissions.[85] The US continues to advocate market-based approaches as the foundation for climate change litigation. The most recent climate change proposals to emerge from the US Senate, as well as the House Bill, ACES, rely heavily on market-based mechanisms and recent efforts to use the CAA to limit GHG emissions using traditional command and control style regulation have been critiqued widely, including criticism from the Obama Administration's Office of Management and Budget (OMB).

As recently as April 2009, the EPA's proposed endangerment finding which, when ultimately released, gave the Agency power to regulate GHG emissions as a pollutant under the CAA, was criticized by the OMB. In a leaked, confidential document, the OMB report criticized the EPA's proposed finding as not giving sufficient weight to the potential costs associated with reducing GHG emissions, revealing the continuing emphasis on cost-benefit-analysis in environmental decision-making.[86] The OMB report included a mixture of comments from various governmental agencies which the OMB amalgamated and then sent to the EPA in April 2009. The report emphasized the 'serious economic consequences' of regulating GHG emissions and questioned the scientific foundations of the EPA's proposed finding.[87] Although an anonymous OMB staff member qualified the statements in the document by saying that '[i]t's a conglomeration of counsel we've received from various agencies...and it's not indicative of an OMB or administration-wide position', the anonymous staff member also confirmed that it was, in fact, prepared by Obama administration staff.[88] The Director of the OMB further emphasized that it had concluded review of the EPA's preliminary

---

[85] Ecological modernization emphasizes a '[d]ecentralized liberal market order that aims to provide flexible and cost-optimal solutions to climate problems'. K Bäckstrand and E Lövbrand, 'Climate Governance Beyond 2012: Competing Discourses of Green Governmentality, Ecological Modernization and Civic Environmentalism' in M Pettenger (ed), *The Social Construction of Climate Change* (Ashgate Publishing, Hampshire, 2007) 124.

[86] See, eg, Rena Steinzor, 'Cass Sunstein Hits the Senate and Climate Change Hits the Media Fan' (Center for Progressive Reform Blog, 13 May 2009) <http://www.progressivereform.org/CPRBlog.cfm?idBlog=38155B38-1E0B-E803-CA91A6BBC1F011CD> accessed 2 January 2010.

[87] Ian Talley, 'White House: CO2 rules to seriously impact economy' (Dow Jones Newswires Release, 12 May 2009) <http://www.speroforum.com/site/article.asp?idCategory=33&idsub=134&id=19275&t=White+House%3A+CO2+rules+to+seriously+impact+economy> accessed 2 January 2010.

[88] Ibid.

finding, thus allowing the EPA to move forward with the proposed finding and noted that:

> OMB simply collated and collected disparate comments from various agencies during the inter-agency review process of the proposed finding. These collected comments were not necessarily internally consistent, since they came from multiple sources, and they do not necessarily represent the views of either OMB or the Administration.[89]

Despite these qualifications, the report highlights how, even with a new climate-friendly presidential administration, economic interests continue to dominate US climate change policymaking processes.

The primacy of economic interest on the US climate agenda is nothing new. During the Clinton Administration, the Byrd–Hagel Resolution early on laid out the US's opposition to the Kyoto Protocol on economic grounds and President George W. Bush was widely faulted in the international community for his 'unashamed declaration that agreeing to implement the US Kyoto emission reduction targets did not suit the economic interests of the US'.[90] And, while the Obama Administration has been much more sympathetic to climate change in both rhetoric and action, economic well-being continues to be the central pillar in climate decision-making, as evidenced by both the leaked OMB report and the President's own words. When encapsulating his position on climate change, President Obama consistently frames actions to address climate change as essential to economic well-being. For example, during his campaign for the US Presidency in 2008, then Senator Obama's official policy statement on climate change did not actually use the words climate change, but instead framed the issue in terms of the economic, security, and environmental threats posed by the US's 'addiction to foreign oil':

> The energy challenges our country faces are severe and have gone unaddressed for far too long. Our addiction to foreign oil doesn't just undermine our national security and wreak havoc on our environment—it cripples our economy and strains the budgets of working families all across America.[91]

Similarly, in a speech given in March 2009, President Obama continued his technique of always presenting climate change as a subset of a larger question of energy security, thereby linking climate change directly with economic prosperity and indirectly with national security:

> So we have a choice to make. We can remain one of the world's leading importers of foreign oil, or we can make the investments that would allow us to become the world's

---

[89] Jesse Lee, 'OMB Director Orszag Corrects the Record on the OMB & EPA' (The White House Blog, 12 May 2009) <http://www.whitehouse.gov/blog/OMB-Director-Orszag-Corrects-the-Record-on-the-OMB-and-EPA/> accessed 2 January 2010.
[90] Christian Reus-Smit (ed), *The Politics of International Law* (CUP, New York, 2004) 86.
[91] Joe Biden and Barack Obama, 'The Obama–Biden Plan' (Statement of their new energy and environment plan prior taking office on 20 January 2009) <http://change.gov/agenda/energy_and_environment_agenda/> accessed 2 January 2010.

leading exporter of renewable energy. We can let climate change continue to go unchecked, or we can help stop it. We can let the jobs of tomorrow be created abroad, or we can create those jobs right here in America and lay the foundation for lasting prosperity.[92]

The critical difference between Presidents Obama's and President George W. Bush's approach is the end goal in linking economic metrics to climate change. While Bush used the power of economics to convince the American public of the dangers inherent in joining international efforts to address climate change, President Obama wields the sword of economics as his most powerful tool in convincing the public of the need for domestic climate change legislation and international re-engagement. That is, President Bush represented legally enforceable emissions reduction obligations—at the domestic or international level—as a threat to US economic well-being and as a direct route to placing the US at an economic disadvantage rapidly to the developing economies of China, India, and Brazil. President Obama, however, has shifted the focus of the debate. Instead of accentuating the economic consequences of measures to reduce GHG emissions, he stresses that continuing economic and political stability requires the US to take on the related tasks of improving energy security and reducing climate-related threats. The Obama Administration spins the economic–energy–climate relationship to reveal the mutual dependency of energy security and environmental protection, stating that new domestic legislation is needed 'to protect our nation from the serious economic and strategic risks associated with our reliance on foreign oil and the destabilizing effects of a changing climate'.[93] He further uses economics to his advantage by suggesting that the US is losing ground to its European and Chinese counterparts in cornering the market in new clean energy technologies and by linking the US's economic woes to our continuing dependence on foreign oil.[94]

Presidents Bush and Obama disagree over the economic necessity of climate change legislation. They agree, however, that economics reigns supreme in US environmental decision-making. Regardless of presidential or congressional politics, the US climate debate is driven first and foremost by near-term economic concerns.

The EU similarly emphasizes economics as a central feature of climate policy-making. The nuances of how economic factors influence transatlantic decision-making vary, however. Historically, the US has focused on short-term costs and benefits and the economic value of flexibility while the EU has concentrated on balancing the costs of early action with the long-term costs associated with climate change.[95] Emphasis on long-term social and economic consequences translated to the EU setting great store in mandatory emissions reductions while questioning the

---

[92] President Barack Obama, 'Energy & Environment' (19 March 2009) <http://www.whitehouse.gov/issues/energy_and_environment/> accessed 2 January 2010.
[93] The White House—President Barack Obama, 'Energy & Environment: Progress' <http://www.whitehouse.gov/issues/energy_and_environment/> accessed 2 January 2010.
[94] Ibid.
[95] Bernd Hansjurgens, 'Introduction' in Bernd Hansjurgens (ed), *Emissions Trading for Climate Policy: US and European Perspectives* (CUP, New York, 2005).

use of flexible market-based mechanisms for reducing GHG emissions. The last ten years, however, has witnessed increasing convergence between US and EU policy choices.

The EU, paradoxically once one of the most vocal sceptics of using emissions trading and flexibility mechanisms preceding Protocol negotiations, went on to launch the world's largest GHG emissions trading regime.[96] Moreover, as of May 2009, EU member states were responsible for the vast majority of ongoing CDM projects (58%) with only Switzerland (21%) and Japan (11%) nearing the EU's high level of investment in CDM projects.[97] While parties have been slow to invest in Joint Implementation projects due to delays in rule-making, EU member states were among the first countries to host JI projects and EU member states dominated the list of project investors.

EU climate policy fully embeds the flexibility mechanisms in regional efforts to reduce GHG emissions. Since 2005, the EU ETS has been the centrepiece of Europe's climate change strategy and the most looked to example for other nations developing emissions trading schemes. The EU Linking Directive further embeds Kyoto's flexibility mechanisms within the EU scheme by allowing operators in the ETS to use credits generated through CDM and JI projects to fulfil regional obligations.[98] The importance of the flexibility mechanisms has become increasingly apparent as the EU begins assessing member states' progress in meeting emissions reduction obligations. With only a handful of EU-15 member states ensured of meeting their emissions reduction obligations under the EU bubble scheme based primarily on domestic policies and measures, the EEA has determined that only if member states increase their use of the Kyoto Protocol flexibility mechanisms will the EU be able to meet its quantified emissions reduction obligations.

The EU's dramatic transition from critic to advocate ensures that the flexibility mechanisms will be a continuing part of any long-term international climate change agreement. In this way, the EU, even more so than the US, has framed the international climate change debate around notions of economic efficiency and regulatory flexibility.

Beyond the flexibility mechanisms, the EU integrates questions of economic competitiveness into its climate policies in various ways. In contrast to the US, which has steadily viewed climate change policy as an economic 'cost', the EU has approached climate change policy not only as a way to avoid long-term costs but also as a way to maximize short-term investments. The EU and the US entered negotiations for the Kyoto Protocol with very different perspectives on the costs and benefits associated with overhauling domestic energy sectors; these

---

[96] K Bäckstrand and E Lövbrand (n 85) 130.
[97] UNFCCC, CDM Statistics, 'Registered projects by AI and NAI investor parties' <http://cdm.unfccc.int/Statistics/Registration/RegisteredProjAnnex1PartiesPieChart.html> accessed 13 May 2009.
[98] Council Directive (EC) 2004/101 of 27 October 2004 amending Directive 2003/87/EC establishing a scheme for greenhouse gas emission allowance trading within the Community, in respect of the Kyoto Protocol's project mechanisms [2004] OJ L338/18.

perspectives have endured. The EU approached international climate negotiations anticipating the economic benefits associated with being a 'first mover' in efforts to overhaul domestic energy sectors by improving energy efficiency and investing in clean energy technologies.[99] The US, in contrast, approached negotiations with little interest or groundwork laid for taking advantage of new market opportunities. The fundamental entry points of the two entities directed economic cost-benefit calculations down very different routes. These initial assessments shaped public perceptions of the costs of climate change, with enduring impact on the willingness of the public to support climate change initiatives.

Framing ambitious climate policymaking as a way for Europe to gain economic advantage has allowed the EU to sustain its climate change law and policymaking efforts, even when those efforts have been unilateral and out-of-step with countries such as the US and Australia. The EU's strategy, however, has not escaped economic critiques. The EU ETS has been heavily criticized as inefficient and, as of yet, dysfunctional, and there remain considerable concerns about its long-term economic viability. The continuing instability of the price for a tonne of carbon creates underlying market instability, especially in times of economic recession. As the EU moves into the midst of Phase II and solidifies auctioning and sector expansion plans for Phase III of the ETS, it is seeking to cure underlying instabilities. The EU ETS offers the grand experiment in marrying market solutions to GHG mitigation. It is still too early to determine the outcome of the experiment.

Many member states and industry lobbyists have criticized the EU ETS and supplemental EU climate policies as jeopardizing industrial competitiveness and potentially driving energy-intensive and export-oriented strategies out of the EU, creating the conditions for a classic race to the bottom. A representative for the cement industry—a GHG intensive sector—has declared that '[a]t current CO2 prices of 25 per tonne, approximately 80 percent of clinker production will be offshored if no free allowances are allocated'.[100] The phenomena of carbon-intensive industries fleeing countries with GHG emission regulations has been referred to as 'carbon leakage' and has become the cause *du jour* of lobbyists opposing stringent EU climate policies.[101] Concern over carbon leakage is rife in the EU, especially as the EU ETS transitions to an allowance auctioning system. Member states—led by Germany, a self-proclaimed climate leader—fought to include widespread exemptions for industries deemed to be at risk as a result of the move to allowance auctioning.[102] The aluminium, steel, iron, and cement industries

---

[99] Reus-Smit (n 90) 88–89.
[100] Thomas Legge, 'EU Climate Policy & Economic Competitiveness' (Presentation at the Peterson Institute for International Economics, Washington DC, 4 March 2009) <http://www.iie.com/publications/papers/legge0309.pdf> accessed 2 January 2010.
[101] EurActiv, '"Carbon leakage": A challenge for EU industry' (27 January 2009) <http://www.euractiv.com/en/climate-change/carbon-leakage-challenge-eu-industry/article-176591> accessed 2 January 2010; see also Jonathan Wiener, 'Protecting the Global Environment' in John Graham and Jonathan Wiener (eds), *Risk vs Risk: Tradeoffs in Protecting Health and the Environment* (Harvard University Press, Cambridge MA, 1995) 193, 214.   [102] Ibid.

are among the sectors expected to receive preferential treatment under Phase III of the EU ETS.

Despite these exemptions, fears of carbon leakage have as of yet proved largely unfounded and there remains considerable disagreement over the extent of the threat. The European think tank, Bruegel, released a report in 2007, arguing that Europe stood to be at a serious economic disadvantage to China and the US in the switch to a low-carbon economy due to Europe's heavy reliance on carbon intensive industries.[103] In contrast, both the International Energy Agency and the OECD have downplayed the risks of carbon leakage in Europe. The International Energy Agency released a report arguing that the negative economic effects of Europe's climate policy would be limited to a handful of manufacturing sectors and dismissed reports that the EU ETS has forced industry to move outside of the EU. The OECD, while confirming the economic threats posed by GHG limitations, sided with the International Energy Agency in downplaying the risks of carbon leakage as limited and manageable.[104] Similarly, a recent report examining German and UK industrial sectors indicates that only 1–2% of GDP is associated with activities that face significant cost increases from carbon pricing.[105]

Carbon leakage has proved a contentious point in EU climate policy and has raised widespread concerns about the economic consequences of increasingly restrictive regulatory reform. The proposal to grant sweeping exemptions to heavy industry under the EU ETS has been closely watched by US policymakers and is reflected in US legislative proposals for a cap-and-trade system. The EU's decision to utilize an emissions trading system, to link that system with other market-based mechanisms, and to modify and update the emissions trading system to accommodate geographic needs profoundly influences the shape of evolving US policy proposals.

Heavy reliance on market-based mechanisms and growing concerns about the economic implications of stringent emissions limits are bringing US and EU climate policies closer together. Convergence around the question of economic well-being is to be expected. What is surprising is how the EU has used economic justifications both to support aggressive policy choices and to limit competitive disadvantages. The EU has used economics to its advantage by emphasizing the economic gains to be achieved as a 'first mover'. It has simultaneously taken the flexibility mechanism gauntlet thrown down by the Americans and embedded market-based mechanisms as the cornerstone of its climate policy. It has co-opted and breathed life into ideas originally proposed by the US and, in so doing, influenced American climate policy.

---

[103] Juan Delgado, 'Why Europe is not Carbon Competitive' [November 2007] *Bruegel Policy Brief* 5.
[104] '"Carbon leakage": A challenge for EU industry' (n 101).
[105] K Neuhoff, J Cust, and K Keats Martinez, 'Implications of Intermittency and Transmission Constraints for Renewables Deployment' in M Grubb, T Jamasb, and M Pollitt (eds), *Delivering a Low-Carbon Electricity System: Technology, Economics and Policy* (CUP, New York, 2008).

Tensions between climate policy and economic pressures will indefinitely influence climate policymaking in the US and the EU. The critical questions will be how the US and the EU calculate the economic costs of climate change, especially in times of economic recession, and what weight economic concerns receive in the policymaking process.

Both President Obama and European leaders, including Stavros Dimas—the European Commissioner for the Environment—have depicted the fight against climate change as an economic golden opportunity. President Obama, for example, declared that: 'The choice we face is not between saving our environment and saving our economy—it's a choice between prosperity and decline.... The nation that leads the world in creating new sources of clean energy will be the nation that leads the 21st century global economy'.[106] Recasting energy policy as a route to creating new jobs and spurring new industry encourages positive perceptions of efforts to address climate change. This approach is inherently limited, however, as the social and economic 'costs' of limiting GHG are unevenly distributed and the losers—often industries with powerful lobbies—are inevitably louder than the winners. In the long-term, the question that will either drive EU and US climate policies closer together or move them even further apart is how policymakers choose to calculate and compare the near-term costs of regulating GHG with the long-term costs of responding to the effects of climate change.

The other point of departure for US and EU climate policy will be the comparative weight given to economic factors in relation to considerations of social welfare and inter- and intra-generational justice. As will be discussed in Chapter 8, despite convergences along economic lines, the relative balance between economic factors and non-economic factors is one of the key points of divergence in US and EU climate policymaking.

Security: The New Driver

Climate change politics in Europe and the US increasingly converge around questions of national security. Environmental issues have often been viewed as separate and distinct from questions of national security. Increasingly, however, policymakers are weighing the security dimensions of environmental degradation, including climate change.

While not all environmental harms create security issues, there is growing concern over the direct and indirect threats environmental degradation poses to national security. Research exploring the possibility for environment-based problems to lead to conflict suggests that environment-security risks are significant

---

[106] The White House, Office of the Press Secretary, 'Fact Sheet: President Obama Highlights Vision for Clean Energy Economy' (Press Release, Washington DC, 22 April 2009) <http://www.whitehouse.gov/the_press_office/Clean-Energy-Economy-Fact-Sheet/> accessed 30 April 2009.

and increasing, especially in relation to tensions over scarce water resources.[107] This research, led by Thomas Homer-Dixon and Peter Gleick, indicates that environmental stresses compound underlying political instabilities in vulnerable areas, leading to conflict and insecurity that have ripple effects worldwide.

Climate change highlights causal relationships between environment and security, prompting policymakers to reassess notions of threats and security. Climate change poses risks to human security in a variety of ways. In particular, it functions as a threat multiplier, exacerbating existing pressures. Climate change places new and additional stresses on already vulnerable societies and degraded environments. It poses threats to resource availability, energy security, and border security, and increases the risks associated with severe weather events, air pollution, disease patterns, and land availability. In these various ways climate change threatens to induce 'humanitarian crises, mass migration, armed conflict over scarce natural resources and economic malaise that breeds despair, violence, and terrorism'.[108] An Institute for Public Policy Research report published in 2008 confirmed the security threats climate changes poses and noted that the question of climate security 'is likely to have a major shaping influence on international affairs in the decades ahead'.[109] The IPCC has similarly identified the threats climate change poses to global security. In 2008, in an address to the European Parliament, Rajendra Pachauri, Chair of the IPCC, warned of the dire consequences climate change poses to human security, warning the international community that there is 'no part of the globe that can be immune to the security threat'.[110]

Questions of environmental security are not new, but climate change has raised the profile of environmental security questions and added a new dimension to on-going discussions about whether traditional notions of security need to be expanded to include more than just military threats to the state.

In the specific context of climate change, security and vulnerability experts emphasize the many ways that the direct and indirect effects of climate change stress social, economic, and political systems with implications for regional and international security. Early concerns over climate-security linkages tended to center on the threats climate change posed to vulnerable states, focusing primarily on the least developed countries and a handful of developing countries perceived

---

[107] Richard A Matthew, 'The Environment as a National Security Issue' (2000) 12 *J Political History* 101, 109.

[108] Nigel Purvis, 'U.S. Global Leadership to Safeguard our Climate, Security, and Economy' (Better World Campaign Report, June 2008) 5 <http://www.policyarchive.org/handle/10207/10917> accessed 3 April 2010.

[109] IPPR Commission on National Security in the 21st Century, *Shared Destinies: Security in a Globalised World* (Interim Report, 27 November 2008) 19.

[110] European Parliament, Environment, 'UN Climate Change Panel Chair Pachauri: "We Swim or Sink Together"' (27 March 2008) <http://www.europarl.europa.eu/sides/getDoc.do?language=EN&type=IM-PRESS&reference=20080319STO24704&secondRef=0> accessed 2 January 2010.

to be particularly geographically vulnerable and lacking in adaptive capacity.[111] In these vulnerable regions, the fear is that climate change will overburden existing infrastructure, pushing societies into social, political, and economic decline. These concerns remain.

However, the climate security debate has expanded beyond concern for the most vulnerable states to reach the political agendas of even the most politically and economically stable regions worldwide. As one commentator noted in 2008, '[a]lmost all heads of government now have a basic understanding that without climate security they will be unable to meet their economic or development goals'.[112] The US and the EU converge in integrating questions of climate security within national security debates.

In the US, for example, the Centre for Strategic and International Studies (CSIS)—a bi-partisan, non-profit Washington, DC think tank—has sponsored a number of workshops and research projects examining the links between global energy security, climate change, and national security. In so doing, the organization has warned of the destabilizing effects climate change could have in both the developed and developing worlds. Similarly, the CNA Corporation—a non-profit research organization known for its conservative policy research and proposals—recently established a panel of high ranking retired US military officials to investigate the linkages between climate change and national security.[113] The panel, including former Generals, Lieutenant Generals, Admirals, and Vice-Admirals, prepared a report finding that '[g]lobal climate change presents a serious national security threat which could impact Americans at home, impact US military operations, and heighten global tensions'.[114] After confirming the links between climate change and national security, the panel recommends specific actions that the US military must take in order to respond to existing and anticipated security threats related to climate change.[115] Key recommendations include:

- fully integrating the national security consequences of climate change into national security and national defense strategies by assessing current military capabilities to respond to national disasters, disease outbreak, and anticipated peacekeeping missions and by more fully integrating the scientific community,

---

[111] See, eg, Timothy McKeown, 'Climate Change, Population Movements, and Conflict' in Carolyn Pumphrey (ed), *Global Climate Change: National Security Implications* (Strategic Studies Institute, Carlisle PA, 2008) 99.

[112] Jennifer Morgan, 'Creating a Secure Climate: The G8 Leadership Challenge' (News Release, 29 April 2008) <http://www.e3g.org/programmes/climate-articles/creating-a-secure-climate-the-g8-leadership-challenge/> accessed 3 April 2010.

[113] CNA Corporation, 'National Security and the Threat of Climate Change: Military Advisory Board' (Military Advisory Board Homepage) (2007) <http://securityandclimate.cna.org/mab/> accessed 2 January 2010.

[114] See CNA Corporation, *National Security and the Threat of Climate Change* (Report) (2007) 46 <http://securityandclimate.cna.org/report/SecurityandClimate_Final.pdf> accessed 2 January 2010.   [115] Ibid.

eg, the National Ocean and Atmospheric Administration and the National Air and Space National, in military planning;
- playing a stronger national and international role in helping stabilize climate change at levels that will avoid dangerous disruption of global security and stability;
- committing to global partnerships to assist with adaptive capacity building in the developing world;
- improving the efficiency of military business processes and accelerating the adoption of clean, more energy efficient technologies in order to simultaneously reduce emissions and minimize the military's dependency on fossil fuels;
- assessing the impact of rising sea levels, extreme weather events, and other likely climate change impacts over the next 30 to 40 years on US military installments worldwide to allow the US Department of Defense (DoD) time to either fortify or relocate affected sites.[116]

The CNA conclusions are definite and resounding, stating that '[p]rojected climate change poses a serious threat to America's national security'.[117] It remains unclear, however, whether the Panel's findings or recommendations will be heeded by Congress or the US Department of Defense (DoD) in future decision-making. Neither CSIS nor the CNA Corporation are governmental entities with decision-making authority. While both institutions are regarded as influential in US military and political circles, neither Congress nor the DoD has yet explicitly detailed whether or how climate change will modify national security and national defense strategies.

The DoD involvement in efforts to address climate change remains oblique. The DoD does not directly sponsor climate change research but it participates in the US Climate Change Science Program (CCSP). As part of the CCSP, the DoD funds 50% of the National Polar-Orbiting Operational Environmental Satellite System, which gathers key data for assessing global environmental conditions, is a key participant in ocean observations systems and global ocean prediction systems, runs one of the US's lead laboratories for polar and sub-polar research, and contributes to various other operations that produce key oceanographic and meteorological information.[118] Beyond this the DoD has expressed support for climate initiatives and, in April 2009, Under Secretary of Defense, Michele Flournoy, characterized climate change as a key factor affecting national security challenges.[119] It is unclear, however, whether the DoD is taking direct measures

---

[116] Ibid 46–48.   [117] Ibid 6.
[118] Climate Change Science Program and The Subcommittee on Global Change Research, *Our Changing Planet: The U.S. Climate Change Science Program for Fiscal Year 2009* (Report and supplement to the President's fiscal year 2009 budget, 2009) <http://www.usgcrp.gov/usgcrp/Library/ocp2009/ocp2009-dod.htm> accessed 2 January 2010.
[119] John J Kruzel, 'U.S. Needs "Pragmatic, Clear-Eyed" Defense Strategy, Flournoy Says' (American Forces Press Service News Release, April 29, 2009) <http://www.defense.gov/news/newsarticle.aspx?id=54130> accessed 3 April 2010.

to prepare personnel, modify military practices, or fortify military infrastructure in response to climate threats.

Within the US Congress, policymaking efforts are directed to crafting framework legislation, with climate security issues receiving little direct consideration but functioning as a justification for and a component of legislative proposals. In 2007, for example, one of the key legislative proposals, the Lieberman–Warner Climate Security Act, was introduced to Congress as fundamental to avoiding catastrophic climate change and ensuring domestic energy security.[120] In presenting the bill, Senator Warner stated that '[i]n my 28 years in the Senate, I have focused above all on issues of national security, and I see the problem of global climate change as fitting squarely within that focus'.[121] Similarly, the 2009 American Clean Energy and Security Act—the first bill to clear the House Committee on Energy and Commerce in the 111th Congressional Session following President Obama's occupation of the White House—characterizes climate change as a national security threat.[122] Directly drawing on language found in the CNA *National Security and the Threat of Climate Change Report*, the bill describes global climate change as a 'potentially significant national and global security threat multiplier' that 'is likely to exacerbate competition and conflict over agricultural, vegetative, marine, and water resources and to result in increased displacement of people, poverty, and hunger within developing countries'.[123] In advocating the creation of an international climate change adaptation program, the bill describes the consequences of climate change as 'likely to pose long-term challenges to the national security, foreign policy, and economic interests of the US', and continues by declaring that:

[i]t is in the national security, foreign policy, and economic interests of the United States to recognize, plan for, and mitigate the international strategic, social, political, cultural, environmental, health, and economic effects of climate change and to assist developing countries to increase their resilience to those effects.[124]

The bill also mandates that the US Agency for International Development—the Agency tasked with administering the adaptation fund—submit annual reports that describe 'the ramifications of any potentially destabilizing impacts climate change may have on the national security, foreign policy, and economic interests of the US', by creating environmental migrants, resource-based conflicts, economic and cultural destabilization, resource constraints, natural disasters,

---

[120] S 3036, 110th Cong (2008).
[121] Joe Lieberman, 'Lieberman and Warner Introduce Bipartisan Climate Legislation' (News Release, 18 October 2007) <http://lieberman.senate.gov/index.cfm/news-events/news/2007/10/lieberman-and-warner-introduce-bipartisan-climate-legislation> accessed 3 April 2010.
[122] 'Global warming poses a significant threat to the national security, economy, public health and welfare, and environment of the United States, as well as of other nations'. HR 2454, 111th Cong, tit VII pt A (2009).
[123] Ibid sub-pt B (Public Health and Climate Change, Part II International Climate Change Adaptation Program). [124] Ibid.

climate-related disease, and the intensification of urban migration.[125] The bill's expansive definition of the threats climate change poses to national security reflects concerns expressed by the IPCC, the UN and domestic groups such as CSIS, and the CNA Corporation. What is notable about the bill's inclusion of security-based considerations is that they appear only in the adaptation portion of the bill and are omitted from the mitigation components of the bill. This placement suggests that while Congress recognizes the security dimensions of climate change, they remain secondary to economic concerns in the political debate over the costs and benefits of mitigation.

Regardless of strategic placement, the 2009 House Bill embeds climate security within the political debate and offers environmental sceptics a more conservative justification for supporting climate legislation.

President Obama regularly draws upon the combined power of economic and security-based justifications for climate actions. Rarely will he discuss climate change without simultaneously discussing energy security, economic opportunities, and national security. The melding of climate change first to energy security and second to economics and national security makes domestic and international climate change initiatives more palatable to an often sceptical American citizenry.

The EU, collectively, and many of its individual member states, have similarly elevated climate change to a security issue. As early as 2003, the European Security Strategy identified the security implications of climate change and the 2006 Commission Communication 'Europe in the World' similarly recognized links between climate change and global security.[126] Following on from this, in 2008, the High Commissioner for the Common Foreign and Security Policy—the main coordinator of the EU's Common Foreign and Security Policy—and the European Commission issued a Joint Report on *Climate Change and International Security* to the European Council.[127] The Report repeats the increasingly common characterization of climate change as a threat multiplier and defines the primary threat as the overburdening of already vulnerable States and regions. It then stresses that climate change risks are not limited to humanitarian crises, but also 'include political and security risks that directly affect European interests'.[128] The Report concludes that 'it is in Europe's self interest to address the security implications of climate change with a series of measures: at the level of the EU, in bilateral relations and at the multilateral level, in mutually supportive ways'.[129]

---

[125] Ibid.
[126] *Report on the Implementation of the European Security Strategy: Providing Security in a Changing World* (11 December 2008) S407/08.
[127] European Commission (EC), 'Climate Change and International Security' (Paper from the High Representative and the European Commission to the European Council, 14 March 2008) S113/08. [128] Ibid 2.
[129] Ibid 3.

After drawing this conclusion, the Report defines the specific threats climate change poses to include: conflict over resources; economic damage and risk to coastal cities and critical infrastructure; loss of territory and border disputes; environmentally-induced migration; situations of fragility and radicalization; tension over energy supply; pressure on international governance. It then offers regional examples of climate threats before outlining a series of recommended responses. In contrast to the American Clean Energy and Security Act, the Report suggests Europe's key response begins with aggressive climate change mitigation efforts, coupled with efforts to engage the global community in cooperative efforts to address climate change. Within this regional and global mitigation framework, the Report then recommends a series of actions, beginning with the reassessment of climate security under the European Security Strategy and continuing with efforts to assess the EU's adaptive capacity and improve regional preparedness for early responses to disasters and conflicts.[130] Underlying the proposed strategy is the supposition that the EU will maintain a strong leadership role in international efforts to improve climate security.

Individual member states such as the UK and Germany have played equally pivotal roles in the global debate over climate security. It was the UK, for example, that first raised climate change as an issue for debate in the UN Security Council in April 2007. The Security Council's first ever debate on the impact of climate change on peace and security was prompted by a letter from the Permanent Representative of the UK to the President of the Council (document S/2007/186) and the debate was opened by the President of the Security Council, UK Foreign Secretary Margaret Beckett.[131] The ensuing debate revealed deep divides among member state representatives over the appropriateness of the Security Council as the proper forum for considering questions of climate change. While China and India—speaking on behalf of the G-77—insisted that the climate change debate belonged in the Economic and Social Council and the General Assembly, Papua New Guinea argued that impact of climate change on small islands was 'no less threatening than the dangers guns and bombs posed to large nations', while UN Secretary General Ban-Ki Moon reminded participants that '[w]ar had too often been the means to secure possession of scarce resources' and said that the Security Council has a role to play in addressing an issue that has 'implications for peace and security'.[132] In her capacity as a representative of the UK, Secretary Beckett confirmed that the UK considers climate change to be a security issue and argued that the Security Council has a role to play in 'building a shared understanding of what the effects of climate change would mean to international peace and security'. A record number of delegations and non-members participated in the debate, indicating

---
[130] Ibid.
[131] UNSC 'Security Council Holds First Ever Debate on the Impact of Climate Change on Peace, Security, Hearing over 50 Speakers' (17 April 2007) Press Release SC/9000.
[132] Ibid.

that despite mixed opinions over forum-choice, questions of climate security engender widespread interest. The Security Council debate did not produce any measurable or enforceable outcomes but it raised the profile of the climate security question and embedded it within global security debates.

At the national level, the UK has embedded questions of climate security within its National Security Strategy. The National Security Strategy defines climate change as 'potentially the greatest challenge to global stability and security, and therefore to national security'. References to the interdependency of climate change with other security risks are interspersed throughout the strategy; however, the Strategy also dedicates independent sections to analyzing climate-related security risks and suggesting methods for tackling climate change.[133] Proposed responses are far-reaching, including: erecting physical defenses, modifying agricultural and development strategies, and restructuring energy policy. The Strategy emphasizes the UK's commitment to playing a leadership role in international forums and underlines its commitment to binding emissions reduction targets. It concludes by outlining the steps it is taking at the national level to address climate threats, to include:

undertaking a systematic detailed analysis, region by region, of how the impact of climate change is likely to affect the United Kingdom; analysing our water and food security issues to ensure sustainable and secure supplies; and increasing our overall investment in climate change research to at least £100 million over the next five years to investigate the dynamics of long-term climate change, the links to international poverty and the impact of climate change on conflict and other factors.[134]

The UK is not alone in treating climate change as a security issue. German Chancellor Angela Merkel played a key role in raising the profile of climate-security linkages during her term as European Council President in 2007.[135] In the same year, the well-respected German Advisory Council on Global Change (WBGU) issued a report entitled 'Climate Change as a Security Risk'.[136] The WBGU report provides a detailed analysis of the interface between environmental change and security before undertaking an exhaustive review of how, where and why climate change poses security risks. The Advisory Council report prompted domestic review of climate security issues and was followed by a governmental report 'Climate Change and Security Challenges for German Development Cooperation', examining the potential for climate to increase incidences of

---

[133] Prime Minister, 'The National Security Strategy of the United Kingdom: Security in an Interdependent World' (Cm 7291, 2008) 20.   [134] Ibid 51.
[135] Woodrow Wilson International Center for Scholars, 'Climate Change Solidly on Germany's Security Agenda, Says ECSP Director' (News Release on environmental change and security program,13 January 2009) <http://www.wilsoncenter.org/index.cfm?topic_id=1413&fuseaction=topics.item&news_id=500740> accessed 2 January 2010.
[136] R Schubert, HJ Schellnhuber, and N Buchmann, *Climate Change as a Security Risk* (Earthscan Publications, London, 2007).

conflict and political insecurity.[137] The report, produced by the Federal Ministry for Economic Cooperation and Development, examines ways that climate change exacerbates conflicts and security risks on a sectoral and regional basis and recommends multiple courses of action the German Development Cooperation should follow to better prepare to respond to climate-related security challenges. The report concludes by noting that the international climate regime currently fails to address linkages between climate and security.

Based on the potential for climate change to have a continuing destabilizing effect on the global community, the governmental report and the WBGU report suggest that as a starting point, resolute action is needed to address the root causes of climate change. The reports further recommend that the German government—in consultation with its European counterparts—seek to offset and prepare for climate-related social destabilization by pursuing a multi-part strategy that includes:

- shaping global change by encouraging the EU to play a greater part in shaping global change;
- reforming the United Nations;
- ambitiously pursuing global climate change policy;
- transforming EU energy sectors;
- identifying threats and vulnerabilities;
- reassessing development priorities;
- supporting adaptation efforts in developing states;
- helping stabilize vulnerable states;
- improving and expanding global information and warning systems; and
- managing mitigation and adaptation efforts through improved international coordination and more effective use of international law.[138]

Together, the reports create a road map for the German Federal Government to use in advocating joined-up responses to climate-security linkages.

Concerns over climate-related security risks are increasingly a point of agreement for EU and US policymakers. The juxtaposition of climate change and national security modifies the tenor of the debate in both regions. In Europe, it adds urgency to existing calls for aggressive climate initiatives and spurs increased efforts to develop comprehensive mitigation and adaptation strategies, while in the US it offers a powerful bi-partisan justification for first actions.

Despite transatlantic convergence around questions of climate security, outstanding questions remain over the security dimensions of climate change in

---

[137] German Federal Ministry for Economic Cooperation and Development, *Climate Change and Security: Challenges for German Development Cooperation* (Report) (Deutsche Gesellschaft für Technische Zusammenarbeit (GTZ) GmbH, Eschborn, April 2008).   [138] Ibid 49–50.

terms of what is being secured, what is being secured against, who is responsible for providing security, whose security is of most concern, when security-based measures are justified, and at what cost.[139] These questions could ultimately divide the transatlantic security debate. Both the EU and the US are driven by self-interest in ensuring global stability; however, EU and US foreign policy strategies are likely to differ over the relative weight given to supporting early mitigation efforts, humanitarian aid, military prowess, and domestic security.

Transatlantic views over climate related security may ultimately diverge over questions of resource priority, but there is currently a shared acknowledgement that climate change poses real threats to national security.

In jointly awarding the 2007 Nobel Peace Prize to Al Gore and the Intergovernmental Panel on Climate Change, the Norwegian Nobel Committee stated:

Extensive climate changes may alter and threaten the living conditions of much of mankind. They may induce large-scale migration and lead to greater competition for the earth's resources. Such changes will place particularly heavy burdens on the world's most vulnerable countries. There may be increased danger of violent conflicts and wars, within and between states.

The Committee's warnings resonate in European and American climate change dialogue. It remains to be seen whether security issues will encourage improved cooperation or provide a new point of transatlantic policy departure.

Shared Trouble Spot: Transport Sector

As well as sharing common concerns for on questions of economics and security, EU and US climate policies struggle with a shared point of weakness: transport policy. The transport sector is one of the fastest growing sectors for GHG emissions worldwide and this holds true for trends in the EU and the US.[140]

Efforts to mitigate greenhouse gas emissions from the transport sector present particular challenges relating to lifestyle choices and cultural preferences. Fossil

---

[139] G Dabelko and P Simmons, 'Environment and Security: Core Ideas and US Government Initiatives' (1997) 17 *School Advanced Intl Studies Rev* 127.

[140] See J Sathaye and M Walsh, 'Transportation in Developing Nations: Managing the Institutional and Technological Transition to a Low-Emissions Future' in Irving Mintzer (ed), *Confronting Climate Change: Risks, Implications and Responses* (CUP, Cambridge, 1992); see also US EPA, Office of Transportation and Air Quality, *Greenhouse Gas Emissions from the U.S. Transportation Sector 1990–2003* (Report) (March 2006) <http://www.epa.gov/oms/climate/420r06003.pdf> accessed 2 January 2010. The report confirms that: 'Total U.S. production of greenhouse gases in 2003 was 13 percent greater than in 1990. By comparison, transportation GHGs grew almost 24 percent over the same period. GHG emissions from the transportation sector increased more in absolute terms than any other sector, growing by 357.4 Tg CO2 Eq. from 1990 to 2003 (Figure 2-1). Ibid. The growth rate of transportation GHGs was equal to the residential sector (also 24 percent), slightly above the commercial sector (22 percent), and considerably greater than agriculture (3 percent) and industry (which decreased by 2 percent).' Ibid.

fuels are fundamental to American and European ways of life; nowhere is this more evident than in the transport sector, where evolving notions of personal mobility have become a defining feature of Western lifestyles.[141] The IPCC early on identified transport as posing challenges for GHG mitigation efforts due to the 'unique role that travel and goods movement play in enabling people to meet personal, social, economic, and developmental needs'.[142] Efforts to reduce emissions from the transport sector are further complicated in the US and the EU by existing disconnects between environmental law and policy and planning law and policy. Further, reliance on personal automobiles and notions of freedom of movement have become so embedded in both regions that it is difficult and politically unpopular to initiate any serious debate about modifying existing transport policy. Transport policy is made even more challenging due to the diversity of stakeholders. Unlike laws and policies targeting specific industrial, governmental, or residential actors, transport policy impacts private, commercial, and public users of transport, vehicle manufacturers and parts suppliers, fuel suppliers, planners, transport sector builders, and transport service providers.[143] Finally, governments notoriously struggle to craft effective regulatory regimes for mobile sources of pollution. The combination of lifestyle choices, diversities of stakeholders, and regulatory complexity make transport one of the most intractable sectors in regards to reducing GHG emissions in the EU and the US.

Transport emissions are rising across the EU and the US and, until very recently, transatlantic governmental entities displayed a common reluctance to take the politically unpopular steps necessary to curb transport emissions.[144] In the EU, for example, climate change activists have declared that the transport sector is 'without a doubt Europe's worse climate performer', while in the US, the transport sector 'produces more CO2 emissions than any other nation's entire economy, except China'.[145]

In the EU, transport is the second largest sector of GHG emission and the fastest growing sector for emissions. Based on current growth rates, if domestic transport is not decoupled from economic growth, emissions will increase for the EU-15 by roughly 31% by 2010 and by as much as 50% by 2020, as against 1990

---

[141] John Urry, *Mobilities* (Polity Press, Cambridge, 2007).
[142] Intergovernmental Panel on Climate Change, 'Technologies, Policies and Measures for Mitigating Climate Change' (Technical Paper prepared under the auspices of IPCC Working Group II, November 1996) <http://www.gcrio.org/ipcc/techrepI/index.html> accessed 2 January 2010, at 21.   [143] Ibid.
[144] Loren Cass, *The Failures of American and European Climate Policy: International Norms, Domestic Politics, and Unachievable Commitments* (SUNY Press, New York, 2007) 105.
[145] CAN Europe and others, *Input from Environmental NGOs at the Start of the Next Round of the European Climate Change Program (ECCP)* (Report in Brussels, 24 October 2005) <http://www.transportenvironment.org/Publications/prep_hand_out/lid/386> accessed 2 January 2010 at 16; D Greene and A Schafer, *Reducing GHG Emissions From U.S. Transportation* (Report prepared for Pew Center on Global Climate Change, May 2003) <http://www.pewclimate.org/global-warming-in-depth/all_reports/reduce_ghg_from_transportation> accessed 2 January 2010, at iii.

levels.¹⁴⁶ Although transport policy is one of the pillars of the ECCP, the EU continues to heavily fund growth in road transport, with the effect that between 1990–2002 emissions from the transport sector increased in excess of 150,000 tonnes—the largest sectoral increase by far—and with aviation accounting for the second largest absolute emissions increase and the highest percentage increase.¹⁴⁷ The only transport sector to see a decrease in emissions was rail transport, due in large part to electrification.¹⁴⁸

Levels of transport emissions vary across member states, with emissions increasing in excess of 50% for eight member states and only decreasing in two member states. Future projections suggest that EU transport emissions will continue to accelerate at a fast pace absent new and additional measures to curb emissions. Existing measures fall short of achieving measurable reductions, actually enabling GHG emissions from the transport sector to grow upwards of 51% for the EU-27 and anywhere between 11% and 157% for the EU-15.¹⁴⁹

The sharp growth—and projected continuing growth—from the transport sector counteracts emissions reductions from other EU sectors. Overall, while transport emissions increased by 27% between 1993–2003, non-transport sectors achieved an 8% emissions reduction during the same time period. With the transport sector currently accounting for 34% of EU energy use and with that figure steadily climbing, emissions reductions from non-transport sectors will quickly be offset unless the EU is able to take new and additional steps to curb and reverse the growth in transport emissions.¹⁵⁰

As discussed in Chapter 5, transport—which excludes aviation—is one of the identified pillars of the EU ECCP. Under the umbrella of the ECCP I and II the EU has adopted a three-fold strategy for addressing transport-based GHG emissions, which includes: (1) voluntary commitments by industry; (2) a car labeling scheme; and (3) the promotion of automobile fuel efficiency through fiscal measures, in conjunction with the EU energy tax as well as the Biofuels Directive. Under the first arm of this strategy, in 1998, the European Commission entered

---

[146] ECCP, *The Second European Climate Change Programme Working Group ECCP review—Transport Final Report* (2 May 2006) <http://ec.europa.eu/environment/climat/pdf/eccp/review_transport.pdf> accessed 2 January 2010 (ECCP Transport Final Report). The report specifies that 'transport is the second largest sector of GHG emissions accounting for 19% of EU-25 emissions and 21% of EU-15 emissions in 2003'. Ibid. Between 1990 and 2003, EU-15 emissions from domestic transport increased by 24% due to continuous increases in road transport volumes (passenger and freight). Ibid. Road transport is the biggest transport emission source (74% share). If domestic transport GHG emissions continue to increase with economic growth they would increase for the EU-15 by almost 31% by 2010 (compared to 1990) and up to 50% by 2020'. Ibid.

[147] Matthias Duwe, 'Climate Policy in the European Union: A Brief Overview of EU-Wide Policies and Measures' (Climate Action Network Europe Presentation in Brussels, 30 November 2004) <http://www.aef.org.uk/downloads/EU%20and%20Climate%20-%20Duwe.pdf> accessed 2 January 2010. [148] Ibid 2.

[149] Ibid.

[150] 'Input from Environmental NGOs at the Start of the Next Round of the European Climate Change Program (ECCP)' (n 145) 16.

into voluntary agreements with the European (ACEA), Japanese (JAMA), and Korean (KAMA) car manufacturers. The voluntary agreements defined fleet average CO2 emission targets from new cars sold in the EU, to be reached collectively by the members of each association. The agreements, which only extended to CO2, required the manufacturers to achieve car fleet average CO2 emissions of 140 g CO2/km7 by 2008 (ACEA) and 2009 (JAMA and KAMA). Mid-term review of the voluntary agreements revealed that while the manufacturers had made substantial progress, they were likely to fall short of meeting agreed-upon targets.[151]

In 2007, responding to the reports showing that auto manufacturers were likely to miss their voluntary commitments to reduce CO2 emissions, the European Commission proposed new binding legislation.[152] The proposed regulation was greeted with staunch resistance from the automotive industry, prompting a series of discussions between the Commission, member states, and industry over ways forward.[153] Following a month of debate, agreement was reached on a deal based on a French proposal to phase in limits on CO2 emissions between 2012–2015. The proposed regulation was adopted by the European Parliament on 17 December 2008, marking an important turning point in the EU's approach to tackling greenhouse gas emissions from the transport sector.[154] In addition to establishing phased-in, fleet-based carbon dioxide limits, the regulation establishes a penalty regime for offenders and sets a long-term target of 95g CO2/km for the new car fleet by 2020.[155]

---

[151] Commission (EC), 'Implementing the Community Strategy to Reduce CO2 Emissions from Cars: Sixth Annual Communication on the Effectiveness of the Strategy' (Communication) COM (2006) 463 final, 24 August 2006.
[152] Commission (EC), 'Proposal for a Regulation of the European Parliament and of the Council Setting Emission Performance Standards for New Passenger Cars as Part of the Community's Integrated Approach to Reduce CO2 Emissions from Light-Duty Vehicles' COM (2007) 856 final, 19 December 2007.
[153] EurActiv, 'EU Clinches Deal on CO2 Emissions from Cars' (News Release, 3 December 2008) <http://www.euractiv.com/en/transport/eu-clinches-deal-co2-emissions-cars/article-177675> accessed 2 January 2010.
[154] Council Regulation (EC) 443/2009 of 23 April 2009 setting emission performance standards for new passenger cars as part of the Community's integrated approach to reduce CO2 emissions from light-duty vehicles [2009] OJ L140/1.
[155] The pertinent provisions of this legislation include:
- A fleet-wide average of 130 g/km by 2012;
- A gradual phasing in of the regulation such that 65% of the fleet must meet this average in 2012, 75% in 2013, 80% in 2014, and 100% in 2015 and beyond;
- Through 2018, a step-up penalty scale for fleet averages beyond the allowance of €5 per car registered in Europe for 1 g/km all the way up to €95 beyond 3 g/km. Beyond 2019 a strict penalty of €95 for each g/km;
- A long term target of 95 g/km by 2020, although this is non-binding at this time and the details of which must be established by 2013; and
- An interim allowance of 7 g/km for fleets incorporating innovative technologies, to encourage alternative energy transportation.

Ibid.

While controls on auto manufacturers have received primary attention in the media, the other two components of the ECCP transport pillar are worth reviewing. As a second component of the ECCP, the EU has adopted a labeling scheme requiring new cars to display fuel consumption and CO2 data, which while relevant to raising consumer awareness has had modest environmental impact.[156] Finally, the EU has used fiscal measures to try and improve efficiency in the transport sector. These measures include road charging (July 2003), harmonized EU energy taxation (January 2004), and proposals for an EU-wide harmonized car tax with the introduction of a CO2 element into the tax base.[157] Cumulatively, these measures have fallen far short of achieving desired emissions reductions, actually allowing significant emissions increases sector-wide.

With the highest percentage growth rate in GHG emissions, aviation has also posed challenges for EU climate policy.[158] Although aviation emissions are not currently covered by the Kyoto Protocol, the EU has lobbied to expand global climate policy to cover aviation and has recently taken steps to do so at the regional level. In November 2008, the EP and the Council adopted a directive mandating inclusion of aviation activities in the third phase of the EU ETS.[159] The directive specifies that airlines will be subject to ETS control beginning in 2012, when total emissions will be capped at 97% of 2004–2006 levels.[160] In 2013 and extending outward indefinitely, the cap will be 95% of 2004–2006 levels.[161] The emissions cap applies to intra-EU flights as well as to international flights either arriving in or departing out of Europe.[162] Minimum operational requirements apply to ensure parity.[163] Extending the EU ETS to include aviation is a significant step in expanding the global reach of the ETS, as it creates a regulatory regime reaching well beyond EU actors.

Despite the extension of the EU ETS, transport policies remain the weakest link in EU climate policies. The regional weakness is mirrored in transport trends in member states. In the UK, for example, National Travel Surveys reveal that average journeys have increased in length by 12%, and that over the last 20 years, average mileage by car and rail has risen by 58% and 34% respectively and average speeds by 19% and 13% respectively. Further, the UK government predicts

---

[156] Matthias Duwe (n 147) 21.

[157] Europa, 'Passenger Car Taxes: Commission Proposes to Improve Functioning of the Internal Market and Promote Sustainability' (Press Release, 5 July 2005) IP/05/839.

[158] Commission (EC), 'Commission Staff Working Document: Summary of the Impact Assessment: Inclusion of Aviation in the EU Greenhouse Gas Emissions Trading Scheme (EU ETS)' COM (2006) 818 final, 20 December 2006 (Commission Aviation Summary); see Janello Veno, 'Flying the Unfriendly Skies: The European Union's Proposal to Include Aviation in Their Emissions Trading Scheme' (2007) 72 *J Air L & Commerce* 659, 673–75 (noting that aviation emissions 'will soon account for a substantial amount of emissions, possibly undermining the attempts by the EU to cut emissions').

[159] Council Directive (EC) 2008/101 of 19 November 2008 amending Directive 2003/87/EC so as to include aviation activities in the scheme for greenhouse gas emission allowance trading within the Community [2009] OJ L8/3.   [160] Ibid.

[161] Ibid   [162] Ibid.   [163] Commission Aviation Summary (n 158) 6.

dramatic increases in road and rail travel, with continuing modest increases in air travel.

Similar patterns emerge in Germany, where the federal government has succeeded in reducing emissions from all sectors except transport.[164] Germany has been notoriously reluctant to regulate transport due in large part to the prominent position of the auto industry in Germany.[165] The first automobile was made in Germany and auto manufacturers continue to be significant economic players, employing approximately 750,000 people in Germany.[166] Cultural and economic factors, thus, dissuade German policymakers from regulating its auto industry, with the effect that German cars are notoriously poor environmental performers in relation to the overall European fleet.[167]

The failure to integrate successful transport policies into the EU climate regime threatens to undermine the EU's larger system of climate policy. Moving forward the EU faces contradictory objectives. On the one hand a primary objective of EU transport policy is to remove barriers to EU-wide transport and to harmonize transport policy and infrastructure availability region-wide. This often leads to improvements in infrastructure and lowering of barriers to movement, resulting in increased use of EU transport systems. On the other hand, the EU seeks to reduce GHG emissions region-wide; to do so requires the EU to integrate GHG emissions strategies into transport policy. Existing transport policy is much more geared toward achieving the first objective—free movement and harmonization of standards and infrastructure. The EU has not yet developed effective initiatives for ensuring that regional transport policy does not promote increased traffic and, thus, increases in GHG emissions. Similar challenges plague member state transport policies.[168]

The EU has identified transport policy as a key priority of the ECCP II. It remains to be seen whether Europe will be able to create effective models for decoupling transport-related GHG emissions from economic growth and regional social and political harmonization. At the moment, this is one area where Europe falls short of providing working models for the international community.

The US faces equally daunting challenges in reducing GHG emissions from its transport sector. The US transport sector is a significant source of GHG emissions domestically and internationally. At the international level, the US transport sector creates more carbon dioxide emissions than any other nation's total economy, with the exception of China.[169] At the domestic level, the transport sector is the

---

[164] P Harris (n 9) 288.
[165] Kerry E Rogers, 'Germany's Efforts to Reduce Carbon Dioxide Emissions from Cars: Anticipating a New Regulatory Framework and its Significance for Environmental Policy' (2008) 38 *Envtl L Rep News & Analysis* 10214, 10214–15.
[166] Verband der Automobilindustrie, 'Facts & Figures' (General information 2007) <http://www.vda.de/en/zahlen/jahreszahlen/allgemeines/> accessed 3 May 2009; 'Briefing: Germany's Car Industry—The Big-Car Problem' [24 February 2007] *The Economist* 81. [167] Ibid.
[168] ECCP Transport Final Report (n 146) 16.
[169] D Greene and A Schafer (n 145) iii.

largest and fastest growing end-use sector for carbon dioxide emissions and is second only to electricity generation in terms of the volume and rate of growth of total GHG emissions.[170] Current projections show that the US transport sector's share of end-use carbon dioxide emissions will increase from 33% to 36% by 2020.

The current goals and initiatives for reducing emissions from the US transport sector fall short of achieving significant emissions reductions. As discussed in Chapter 2, under President George W. Bush, transport policy included support for the 'FreedomCAR' initiative, incentives for fuel cell technology research, tax credits for hybrid and fuel cell vehicles, modifications for the corporate average fuel economy standards (CAFE), a tire pressure monitoring system, and new agreements with private industry to develop more efficient automobiles. These measures offered financial support for research and development and created an initial framework for reducing the climate-footprint of automobiles but fell far short of implementing any concrete or measurable strategies for reducing transport-based GHG emissions.

US CAFE standards are one of the weakest links in American transport policy and have proven to be remarkably intractable. CAFE standards offer low-hanging fruit for regulators seeking to limit transport-based GHG emissions. The failure to incrementally increase US CAFE standards is indicative of a larger resistance to regulating or otherwise inhibiting private transport choices in the US. The first CAFE standards were set in 1975; between 1975 and 2007 the fuel economy standards remained largely unchanged despite technological development and advanced fuel standards in other developed countries. Congressional refusal to mandate higher CAFE standards coupled with low gasoline taxes facilitated the emergence of a geographically dispersed, automobile dependent society.[171] Between 1990 and 2006 alone, for example, the number of automobiles on US roads increased in excess of 55 million.[172] After 32 years of essentially frozen standards, with the passage of the Energy Independence and Security Act of 2007, Congress required improvements in fuel economy standards. However, the Act only mandated an increased fleet average of 35 miles per gallon by 2020, roughly equivalent to existing European and Japanese averages.[173] The 2007 modification to CAFE standards was prompted by a desire to improve energy independence and security. But even this modest increase proved politically unpalatable, and the National Highway Traffic Safety Administration (NHTSA)—the Agency delegated authority to establish CAFE standards—failed to take timely steps to implement the change.

---

[170] Ibid.
[171] Thomas Friedman, *Hot, Flat & Crowded: Why We Need a Green Revolution—and How it can Renew America* (Farrar, Straus and Giroux, New York, 2008) 17.
[172] Arnold W Reitze Jr, 'Federal Control of Carbon Dioxide Emissions: What are the Options?' (2009) 36 *Boston College Envtl Affairs L Rev* 1. [173] T Friedman (n 171) 16.

Historical inaction in regards to CAFE standards reveals the sensitivity of questions of transport policy in the US. The infamous American love affair with automobiles, personal property, and freedom of movement make any suggestions relating to regulating vehicle use and land-use planning unpopular across political lines. Questions of transport policy are further complicated by complex questions of legal domain. Jurisdiction over key transport-related decision-making processes are divided among federal, state, and local governments and controlled by an intricate web of federal and sub-federal laws. The federal government, for example, exercises virtually complete control over automobile regulation while state and local governments control the vast majority of planning decisions that impact lifestyle patterns and, thus, transport choices. Federal and state environmental laws are then superimposed on automobile and planning laws with varying impacts. Lack of coordination between transport, planning, and environmental policy further complicates the already intractable debate over transport policy.

As a consequence, US climate-transport debates centre on technological opportunities and shy away from policy proposals related to reducing automobile dependency or overhauling land-use planning law at the federal or state levels. With very few exceptions, over the past 30 years, US policy has promoted rather than curbed growth in the transport sector. Until 2009, there was little indication that the federal government would integrate automobile regulation into an overarching climate strategy. In May 2009, however, driven by California's persistent efforts to receive a waiver under the CAA to regulate GHG tailpipe emissions, President Obama unveiled new national standards for automobile fuel economy and the first-ever federal GHG emissions standards for cars and trucks. The standards were agreed upon by state and federal governmental leaders in conjunction with automobile manufacturers, the United Auto Workers, and environmental NGOs after years of debate and litigation over California's proposed tailpipe regulations. In revealing the new standards in a White House press conference, President Obama noted that '[i]n the past, an agreement such as this would have been considered impossible. That is why this announcement is so important, for it represents not only a change in policy in Washington, but the harbinger of a change in the way business is done in Washington.'[174]

The President's remarks reflect the intransigency characterizing US transport policy and suggest, at a minimum, that inroads are being made into a historically change-resistant policy sphere. Modest fuel economy and GHG emission standards will have a limited effect in transitioning the US away from its automobile-dependency. The standards are significant, nevertheless. With highway vehicles accounting for 72% of total transport emissions and passenger cars and light

[174] The White House, Office of the Press Secretary 'President Obama's Announcement on National Fuel Efficiency Policy' (Press Release, 19 May 2009) <http://www.america.gov/st/texttrans-english/2009/May/20090520171509eaifas0.6000286.html&distid=ucs> accessed 2 January 2010.

trucks accounting for in excess of 50% of total transport emissions, the proposed standards are well targeted.[175]

As a follow-up to the President's announcement, on 15 September 2009, the EPA and the NHTSA proposed a new program for regulating greenhouse gas emissions and improving fuel economy for light duty vehicles sold in the US. The joint program would be the first regulatory program of its kind and with the release of the EPA's final endangerment finding on 7 December 2009, Administrator Jackson created the necessary legal framework for it to proceed.

Following President Obama's announcement of the new automobile standards, the US House of Representatives approved ACES. The bill includes a section on 'Clean Transportation', which includes provisions on electric vehicle infrastructure, large-scale vehicle electrification, plug-in electric drive vehicle manufacturing, and investment in clean vehicles. In relevant part, the bill:

- proposes to amend the Public Utility Regulatory Policies Act to require utilities to consider developing plans to support electric vehicle infrastructure and to consider establishing protocols for integration with smart grid systems;
- authorizes the Secretary of Energy to provide financial assistance for regional deployment and integration of grid-connected vehicles;
- authorizes the Secretary of Energy to provide financial assistance for retooling existing factories for the manufacture of electric vehicles and authorizes the Secretary of Energy to provide financial assistance to help auto manufacturers purchase batteries for first production vehicles;
- provides for distribution of allowances for plug-in electric drive vehicle manufacturing and deployment and advanced technology vehicles.[176]

Although the bill stalled in the Senate, if something similar is ultimately agreed upon by Congress, it will support modest but meaningful steps towards moving beyond CAFE standards in regulatory efforts to improve the environmental footprint of the transport sector.

New automobile standards move US's transport policy forward. Successful implementation of the standards will be critical to attempts to curb domestic GHG emissions. The size and energy intensity of the US transport sector will require a more balanced effort in the long-term to not only curb growth but actually achieve emissions cuts from the transport sector. The measures promoted in ACES would instigate efforts to overhaul the fossil-fuel intensive transport sector. Even these proposed measures are modest in relation to the scale of the US transport sector. In the short-term and absent far-reaching new proposals, the US transport sector will continue to account for a significant percentage of total global GHG emissions.

---

[175] D Greene and A Schafer (n 145) iii.   [176] HR 2454 (n 122).

Transport policy is the Achilles' heel for EU and US climate policies. While there is greater consensus over progressive actions on climate change in Europe, this consensus has not translated to support for climate-based overhauls of transport policy. Growing consensus on climate change in the US similarly stops short of extending to transport policy. In both contexts, policymakers have identified transport as a trouble spot and are beginning to take steps toward integrating transport policy more closely into climate change strategies. Neither actor has yet developed a model worth emulating but skeleton frameworks are emerging.

Other Macro-Level Convergences

European and American climate policy converges around two final central points: the need to expand policy choices to include adaptation strategies and the necessity of China, India, and Brazil adopting emissions reduction commitments in the post-Kyoto agreement.

Political efforts to address climate change in the EU and the US have focused on mitigation strategies. With the release of the Fourth IPCC Assessment in 2007 and subsequent scientific reports confirming that the global community is wedded to a certain amount of warming, both the EU and the US have broadened policy initiatives to consider adaptation needs. While mitigation continues to receive the lion's share of attention, questions of adaptation underpin growing security concerns, influence economic assessments of proposed mitigation strategies, and spawn decentralized policymaking aimed at alleviating local effects.

Adaptation: The European Union Approach

In Europe, adaptation was defined as one of six working groups in the ECCP II. In a final assessment report—'Green Paper on Adapting to climate change in Europe'—released in June 2007, the European Commission laid out a recommended course of action for EU climate change adaptation efforts.[177] The Commission prefaced the paper by emphasizing the preeminence of mitigation measures and framing adaptation measures as a secondary but essential component of a European climate change strategy. In reviewing the relationship between mitigation and adaptation, the Commission noted that 'a certain degree of climate change is inevitable throughout this century and beyond, even if global mitigation efforts over the next decades prove successful. While adaptation action has therefore become an unavoidable and indispensable complement to mitigation action, it is not an alternative to reducing GHG emissions. It has its limits.'[178] The Green Paper approaches adaptation as a regional priority to be achieved in close coordination with member states but stresses that adaptation measures must be considered within a larger international context, as overseas

---

[177] Commission (EC), 'Adapting to climate change in Europe—options for EU action' (Green Paper) COM (2007) 354 final, 29 June 2007.   [178] Ibid 3.

adaptation aid will be an essential component of the EU strategy. The Paper examines the likely impacts of climate change in the key areas of water availability, ecosystem and biodiversity functioning, food availability, coastal zone management, and human and animal health and stresses that 'Europe will not be spared'.[179] The Paper then makes the case for early adaptation efforts and outlines four lines of proposed action:

- Where current knowledge is sufficient, adaptation strategies should be developed in order to identify optimal resource allocation and efficient resource use which will guide actions at the EU level, through the EU sectoral and other policies and through the available Community Funds.
- The EU needs to recognize the external dimension of impacts and adaptation and to build a new alliance with its partners all around the world and particularly in developing countries. Adaptation action should be coordinated with its neighbors and cooperation with international organizations should be further strengthened.
- Where there are still important knowledge gaps, Community research, exchange of information and preparatory actions should further reduce uncertainty and expand the knowledge base. Integration of research results into policy and practice should be reinforced.
- Coordinated strategies and actions should *inter alia* be further analyzed and discussed, in a European Advisory Group on Adaptation to Climate Change under the European Climate Change Programme.[180]

Beyond these generalized suggestions, the Paper recommends adaptation measures to be taken in specific regional sectors, eg: reassessing agricultural practices under the Common Agricultural Plan; climate proofing the transport system; diversification of energy sources to deal with climate fluctuations; integrating climate change adaptation strategies into business plans; adopting a Commission Communication on establishing a framework to address the impacts of climate change on human and animal health; integrating climate-proofing into integrated into the Environmental Impact Assessment (EIA) Directive and the Strategic Environmental Assessment Directive; developing new policy responses that facilitate joined-up responses to vulnerability and risk assessments.

Externally, the Green Paper advocates incorporating climate adaptation questions into EU Common Foreign and Security Policy (CFSP), identifying CFSP as the best avenue for preventing and responding to climate-related conflicts, natural disasters, and forced migration. Importantly, the Green Paper frames external adaptation responses within the context of common but differentiated responsibilities, emphasizing developed countries' historic responsibility for atmospheric GHG accumulation and, thus, their responsibility to take the lead in

---

[179] Ibid 4.   [180] Ibid 14.

not only mitigation but also adaptation measures. The Paper further characterizes EU climate change adaptation responsibilities as necessary components to achieving the UN Millennium Development Goals and stresses that adaptation policies should be linked with poverty reduction policies. The Paper recommends a multi-pronged approach to implementing external adaptation measures that advocates the following: (1) improving adaptation strategies within the context of the UNFCCC system, drawing upon the existing 2004 EU Action Plan on Climate Change and Development to support adaptation measures in developing countries; (2) building a Global Climate Change Alliance to facilitate dialogue and cooperation between the EU and developing countries; (3) creating an integrated research scheme designed to reduce scientific uncertainties and improve knowledge on climate change; and (4) improving measures for more closely involving European society, businesses and the public sector in efforts to develop adaptation strategies.

Throughout, the Green Paper stresses the need for timely, joined-up regional efforts to prepare domestic and external adaptation strategies in order to minimize and avoid long-term social and economic instability.

In addition to compiling the Green Paper, the Commission has established working groups to assess projected impacts and proposed adaptation strategies in specific sectors, including: impacts on water cycle and water resources management and prediction of extreme events; marine resources and coastal zones and tourism; human health; agriculture and forestry; biodiversity; regional planning, built environment, public, and energy infrastructure; structural funds; urban planning and construction; development cooperation; role of insurance industry; building national strategies for adaptation (country reports).

The 2007 Green Paper offered a comprehensive overview of the domestic and international dimensions of adaptation, prompting the Commission to call for the creation of an EU Adaptation Framework. In a 2009 White Paper, 'Adapting to climate change: Towards a European framework for action', the Commission urges the adoption of a comprehensive EU Adaptation Framework 'to improve the EU's resilience to deal with the impact of climate change'.[181] According to the White Paper, scientific findings suggest that the impacts of climate change will be felt much sooner than indicated in 2007 and that Europe must begin to prepare immediately. To this end, the White Paper offers a regional adaptation framework with immediate effect. The framework begins with an initial phase that will run from 2009–2012 and will lay the groundwork for preparing a comprehensive EU adaptation strategy to be implemented during Phase 2, commencing in 2013. During Phase I, the framework focuses on:

1. building a solid knowledge base on the impacts and consequences of climate change for the EU;

---

[181] Commission (EC), 'Adapting to climate change: Towards a European framework for action' (White Paper) COM (2009) 147 final, 1 April 2009.

2. integrating adaptation into EU key policy areas;
3. employing a combination of policy instruments (market-based instruments, guidelines, public-private partnerships) to ensure effective delivery of adaptation and
4. stepping up international cooperation on adaptation.[182]

The White Paper makes the economic case for supporting strong adaptation responses; suggests mechanisms for improving the resilience of different sectors; outlines the need to develop a clearinghouse mechanism to facilitate the exchange of information on climate risks, impacts and responses; advocates using revenue generated from auctioning allowances under the EU ETS for adaptation purposes and proposes the creation of an Impact and Adaptation Steering Group to coordinate regional and member state initiatives. Expanding upon findings in the 2007 Green Paper, the White Paper creates mechanisms for moving forward with a more integrated adaptation strategy.

Adaptation: The US Approach

The US has not developed a similar proposal for a coordinated federal adaptation strategy. Adaptation concerns continue to take a back-seat to the more pressing Congressional debate over GHG mitigation initiatives. As discussed in reference to climate security concerns, however, Congressional climate change proposals increasingly integrate adaptation and mitigation measures into climate change bills. ACES, for example, proposes a detailed two-part plan of action for adapting to climate change. The first part of the bill focuses on domestic adaptation. The bill proposes creating a National Climate Change Adaptation Program—akin to the EU adaptation framework—and establishes separate strategies for addressing questions of public health and natural resource adaptation. The second part of the bill proposes the creation of an International Climate Change Adaptation Program.

Domestic Adaptation

The National Climate Change Adaptation Program would function as the heart of adaptation efforts and would expand the US Global Change Research Program to include a climate change adaptation program, establish a National Climate Service within the National Ocean and Atmospheric Administration, and distribute a limited number of emissions allowances generated through the proposed emissions trading regime to assist states in implementing adaptation measures, contingent on the completion of an approved State Adaptation Plan—thus, encouraging decentralized development of adaptation strategies.

The bill also proposes formally recognizing the public health impacts of climate change, mandating actions by the Secretary of Health and Human Services

[182] Ibid 7.

to prepare and implement programs for responding to health-based threats, creating a science advisory board to advise the Secretary on science related to the health effects of climate change, and establishing a supplemental Treasury fund to help finance health-based adaptation strategies.

As a final component of the domestic adaptation program, the bill provides for the creation of a Natural Resources Climate Change Adaptation Panel, chaired by the White House Council on Environmental Quality, to facilitate interagency coordination on natural resources adaptation. It otherwise provides for improved natural resource adaptation strategies by requiring federal agencies and states to develop natural resource adaptation plans and establishing a Natural Resources Climate Change Adaptation Fund.

International Adaptation
The second part of the adaptation plan focuses on international measures. As discussed in relation to climate security concerns, the bill calls for the establishment of an International Climate Change Adaptation Program within the US Agency for International Development, with the objective of assisting the most vulnerable developing countries in adapting to climate change. The bill instructs that 40 to 60% of all allowances be directed to multilateral institutions for dispersal and dictates that aid should be funneled to those developing countries most vulnerable to climate change. The details of funding levels, project priorities, vulnerability assessments, and adaptation objectives are left for future development.

If passed, ACES would have created the legal backbone for the development of domestic and international adaptation programs. The APA, the bill being debated in the Senate in Summer 2010, contains similar domestic and international adaptation provisions but while the domestic adaptation provisions more closely mirror their ACES counterparts, the international provisions fall short of creating anything resembling a comprehensive adaptation strategy. While adaptation initiatives continue to be overshadowed by mitigation debates in Congress, growing awareness of the impacts of climate change prompt widespread public concerns about the feasibility and costs of climate change and encourage increasingly bi-partisan support for adaptation planning.

The leveling effect of adaptation planning, for example, is visible amongst US coastal communities, where most:

existing beaches and about half our existing coastal wetlands could be eroded or inundated by even a two-foot rise. Seventy to ninety percent of our wetlands could be eliminated by a seven-foot rise.[183]

---

[183] James G Titus, 'Does the U.S. Government Realize that the Sea is Rising? How to Restructure Federal Programs so that Wetlands and Beaches Survive' (2000) 30 *Golden Gate U L Rev* 717.

Current federal and state laws—eg, the US Clean Water Act and Coastal Zone Management Act—fall far short of providing adequate options for mitigating or adapting to sea level rise or its concomitant effects on beaches, wetlands, urban development, fisheries etc.[184] As evidence mounts of the relationship between global warming and sea level rise, and projections are developed showing the impacts on specific domestic communities—eg, revealing that sea levels around Charleston, SC will raise 1.03 feet in 100 years—local, state, and federal politicians are increasingly called upon to create coastal adaptation strategies.[185] In this way, calls for adaptation precede calls for mitigation among certain constituencies. Mounting focus on adaptation, thus, widens the base of the climate change debate.

In both the EU and the US adaptation measures lag behind mitigation initiatives. Yet, in both contexts, the balance is rapidly shifting towards greater focus on adaptation as the internal and external consequences of climate change become more evident.

The shifting nature of EU and US debate mirrors a similar change at the international level, where the UNFCCC Bali Action Plan prioritized enhanced action leading up to the negotiation of a post-Kyoto agreement. Desire to improve adaptation planning for purposes of domestic well-being and international peace and security thus emerge as a point of convergence both in EU and US policy and within wider international debate.

Broadening International Emission Reduction Commitments

The US has persistently objected to the Kyoto Protocol based on the Protocol's disparate treatment of developed and developing countries. More specifically, the US has historically opposed any international agreement that creates legally binding emissions obligations for the US while omitting major developing economies, such as China, India, and Brazil, from some form of emissions reduction obligation. Yet, differentiation of commitment lies at the heart of the international climate regime. The UNFCCC and the Kyoto Protocol are premised on the notion that legal responsibilities to address climate change are assigned 'on the basis of equity and in accordance to their common but differentiated responsibilities'.[186] The principle of common but differentiated responsibilities has been interpreted to include two primary considerations: historic contribution to greenhouse gases and present capacity to respond, that is, level of economic development. The US has steadfastly opposed the principle as it has been interpreted by the UNFCCC,

---

[184] Ibid.
[185] IPCC, '2007 Summary for Policy Makers' in S Solomon and others (eds), *Climate Change 2007: The Physical Science Basis* (Contribution of Working Group I to the Fourth Assessment Report of the Intergovernmental Panel on Climate Change) (CUP, New York, 2007); National Oceanic and Atmospheric Administration, 'Mean Sea Level Trend for Charleston, SC' (Chart, revised 9 December 2008) <http://tidesandcurrents.noaa.gov/sltrends/sltrends_station.shtml?stnid=8665530> accessed 6 May 2009.
[186] United Nations Framework Convention on Climate Change (adopted 9 May 1992, entered into force 21 March 1994) 31 ILM 849 art 3.1.

both in regards to its assignment of responsibility for past harms to present generations of US citizens as well as for its failure to recognize that responsibilities should evolve as countries experience economic growth.[187] The exclusion of China, India, and Brazil made US participation in the Kyoto Protocol domestically untenable based primarily on concerns of economic disadvantage.

In contrast, the EU has been a staunch proponent of the principle of common but differentiated responsibilities based on notions of equity. As the Copenhagen negotiations demonstrated, however, the US and the EU are drawing closer together on the necessary scope of emissions reduction obligations in a post-2012 world. While the US inches closer to participating in a legally—or, at least, politically—binding emissions reduction regime, the EU is increasingly siding with the US in insisting that emissions reduction obligations be extended based on present levels of economic development and present levels of GHG emissions.

The changing nature of global economics and, thus, global pollution has prompted transatlantic reassessment of abatement responsibilities. The 2007 IPCC report, for example, revealed that, of the projected 40–110% increase in CO_2 levels by 2030, two thirds to three quarters of this amount will be generated in developing countries. Similarly, China has already surpassed the US as the largest net global GHG emitter and Chinese emissions continue to steadily increasing as it seeks to quadruple its 2000 GDP level by 2020. Consequently, the EU and the US increasingly converge around the notion that any effective international climate regime necessarily must include the world's biggest emitters and that notions of common but differentiated responsibilities must evolve to more closely attribute responsibilities based not only on past responsibilities but also on present contributions and capacities.

The extent of convergence on this point remains unknown. Two key issues underpin successful extension of legally binding emission reduction obligations in a post-Kyoto agreement: (1) the US's acceptance of legally binding emissions reduction obligations; (2) the ability of the developed and developing countries to agree upon an equitable framework for enlarging the legally binding regime.[188] The first component is a necessary predecessor to the resolution of the second component and, as the Copenhagen Climate Change Conference in December 2009 revealed, there exists very little effective dialogue among developed countries, such as the EU and the US, or between developed and developing countries on an agreeable framework for successful resolution of the second component. Reassessment of the principle of common but differentiated responsibilities remains one of the most contentious points in international climate negotiations.

---

[187] Lavanya Rajamani, 'From Berlin to Bali and Beyond: Killing Kyoto Softly?' (2008) 57 *Intl & Comparative LQ* 909.

[188] H Winkler, B Brouns, and S Kartha, 'Future Mitigation Commitments: Differentiating among Non-Annex I Countries' (2006) 5 *Climate Policy* 469.

Conclusion

Upon coming into office in 2009, US President Barrack Obama declared:

We must be honest with ourselves. In recent years, we've allowed our alliance to drift... In America, there's a failure to appreciate Europe's leading role in the world.... Together we must confront climate change by ending the world's dependency on fossil fuels by tapping the power from the sources of energy like the wind and the sun and calling upon all nations to do their part. And I pledge to you that in this global effort the U.S. is now ready to lead.[189]

Responding to the new President's position on climate change, European Commission President José Manuel Barroso commented that the new US administration is 'much clearer and more ambitious' on climate change and that '[w]e have welcomed very positive changes the U.S. is making.... Only together can we convince others in our common efforts to fight climate change.'[190] The statements of these two leaders reflect the schisms in EU–US climate relations but they also reflect continuing recognition of the 'indispensable partnership' between the EU and the US on matters of global importance.

Within the context of this globally significant alliance, variations in regional approaches to climate change have symbolic and substantive importance that extends well beyond the impact of individual policy choices. Examining variations in transatlantic climate policy offers insight into the structural differences driving policy choices, policy performance, and the likely shape and success of future climate policy regimes in the US, the EU and beyond in a post-Kyoto world. The simplistic image of the EU as the leader and the US as the laggard in international climate politics ignores layers of transatlantic policy successes and failures and macro-level points of convergence and impedes constructive analyses.

As the international community works to develop a more effective and lasting global climate regime, the common rhetoric of transatlantic divergence on climate policy offers little direction in moving forward and masks very real lessons to be learned by more closely examining US and EU climate laws and policies as the global community seeks to negotiate a post-Kyoto legal framework.

---

[189] The White House, Office of the Press Secretary, 'Remarks by President Barack Obama: Hradcany Square, Prague, Czech Republic' (Press Release, 5 April 2009) <http://www.whitehouse.gov/the_press_office/Remarks-By-President-Barack-Obama-In-Prague-As-Delivered/> accessed 30 December 2009.

[190] Mark John, 'Obama Climate Pledge gets Cautious EU Welcome' *Reuters* (5 April 2009) <http://www.reuters.com/article/idUSTRE5340Z820090405> accessed 30 December 2009.

8
Socio-Legal Factors Influencing US and EU Law and Policymaking

Introduction

After exploring the fundamental convergences and divergences between US and EU climate change regimes, the key question that remains is why do they differ? European and American approaches to climate change are as much a product of culture as a response to scientific inquiry. Transatlantic cultural differences give rise to location specific ideas, politics, and dialogue, which in turn, generate distinct legal and political responses to social problems.[1]

Environmental law and policy inevitably draws upon a complicated mixture of established legal culture, *sui generis* reforms, non-legal regulatory ideals, and social, political, and legal norms.[2] It is inseparable from and defined by the culturally-specific economic and ethical foundations that underpin its development. Climate change law and policy is no different. It is impossible to comparatively analyze climate laws and policies without examining the socio-legal context within which they arise.[3]

The EU and the US are both supreme allies and supreme competitors. They frequently support one another on questions of international relations, while simultaneously conflicting on others. Accordingly, the diverging US and EU climate change regimes come as no surprise to international relations and international law experts. Yet, the questions of how and why the EU and the US are responding to climate change and what socio-legal factors drive these choices remain underexplored.

---

[1] Marc R Poirier, 'Property, Environment, Community' (1997) 12 *J Envtl L & Litigation* 43, 64.
[2] Cinnamon Carlarne and others, 'Maturity and Methodology: Starting a Debate about Environmental Law Scholarship' (2009) 21 *J Envtl L* 213.
[3] Environmental law is best understood in its larger socio-legal context. See, eg, B Hutter, 'Socio-legal Perspectives on Environmental Law: An Overview' in B Hutter (ed), *A Reader in Environmental Law* (OUP, Oxford, 1999); K Hawkins, *Environment and Enforcement* (OUP, Oxford, 1984).

To better understand convergences and divergences in EU and US climate change strategies, this chapter examines five categories of factors that influence climate change law and policymaking in Europe and America, including: (1) systems of governance; (2) risk perception and notions of precaution; (3) the roles of media and civil society; (4) modes of capitalism; (5) notions of equity. It is not possible here to undertake an exhaustive review of each of these topics, all of which have generated a rich body of literature. Rather, the goal of this chapter is to outline how each of these fundamental factors shapes transatlantic climate debate and highlight points requiring further enquiry.

Giving Context to Transatlantic Policy Choices

The EU and the US share many socio-legal similarities. As two of the largest global economic entities, they demonstrate similarly high levels of industrialization and social development. In common with other wealthy democracies, European and American laws reveal remarkable similarities.[4] In both contexts, environmental protection ranks high on the political agenda, and both the normative foundations and the scope of environmental protection regimes overlap to a significant degree.[5]

While the EU and the US share many commonalities in terms of basic social, economic, and political foundations, the nuances of how these systems interact can lead to profoundly different legal and political responses. Further, the US possesses a distinct legal system that differentiates it from its European counterparts. There is a rich body of literature exploring the unique qualities of the US legal and regulatory system. This literature identifies eight key features that distinguish the US system, including: (1) more complex and detailed bodies of rules; (2) more frequent recourse to formal legal methods of implementing policy and resolving disputes; (3) more adversarial and expensive forms of legal contestation; (4) more punitive legal sanctions (including larger civil damage awards); (5) more frequent judicial review, revision, and delay of administrative decision-making; (6) more legal uncertainty, malleability, and unpredictability; (7) more political controversy about legal rules and institutions and processes; (8) more legal uncertainty and instability.[6]

Beyond systemic differences, in the environmental arena there is a lively debate over the varying roles of the EU and the US as innovators and drivers of legal and

---

[4] Robert A Kagan, *Adversarial Legalism: The American Way of Law* (Harvard University Press, Cambridge MA, 2001) (*Adversarial Legalism*); Robert A Kagan and Lee Axelrad (eds), *Regulatory Encounters: Multinational Corporations and American Adversarial Legalism* (University of California Press, Berkeley CA, 2000) (*Regulatory Encounters*).
[5] *Adversarial Legalism* (n 4); *Regulatory Encounters* (n 4); Cliona Kimber, 'Environmental Federalism: A Comparison of Environmental Federalism in the United States and the European Union' (1995) 54 *Maryland L Rev* 1658, 1659.   [6] *Adversarial Legalism* (n 4).

political change.[7] Many commentators argue that the US has ceded its role as environmental policy innovator to the EU.[8] Climate policy is frequently highlighted as a case in point. The realities of domestic and international environmental regulation, including climate regulation, are much more complex than this categorization would suggest.[9]

Systemic legal differences coupled with shifting notions of environmental law leadership, however, define the contours of the debate over how and why EU and US legal and political responses to climate change diverge. Within these contours, it is necessary to explore how systems of governance, societal notions of risk and precaution, media and civil society involvement, and economic and ethical reasoning interact to form culturally specific decision-making frameworks.

Governance

Analyses of climate law and policy are frequently framed within the larger context of systems of climate governance.[10] Discussions of climate governance appear frequently in academic and popular literature but, as a conceptual term, 'climate governance' remains ill-defined. Consequently, the first question that must be posed is what do we mean by climate governance, and what do examinations of climate governance reveal about the socio-legal context shaping climate strategies in the EU and the US.

Exploring climate law and policy within the larger framework of governance is appealing because it offers a wider lens within which to examine decision-making. Notions of governance enable commentators to look beyond the instrumental value of law and beyond narrow and outdated notions of governmental functioning to consider 'the multiplicity of actors, institutions and relationships involved in the process of governing'.[11] Governance studies recognize interdependencies among governmental, civil society, and economic actors and look beyond traditional notions of the state as the sole provider of services to re-frame legal and political analysis within a globalized political economy.

---

[7] David Vogel, 'The Hare and the Tortoise Revisited: The New Political of Consumer and Environmental Regulation in Europe' (2003) 33 *British J Political Science* 557.
[8] Ibid. But see Jonathan Wiener, 'On the Political Economy of Global Environmental Regulation' (1999) 87 *Georgetown LJ* 749.   [9] J Wiener (n 8).
[10] Cinnamon Carlarne, 'Good Climate Governance: Only a Fragmented System of International Law Away?' (2008) 30 *L & Policy* 4 (Special Issue on Global Warming).
[11] European Commission, *EU Enlargement and Multi level Governance in European Regional and Environment Policies: Patterns of Institutional Learning, Adaptation and Europeanization Among Cohesion Countries (Greece, Ireland and Portugal) and Lessons for New Members (Hungary and Poland)* (Final Report on Research of Socio-economic Sciences and Humanities 2007) 21–22 (EC Environmental Study); see also J Scott and J Holder, 'Law and New Environmental Governance in the EU' in G de Burca and J Scott (eds), *Law and Governance in the EU and US* (Hart Publishing, Oxford, 2006).

At the supranational level, for example, Dan Esty defines governance as:

refer[ing] to any number of policymaking processes and institutions that help to manage international interdependence, including (1) negotiation by nation-states leading to a treaty; (2) dispute settlement within an international organization; (3) rulemaking by international bodies in support of treaty implementation; (4) development of government-backed codes of conduct, guidelines, and norms; (5) pre-negotiation agenda-setting and issue analysis in support of treaty-making; (6) technical standard-setting to facilitate trade; (7) networking and policy coordination by regulators; (8) structured public-private efforts at norm creation; (9) informal workshops at which policymakers, NGOs, business leaders and academics exchange ideas; and (10) private sector policymaking activities.[12]

This definition of governance accounts for the processes, institutions, and outcomes of law, while also including the peripheral public, private, and intergovernmental activities that shape and support lawmaking processes. Notions of governance, thus defined, reflect the realities of modern political systems, where law is a fundamental, but nevertheless, component part of a larger system that creates the parameters of legal, political, and economic decision-making.

In considering EU and US governance structures, one must first consider what level of governance to analyze. This analysis will focus on supranational/federal level governance structures, recognizing that this is the loci for primary decision-making with far-reaching regional and global effect.

Before considering distinctive elements in EU and US domestic governance systems, however, there is an *a priori* political distinction that must be explored. That is, the extent to which the EU and the US engage with international law and policymaking, particularly within the field of international environmental law, differs with concomitant effect on domestic systems of governance.

At a very basic level, the US is a more reluctant player in multilateral law and policymaking than is the EU. For example, in the wider field of international law, the US has refused to participate in the International Criminal Court, treated International Court of Justice decisions with considerable scepticism, ignored UN Security Council protocol, and abstained from various International Human Rights treaties. Within the narrower field of international environmental law, the US has refused to ratify the Kyoto Protocol, the Convention on Biological Diversity (CBD), and the International Convention on the Law of the Sea.[13]

---

[12] Daniel Esty, 'Good Governance at the Supranational Scale: Globalizing Administrative Law' (2006) 115 *Yale LJ* 1497.

[13] Letter by President George W. Bush (Letter to Senators Hagel, Helms, Craig, and Roberts, 13 March 2001). In an effort to avoid participating in an international agreement that would bind the US to specific goals, the US is instead negotiating with other like-minded nations to develop a multilateral climate change agreement based on non-binding commitments. US Department of State, 'Fact Sheet: President Bush and the Asia-Pacific Partnership on clean Development' (Press Release, 27 July 2005) <http://georgewbush-whitehouse.archives.gov/news/releases/2005/07/20050727-11.html> accessed 1 June 2010.

Even when directly engaged in international conflicts, US citizens have historically retained primary focus on matters of domestic politics.[14] This inward-looking focus creates a feedback loop whereby the more American citizens focus on domestic issues, the less sustained coverage the media provides of international politics and the less likely the domestic citizenry is to engage with issues of international affairs. As a consequence, Congressional handling of international issues has traditionally received less popular attention and critique than have domestic issues, leaving the politics of international affairs vulnerable to excessive influence by special interest groups.[15]

Domestic pressure groups, thus, exercise considerable influence over American foreign policy, with the effect that much post-Cold War American foreign policy has tended toward 'unilateral and bullying conduct'.[16] As America has sought to solidify its position in the global economic and political world order, unilateral decision-making and prevalent notions of the US as a modern empire have 'smack[ed] of hubris and arrogance' with the effect of alienating global allies, such as the EU.[17]

Following the terrorist attacks on 11 September 2001, public interest in international affairs increased in the US, but with mixed effect for multilateralism. While the American public initially expressed interest in and support for the military campaigns in Afghanistan and Iraq, the support was directed towards unilateral efforts to protect US security rather than towards promoting sustained engagement in multilateral law and policymaking. Renewed interest in international affairs has not correlated with increased regard for international law and the United Nations and, in fact, has tended to imbed disregard for multilateralism.[18] The continuing trend within the US to repudiate and undermine multilateral agreements and negotiations has been characterized as a pattern of 'lawlessness'.[19] Further, renewed interest in matters of international security did not translate to greater levels of interest in international environmental law and, in fact, diverted public and political support away from these issues.

Inadequate media coverage coupled with cartelization of international politics has created a culture of indifference around international environmental affairs among the larger American public that recent events have done little to allay. Public complacency frequently translates to disinterest in engaging in

---

[14] Joseph Nye Jr, 'The American National Interest and Global Public Goods' (2002) 78 *Intl Affairs* 233; Glen Sussman, 'The USA and Global Environmental Policy: Domestic Constraints on Effective Leadership' (2004) 25 *Intl Political Science Rev* 349.   [15] Nye (n 14).
[16] Ibid 234, quoting Henry Kissinger, *Does America Need a Foreign Policy?: Toward a Diplomacy for the 21st Century* (Simon & Schuster, New York, 2001) 15.   [17] Ibid 235.
[18] Disregard for international law and multilateralism is evidenced by persistent efforts on the part of the US to repudiate the International Criminal Court. Eg, J Vogler and C Bretherton, 'The European Union as a Protagonist to the United States on Climate Change' (2006) 9 *Intl Studies Perspectives* 1.
[19] Phillipe Sands, *Lawless World: America and the Making and Breaking of Global Rules* (Penguin Books Ltd, London, 2005).

international environmental affairs that the public perceives as costly, remote, and needless. Larger debates over the appropriate principles guiding America involvement in international affairs—ie, isolationism, unilateralism, multilateralism—similarly pervade and undermine US participation in international environmental politics.

In the politics and international relations literature, there is a lively debate over the question of whether domestic politics or inter-state relations are the key driver in formulating state positions on matters of foreign relations.[20] Within the more specific context of international environmental politics, there is increasing support for the argument that domestic politics drive the level and character of US involvement in international environmental politics.[21] As discussed in Chapter 7, economic considerations have been a primary driver in US engagement with global climate change policymaking. Economics is a primary but not exclusive driver. The US's failure to engage in several key multilateral agreements, eg, the CBD, the Kyoto Protocol, and the United Nations Convention on the Law of the Sea, reflects not only economic concerns but also 'a deeper opposition to international constraints on US sovereignty'.[22] Domestic politics and widespread scepticism towards international law and, more importantly, international constraints on US law and policy, thus, heavily influence the role the US plays in international environmental law.

Looking back to the inception of the climate debate, the mode of US participation in the UN Earth Summit in 1992 plotted the role the US would play over the next 15 years of international environmental law and policy. At the 1992 Earth Summit, the international community finalized two of the most important and far-reaching multilateral environmental agreements, the UNFCCC and the CBD. Leading up to the Earth Summit, the US had been an early innovator and leader in the field of environmental law. US leadership had stalled during the 1980s, however, largely due to the increasingly partisan nature of environmental politics.[23] By the 1992 Earth Summit, partisan politics impeded environmental law-making at the federal and international levels. At Rio, President George H.W. Bush was the only

---

[20] N Carter, *The Politics of the Environment: Ideas, Activism, Policy* (Harvard University Press, Cambridge MA, 2001). PS Chasek 'The Global Environment in the Twenty-First Century: Prospects for International Cooperation' in PS Chasek (ed), *The Global Environment in the Twenty-First Century* (United Nations University Press, New York, 2000). Cf, DF Sprinz and M Weifrs, 'Domestic Politics and Global Climate Policy' in U Luerbacher and DF Sprinz (eds), *International Relations and Global Climate Change*. (MIT Press, Cambridge MA, 2001); ER DeSombre, *Domestic Sources of International Environmental Policy: Industry, Environmentalists, and US Power* (MIT Press, Cambridge MA, 2000).

[21] P Holtrup, 'The Lack of U.S. Leadership in Climate Change Diplomacy' in B May and MH Moore (eds), *The Uncertain Superpower: Domestic Dimensions of US Foreign Policy after the Cold War* (Leske & Budrich, Opladen, 2003); R Paarlberg, 'Lapsed Leadership: U.S. International Environmental Policy Since Rio' in NJ Vig and RS Axelrod (eds), *The Global Environment: Institutions, Law, and Policy* (CQ Press, Washington DC, 1999).

[22] Daniel Bodansky, 'Targets and Timetables: Good Policy But Bad Politics?' in J Aldy and R Stavins (eds), *Architectures for Agreement: Addressing Global Climate Change in the Post-Kyoto World* (CUP, New York, 2007).   [23] Sussman (n 14) 358.

state delegate who refused to sign the CBD and he agreed to sign the UNFCCC only after mandatory objectives were replaced by voluntary goals and timetables.

President George H.W. Bush's initial opposition to both agreements turned on issues of economics and sovereignty. As concerned the CBD, unanswered questions over the inviolability of intellectual property rights proved to be one of the major obstacles to US participation, while fears that 'limits on emissions would require major changes in Americans' way of life and would threaten an already weak economy' prevented the US from signing the UNFCCC until the framework was modified to remove mandatory commitments.[24]

For almost 17 years, the US position on the CBD and the UNFCCC remained largely unchanged. During the late 1990s, President Clinton endorsed the CBD and the Kyoto Protocol to the UNFCCC, but a Republican-controlled Senate blocked any possibility of either treaty being ratified. Upon President George W. Bush's ascendancy to the US Presidency in 2001, executive support for the CBD and the Kyoto Protocol dissipated.

The years since the Rio Summit have been marked by the absence of US leadership in key domains of international environmental lawmaking. The absence of US leadership during this critical period for international climate change law is largely attributable to the increasingly partisan nature of environmental politics, which 'has had a direct impact on US environmental policy-making, both domestically and globally'.[25] Partisan politics at the federal level coupled with uneven media coverage of global environmental issues and institutions has created an atmosphere of scepticism and uncertainty among the US public over the purpose and practicality of international environmental agreements, compounding public reluctance to support US involvement in these venues.

In contrast to the US cautious and selective engagement in international law and policymaking, the EU has adopted a more active role on the international scene, particularly with regard to human rights, international development, and international environmental law.[26] The EU has engaged more widely with international institutions generally, participating in the International Criminal Court, the European Court of Human Rights, and the United Nations Convention on the Law of the Sea.[27]

---

[24] Ibid 362, quoting G Bryner, *Gaia's Wager: Environmental Movements and the Challenge of Sustainability* (Rowman & Littlefield, Lanham MD, 2001) 142.   [25] Ibid.

[26] Robert Falkner, 'The Political Economy of "Normative Power" Europe: EU Environmental Leadership in International Biotechnology Regulation' (2007) 14 *J Eur Public Policy* 507; J Vogler and H Stephan, 'The European Union in Global Environmental Governance: Leadership in the Making?' (2007) 7 *Intl Envtl Agreements* 389. There are clear exceptions, eg, early opposition on the part of the EU to negotiations for the Vienna Convention on Substances that Deplete the Ozone Layer. See G Porter, JW Brown, and PS Chasek, *Global Environmental Policy* (Westview Press, Boulder CO, 2000); I Manners, 'Normative Power Europe: A Contradiction in Terms?' (2002) 40 *J Common Market Studies* 235.

[27] KE Smith, *European Union Foreign Policy in a Changing World* (Polity, Cambridge, 2003); ME Smith, *Europe's Foreign and Security Policy: The Institutionalization of Cooperation* (CUP, Cambridge, 2004).

The role of the EU in the international arena has expanded exponentially over the past three decades as the EU has acquired additional competencies, particularly in the area of environmental policy. The European Commission has gradually acquired broad authority to negotiate in international environmental arenas alongside EU member states. With the end of the Cold War, declining US economic domination, and growing regional competencies, the 1990s witnessed the EU taking an increasingly active role in international politics. Drawing upon the underlying objective of 'assert[ing] its identity on the international scene', as articulated in article 2 of the TEU, the EU has made a concerted effort to strengthen the Community's role in international law and policy since the early 1990s.[28]

Following on from the negotiation of the TEU, at the 1992 Rio Earth Summit, the EU assumed a more prominent leadership role in global environmental governance, especially in respect to 'the protection of the ozone layer, biotechnology, biodiversity and related UN reform'.[29] Within the broad field of global environmental governance, the EU continues to play an active role in negotiating, ratifying, and implementing multilateral environmental agreements. The decision by the EU to espouse a strong leadership role in international climate change negotiations reflects both fundamental concerns about the impacts of climate change and a continuing interest in acting as a leader on issues of global environmental governance. Widespread public concern prompted European leaders to champion international and regional climate change measures. Domestic interest in increasing the prominence of the EU's international personality also motivated Europe's climate leadership. Climate change created an opportunity for the EU to take up the mantle of global leadership that the US had so precipitously dropped.

Much of the EU's success in this regard is attributable to greater intermingling of domestic, regional, and international politics. While interest and support for international engagement varies considerably across the Union and across issue lines, the geographic positioning, historical interdependency, and political configuration of the EU predispositions European citizens and politicians to be more attuned to issues beyond domestic politics.

The degree to which domestic issues dominate US politics should not be underestimated. In contrast, European politics whether at the State or supranational level reflect a more integrated melding of local, regional, and international issues. In the context of climate change, variations in internal political dialogue have prompted diverging responses in the US and the EU. Rather than deterring

---

[28] European Union, 'Consolidated Versions of the Treaty on European Union and of the Treaty Establishing the European Community' (Consolidated versions with amendments, 29 December 2006) <http://eur-lex.europa.eu/LexUriServ/LexUriServ.do?uri=OJ:C:2006:321E:0001:0331:EN:pdf> accessed 30 December 2009 (Maastricht Treaty and EC Treaty).

[29] S Oberthur and CR Kelly, 'EU Leadership in International Climate Policy: Achievements and Challenges' (2008) 43(3) *The Intl Spectator* 35.

European participation in the international climate regime, the US's political aloofness has deepened EU resolve to lead global climate politics.

The US's failure to take the helm of international climate change politics at the start weakened global efforts and created a leadership void. Europe has sought to fill the leadership void. However, even with European leadership, global climate politics suffer from the US's continuing refusal to fully engage in multilateral negotiations. Anything but full engagement on the part of the US creates substantive and procedural obstacles that disadvantage international climate change lawmaking. The US's response reflects issue specific concerns as well as deep-seated apprehension about the dangers of limiting domestic sovereignty by engaging in binding multilateral frameworks. Over time, issue specific concerns have lessened as the science behind climate change has improved and been more widely disseminated; it is not yet clear, however, whether underlying opposition to constraints on State sovereignty—particularly with regard to economic activity – will prove the ultimate stumbling block to US participation in the international climate regime.

Polity

The extent to which the EU and the US engage with international environmental law and policymaking reflects important differences in underlying social, legal, and political systems and ultimately plays a seminal role in shaping climate governance choices. The vagaries of international politics, however, vary with time and political administration, making it essential to look more closely at underlying differences in US and EU governance.

There is a growing body of literature exploring how the US and the EU differ in their respective political competencies and placements of power. In the context of climate change, several key points deserve consideration. Both the US and the EU share legal authority over environmental policy with their constituent states. In the case of the US, however, only the federal government has the ultimate ability to negotiate, adopt, and implement international environmental law in line with Constitutional constraints. EU authority, by contrast, is far more fragmented. Power is shared horizontally among the institutions of the EU and vertically with member state governments. The Council, Commission, Parliament, and Court share legal authority over environmental decision-making at the regional level.

At the international level, the Commission negotiates alongside but not on behalf of member states. In order to negotiate a collective EU position on a question of international environmental law or policy, that position must be debated and agreed upon by member states prior to the commencement of international negotiations. Further, because the EU is represented in international climate negotiations by the current President of the European Council, 'much may depend upon the willingness and capability of the "President in Office" to pursue the climate-change agenda'.[30] Thus, for example Chancellor Merkel's Presidency was marked by aggressive EU leadership on climate change while Czech President Klaus's term in office was marked by noticeable internal divisions and less decisive European leadership. Yet, the authority of the President is directly constrained by Council decisions and he/she has no independent authority to enter agreements or modify Council positions during negotiations. The EU, thus, shares negotiation and representation powers both among Community institutions and between the Community and the member states. There is no one governmental institution or head of state who wields the power to determine Community climate policy. Climate change policymaking authority is shared horizontally and vertically, creating numerous access points for policy formation. The 'open-ended and competitive governance structure of the EU' creates a climate change policymaking environment that is accessible to a wide range of voices representing diverse constituencies and political perspectives.[31] The breadth of the debate allows diverse perspectives to be voiced, encourages internal competition for policy leadership, and prevents partisanship or special interests from being able to control the direction of decision-making. The breadth of the debate, however, also means that negotiations can be time-consuming and require great compromise to reach consensus.

In contrast, the US President wields great power in international relations, retaining the power to negotiate and enter into international agreements, constrained only by Constitutional limits and the US Senate's powers of 'advice and consent'.[32] US participation in international law and policymaking, thus, is characterized by narrower and more closely defined chains of authority. The President, as the Head of States, is vested with primary authority in entering into

---

[30] Vogler and Bretherton (n 18) 13.

[31] M Schreurs and Y Tiberghien, 'Multi-Level Reinforcement: Explaining European Union Leadership in Climate Change Mitigation' (2007) 7(4) *Global Envtl Politics* 19, 24.

[32] The US president more directly shapes US foreign policy, although on the basis of a very different system of checks and balances. The US President has extensive authority to negotiate on international manners. Unlike the President of the Council, the US President is not directly constrained by state opinions or preferences. The President of the European Council may only represent Council positions, whereas the US President has extensive powers to enter into international negotiations. The President's Constitutional powers in this regard are constrained by the US Senate's authority of 'advice and consent' on the treaty-making process, but only modestly so. Neither the President of the Council nor the Commission have similar powers to singularly and fluidly speak on behalf of the member states. This is relevant for two reasons. First, if the US President chooses not to participate in an international environmental agreement, there is very little the country can due to compel participation. Second, however, if the President does choose to sign an agreement, he then relies on legislative support to ratify and implement the Treaty. In contrast, the decision to participate or not to participate in an international environmental agreement is not linearly invested in one authority or governmental institution in the EU. Decisions over participation and follow up negotiations on ratification and implementation are funnelled through more dispersed chains of power. Narrower and closely defined chains of authority in the US create greater room for flexibility in negotiations while simultaneously increasing opportunities for partisan politics to determine outcomes. Variations in channels of authority shape EU and US politics at the Community and international levels.

international agreements; the Senate maintains sole authority to ratify treaties; the judiciary retains powers of oversight and enforcement. In contrast, the EU lacks a single Head of State who is vested with significant authority at the international level; similarly, ratification, implementation, and enforcement authorities are shared among the Council, the EP, the Commission, and member states.

The US's triadic system of checks and balances affords greater room for flexibility in international negotiations while simultaneously increasing opportunities for partisan politics to determine outcomes. The narrowness of political channels is particularly relevant in the case of climate change, where it affords greater leeway for powerful special interest groups such as the oil and gas lobbies to influence executive and legislative officials at critical decision-making funnel points, eg, Presidential advisers, House Energy and Commerce Committee members. The EU's system of shared powers, in contrast, requires more extensive and time-consuming domestic negotiating and compromising and creates rigidity in international negotiations, but it also affords greater opportunities to overcome partisan politics and to minimize the influence of special interest groups at the negotiating and implementation stages.

Variations in channels of authority indelibly shape EU and US participation in international climate change law and policymaking by creating different filters for determining whose preferences determine negotiating positions. Thus, for example, while a Republican Congress or a Republican President can effectively block US participation in the international climate regime, the political party of the Council President, the German Chancellor, or the UK Prime Minister does not similarly determine EU participation. The multiplicity of perspectives voiced in the EU debate and the variety of political parties involved in EU politics requires regional negotiations to move beyond party politics and encourages issue-specific dialogue. EU climate debates, like US climate debates, are laden with interest-based politics. More dispersed channels of decision-making and representation in the EU, however, facilitate a more comprehensive debate and prevent interest-based and partisan politics from being determinative.

Domestic environmental politics reflect parallel variations in legal authority. Both the US and the EU share power between a centralized government and constituent political units. While the modes of power-sharing differ, efforts to allocate roles and responsibilities complicate environmental lawmaking in both contexts. In the US, the federal government is vested with more expansive environmental lawmaking, implementation, and enforcement powers and has taken advantage of these powers to craft a more 'comprehensive and coherent body of environmental law'.[33] The power of the EU in this regard remains more tightly constrained. Although environmental competences have expanded significantly over time, the EU continues to be constrained by questions of subsidiarity and dispersed implementation and enforcement authority. Due to the highly fragmented

---

[33] The EU Kimber (n 5).

nature of law and policymaking, the EU relies less on complex and detailed central regulation and more on central regulation that creates overarching objectives within which member states retain substantial implementation and enforcement authority. Continuing EU expansion has further complicated environmental law and policymaking by making 'policies that would benefit all...harder to find, and reaching decisions...correspondingly more difficult'.[34]

Extensive centralized environmental powers, however, do not always translate to greater willingness or success in enacting environmental law in the US, as the case of climate change demonstrates. Until 2009, for example, both the US Congress and US EPA refused to exercise federal powers to regulate greenhouse gases. The refusal of federal level institutions to regulate on questions of climate change impeded state and regional climate programs and created regulatory disarray. Equally, more decentralized power sharing has enabled climate law and policymaking to progress more rapidly by allowing the EU institutions to agree on framework goals at the central level while leaving contentious details to member states. The decentralized framework, however has also resulted in patchy implementation and enforcement efforts Community-wide. Power-sharing relationships, thus, challenge both US and EU efforts to create comprehensive GHG reduction schemes.

An added point of departure in domestic environmental politics revolves around regulatory capacity and the powerful role of administrative law in the US. It is impossible to conceive of US environmental law in the absence of the EPA. The EPA serves as rule-maker, rule-implementer, and rule-enforcer. It is the architect of existing environmental regulatory regimes. The course and effectiveness of US environmental law can be charted through EPA's successes and failures. Without an equivalent organization, the development of EU environmental law relies on dissimilar patterns of rule-making and implementation.[35]

The absence of a competent EU environmental administrative system means that member states retain more discretion in interpreting, implementing, and enforcing EU environmental rules. While this system comports with EU notions of subsidiarity, it creates greater challenges to harmonizing environmental law and to ensuring the effective implementation of existing measures. In the field of biodiversity protection, for example, great disparities continue to exist among member state compliance with the EU Natura 2000 program. Thus, while the EU has negotiated an overarching biodiversity framework, it continues to struggle with implementation and enforcement of the provisions.

Similar challenges now confront EU efforts to implement and enforce the complex web of climate-related provisions. The absence of an empowered environmental agency, thus, challenges efforts to implement the increasingly complex

---

[34] Maria Lee, *EU Environmental Law: Challenges, Change and Decision-Making* (Hart Publishing, Oxford, 2005) 21.
[35] Jonathan Wiener, 'Better Regulation in Europe' in J Holder and C O'Cinneide (eds), *Current Legal Problems 2006: Vol 59* (OUP, New York, 2006).

climate regulatory regimes. However, it also creates greater opportunities for innovation at the higher levels of government and fewer constraints on member states in crafting location-specific regulatory strategies. Thus, for example, while the US EPA facilitates implementation and enforcement of environmental rules, until 2009, the failure of the EPA to exercise its rule-making, implementation, and enforcement powers proved an impervious impediment to sub-federal efforts to address climate change.

The extent of the EPA's ability to influence climate change lawmaking was demonstrated in 2009, when the agency issued a finding that GHGs posed a danger to human health. This finding, in essence, created regulatory authority for the EPA to regulate GHGs. The EPA's decision not only created the conditions for a complete reframing of climate change regulation in the US, but it also prompted the US Congress to press forward with proposed climate change legislation. The possibility of EPA stepping in and filling the climate change regulation void changed the dynamic of the entire US climate change governance debate.

Regulatory distinctions extend beyond the source of administrative authority, however. The literature on environmental regulation is rich with analyses of how, why, and to what effect US and EU regulatory approaches differ.[36] In both the US and the EU, there is increasing recognition that patterns of regulations reflect complex interactions not only between government and well-organized interest groups, but also between 'ideologically-motivated policy entrepreneurs, social movements that generate 'norm cascades', and disasters, scandals, or analyses that suddenly shift the politics of regulation'.[37] The dynamics of regulation are shaped by context-specific social norms, including evolving views on environmentalism, precaution, and self-regulation that prompt shifting responses from regulators and regulated communities, alike.[38] In the US, for example, powerful special interest groups were fuelled in early efforts to impede climate-based regulation based on social norms favoring 'sound science', self-regulation, and market-based solutions to climate-based problems.

---

[36] D Fiorino, *The New Environmental Regulation* (MIT Press, Cambridge, 2006); J McEldowney and S McEldowney, *Environmental Law and Regulation* (Blackstone Press, London, 2001) 89; B Hutter, 'Socio-legal Perspectives on Environmental Law: An Overview' in B Hutter (ed), *A Reader in Environmental Law* (OUP, Oxford, 1999); K Hawkins, *Environment and Enforcement* (OUP, Oxford, 1984); E Fisher, *Risk Regulation and Administrative Constitutionalism* (Hart Publishing, Oxford, 2007) 669; R Stewart, 'A New Generation of Environmental Regulation?' (2001) 29 *Capital U L Rev* 21.

[37] Dorothy Thornton and others, 'The Persistence of Economic Factors in Shaping Regulation and Environmental Performance: The Limits of Regulation and Social License Pressures' (Paper prepared for the Annual Meeting of the Law and Society Association, Berlin, Germany, 26–29 July 2007).

[38] C Coglianese and J Nash, *Leveraging the Private Sector: Management-Based Strategies for Improving Environmental Performance* (Resources for the Future Press, Washington DC, 2006); C Coglianese and D Lazer, 'Management-Based Regulation: Prescribing Private Management to Achieve Public Goals' (2003) 37 *L & Society Rev* 691.

Over time, however, the regulatory debate has changed as climate science has evolved, social and economic concerns over climate change have grown, and dominant political ideologies have shifted, to create a more regulation-friendly social and political environment. In the EU, regulatory practices favoring cooperative approaches to negotiating solutions between bureaucrats, business, and civil society coupled with dominate social norms favoring precautionary, rules-based approaches to environmental problems facilitated early efforts to regulate in the climate arena. Yet, the regulation-friendly environment in Europe contrasts with the ability of the EU to craft regulatory regimes that follow-through on questions of implementation and enforcement. That is, while the European regulatory environment has proven more conducive to climate change rule-making (eg, the myriad of climate-related directives, regulations and decisions), it has proven to be a more difficult terrain when it comes to overseeing the harmonized implementation, evaluation, and enforcement of these measures. The EU's continuing lack of administrative capacity is likely to prove one of the most difficult obstacles to overcome as it moves forward in efforts to implement its increasingly complex climate change governance regime.

Variations in governmental structures and regulatory capacity are compounded by fundamental differences in the modes of law implementation and adjudication in the US and the EU. As previously introduced, the US's unique legal and regulatory style is distinguished from foreign systems in eight key ways:

1. more complex and detailed bodies of rules;
2. more frequent recourse to formal legal methods of implementing policy and resolving disputes;
3. more adversarial and expensive forms of legal contestation;
4. more punitive legal sanctions (including larger civil damage awards);
5. more frequent judicial review, revision, and delay of administrative decision-making;
6. more legal uncertainty, malleability, and unpredictability;
7. more political controversy about legal rules and institutions and processes; and
8. more legal uncertainty and instability.[39]

These distinctive features create a legal style, often termed 'adversarial legalism', in which the courts are actively involved in the on-going interpretation and enforcement of a complex and evolving set of primary and secondary rules. This legal style 'reflects deliberate government encouragement of private litigation and judicial action to help implement public policy'.[40] It is fuelled by a 'politically selected judiciary, armed with significant lawmaking and remedial powers; a highly entrepreneurial legal profession (including public interest law firms)

---

[39] *Adversarial Legalism* (n 4); *Regulatory Encounters* (n 4).   [40] *Adversarial Legalism* (n 4).

empowered by wide-ranging rights to pre-trial discovery and the right to bring remunerative, potentially devastating class actions'.[41]

The US's brand of adversarial legalism is set against a background of public scepticism of big government, demand for personalized judgments, and preference for decentralization and minimal market regulation. In contrast, the EU's legal system is characterized by a more bureaucratic, impersonal form of legalism, wherein the courts are less intimately involved in the minutiae of law-making. There is greater deference to government and underlying concerns for social welfare preclude conversion to US-style adversarialism.[42]

Differing notions of the role of law underlie systemic and stylistic governance differences in the US and the EU. The US's wariness of big government and penchant for market-based solutions to social problems contrasts with greater respect for bureaucracy, social welfare, and command-and-control style laws in the EU. In the US, detailed primary rules give rise to even more detailed sets of secondary rules that provide ample opportunity for the regulated community and civil society to challenge the scope, legality, and applicability of relevant rules. Recourse to judicial review, thus, has become increasingly common. In the EU, while the judiciary is one of the most powerful regional institutions, far more deference is accorded to experts and bureaucratic decision-makers, with lawyers playing a much less pronounced role throughout the lawmaking and implementation process. Similarly, final policy decisions are accorded more weight and are less often subject to challenge and judicial review.[43]

The peculiarly American legal style introduces high levels of inefficiency and uncertainty into the legal system. However, during periods where narrow channels of deliberation and representation have precluded federal action on climate change within the executive and legislative branches—as well as within the US system of administrative law—adversarial legalism has created pressure points for legal progress on climate change. The highly adversarial nature of the American legal system proved to be an advantageous feature for challenging public and private actions and omissions on the question of climate change. As discussed in Chapters 4 and 5, climate change litigation is rife in the US and has prompted high-level judicial, executive, and regulatory responses.

US and EU governance styles diverge over questions of the proper loci of decision-making, preferred modes of regulation, underlying rationales for governmental interference in the economy, and authority accorded to policy decisions.[44] Yet, in both contexts there are converging trends towards adversarialism

---

[41] Ibid.
[42] Eg, David Vogel, *National Styles of Regulation: Environmental Policy in Great Britain and the United States* (Cornell University Press, New York, 1986) (arguing that the American business community is less deferential to government than its counterparts in Western Europe and is more likely to challenge laws and regulation in the courts). [43] Ibid.
[44] See, eg, Jaggard's discussion of the consensual nature of German decision-making, which has influenced EU modes of decision-making. Lyn Jaggard, 'The Reflexivity of Ideas in Climate Change Policy: German, European and International Politics' in Paul G Harris (ed), *Europe and*

and privatization prompted by globalization of the world economy and mobilization of increasingly cosmopolitanist civil societies.[45] This trend towards the 'Americanization' of transatlantic politics brings European and American governance systems closer together.[46]

Yet significant differences remain. First, Europe continues to maintain more expansive social welfare benefits, health care, and employee protections and benefits as well as a more demanding system of tax law.[47] Second, the US continues to operate a more rigid and punitive system of sanctions in respect to criminal law, business regulations, and tort law.[48] Third, the US continues to rely on more detailed, complex, and prescriptive rules. Thus, while there has been a trend towards minimizing the role of State governments as service providers and augmented roles for the judiciary in the EU, these changes have been accompanied by a simultaneous trend towards greater 'Europeanization', whereby efforts to modernize regional policy and encourage global economic competiveness have been accomplished in line with continuing European reliance on cooperative decision-making and civil and social citizenship rights.[49]

Despite increasing convergence around legal styles conducive to global economic competitiveness, institutional and cultural differences continue to distinguish US and EU legal landscapes. Climate change law and policy continues to reflect these differences. In Europe, climate change law and policy emanates from centralized sources; climate law and policy is plentiful but is fragmented and relies on decentralized modes of implementation and enforcement. Litigation plays a modest role in the evolution of climate policy. In the US, distrust of big government is running headlong into adversarial legalism. The ability of the federal government, either through acts or omissions, to impede the development of domestic and international climate law and policy are being chipped away by judicial decisions, evolving social norms and shifting balances in partisan politics. The effect of these shifts is that the powerful, albeit distrusted, institutions of the federal government are poised to centralize the

---

Global Climate Change: Politics, Foreign Policy and Regional Cooperation (Edward Elgar Publishing, Northampton MA, 2007) 323. But see Susan Rose-Ackerman, *Controlling Environmental Policy: The Limits of Public Law in Germany and the United States* (Yale University Press, New Haven CT, 1995) (comparison of US and German approaches to environmental policymaking that suggests that the German regulatory process lacks sufficient public participation and judicial oversight and vests too much power with bureaucrats and technical experts).

[45] EC Environmental Study (n 11) 21–22.
[46] 'The 'Americanization' of European law might be defined as the adoption by Western European countries of laws, legal practices, and legal frameworks that had been entrenched in the US significantly earlier and that represent significant departures from longstanding European legal traditions'. Robert Kagan, 'Questioning the "Americanization" of European Law' (2007) 6(1) *Newsletter Eur Politics & Society Section of the American Political Science Association* 5.
[47] Ibid.   [48] Ibid.
[49] Eg, EC Environmental Study (n 11); Lisa Conant, 'Courts and the Americanization of Social Rights in Europe?' (2007) 6(1) *Newsletter Eur Politics & Society Section, American Political Science Association* 8.

existing patchwork of climate policy. Mobilization of the US system of administrative law, coupled with the possibility of Congressional action will alter the entire context of US climate politics to unknown effect.

Scientific Uncertainty and Risk

Social and political reactions to scientific uncertainty and risk influence the scope and pace of development of US and EU climate change policy. In particular, questions over the application of the precautionary principle, depictions of risk, and the validity of different forms of risk assessment highlight rifts in US and EU approaches to climate change.

The Precautionary Principle

Disagreements over the meaning and application of a precautionary approach to climate change have generated great discord in transatlantic climate debate. The precautionary principle, a commonly applied principle of international environmental law, states that '[w]here there are threats of serious or irreversible damage, lack of full scientific certainty shall not be used as a reason for postponing cost-effective measures to prevent environmental degradation'.[50] Since the mid-1980s, the precautionary principle has been extensively referenced and relied on in multilateral environmental agreements.

Both the US and the EU have incorporated the concept of the precautionary approach into their domestic environmental legislation.[51] They have, however, often disagreed over the use and meaning of the precautionary principle in international forums. The EU has been the most vocal proponent of the precautionary principle, arguing for its inclusion in the Biosafety Protocol to the CBD and for its inclusion in WTO decision-making. The US, in contrast, has vocally opposed incorporating the precautionary principle into certain multilateral environmental agreements and international decision-making processes, eg, the Biosafety Protocol and the WTO dispute settlement body.[52] Conflicting notions of the role and meaning of precaution proved fundamental to early and continuing disagreements between the US and the EU over the development of the international climate change regime.

---

[50] *Report of the United Nations Conference on Environment and Development* (Rio de Janeiro, 3–14 June 1992) (12 August 1992) UN Doc A/CONF.151/26 (Vol 1) (Rio Declaration).

[51] See, eg, US Clean Air Act, 42 USC §§ 1251-1376 (2000); US Endangered Species Act, 16 USC §§ 1531–44 (2000); see also Commission (EC), 'On the Precautionary Principle' (Communication) COM (2000) 1 final, 2 February 2000 (Communication from the Commission on the Precautionary Principle).

[52] WTO, *EC—Measures Concerning Meat and Meat Products (Hormones)* (16 January 1998) WT/DS26/AB/R and WT/DS48/AB/R at 41–42 (EC—Beef Hormones); see also Anais Laidlaw, 'Is it Better to be Safe than Sorry?: The Cartagena Protocol Versus the World Trade Organisation' (2005) 36 *Victoria U Wellington L Rev* 427.

Since the inception of international climate talks, the EU has pushed for a climate regime centered on precaution.[53] The US, in contrast, has advocated a 'sound science' approach to climate change.[54] The term 'sound science', while possessing no technical meaning, was adopted by climate change skeptics as an alternative framework to the precautionary principle for guiding climate decision-making.

The term 'sound science' became a popular catch phrase during transatlantic trade debates over the use of genetically modified organisms (GMOs).[55] Debates between the US and the EU over the legality of trade in GMOs witnessed the US employing the term 'sound science' to counter EU reliance on precaution as the defining principle for regulating GMOs. In this context, the US supported widespread use, release, and trade in GMOs as a policy decision based on principles of 'sound science', as required by the WTO.[56] The EU, in contrast, adopted a more cautious approach to the use and release of GMOs, basing its regulatory strategy on the precautionary principle.[57] The different regulatory approaches adopted by the US and the EU sparked an international trade dispute that ultimately resulted in a decision in favor of the US. In this conflict, the WTO dispute settlement panel found that an EU *de facto* ban on GM products violated the WTO Sanitary and Phytosanitary Agreement because the 'measures were not based on risk assessments satisfying the definition of the SPS Agreement and hence could be presumed to be maintained without sufficient scientific evidence'.[58]

Transatlantic trade disputes have frequently arisen over differing responses to scientific uncertainty, with the EU more often advocating precautionary approaches while the US resists market interference in the absence of a higher level of certainty over predicted risks. The precaution–sound science dichotomy spilled over into the climate change debate. Advocates of a sound science approach demanded higher standards of proof and greater levels of certainty as well as more rigorous cost-benefit analyses as prerequisites for action on climate change.[59] Calls for a sound science approach became synonymous with US resistance to early action to reduce GHG emissions. The US relied on a perceived lack of scientific certainty—ie, the absence of sound science—to postpone adopting domestic GHG reduction measures.

---

[53] Communication from the Commission on the Precautionary Principle (n 51).

[54] Philippe Sands, *Transnational Environmental Law* (Kluwer Law Intl, London, 1999) 134; see also J Murphy, L Levidow, and S Carr, 'Regulatory Standards for Environmental Risks: Understanding the US–European Union Conflict over Genetically Modified Crops' (2006) 36 *Social Studies Science* 1.

[55] Cinnamon Carlarne, 'From the USA with Love: Sharing Home-Grown Hormones, GMOs, and Clones with a Reluctant Europe' (2007) 37 *Envtl L* 2; Murphy, Levidow, and Carr (n 54).

[56] Murphy, Levidow, and Carr (n 54). [57] Ibid.

[58] WTO, European Communities—Measures Affecting the Approval and Marketing of Biotech Products (29 September 2006) WT/DS291, 292, 293/R.

[59] Myanna Lahsen, 'Technocracy, Democracy, and U.S. Climate Politics: The Need for Demarcations' Special (2005) 30 *Science Technology & Human Values* 137.

The sound science–precaution debate wrongly suggested to the public that the precautionary principle, as advanced by EU climate policies, and scientific rigor, as demanded by US policymakers, are antithetical.[60] The debate actually turned on political responses to scientific uncertainty. Public perceptions of climate change radiate from the dominant frame of reference used in political debate. By framing the debate in terms of science rather than politics, decision-makers sought to attach more credibility to their respective positions.

For example, for years, whenever President George W. Bush discussed climate change he repeatedly referred to 'uncertainty' and a 'lack of scientific consensus' in relation to the debate over human induced global warming, despite the fact that most scientists contested his use of these terms.[61] His use of language was strategic. By framing climate change in this way, President Bush disseminated the ideas that uncertainty and confusion dominated climate change science, making it more politically palatable for his administration to promote a lenient approach towards climate change.[62] In contrast, EU politicians have engaged in a more vigorous debate over the science of climate change wherein uncertainty has not been denied but has been more precisely analyzed, eg, the uncertainties involved in using different climate models rather than generalized uncertainties relating to the entire phenomenon of climate change.[63]

The sound science–precaution debate became less pronounced in 2007, when the Bush Administration's stance towards climate change softened, and has continued to lose force as President Obama has appointed scientists and science-literate actors into prominent cabinet positions. Despite converging opinions on the science of climate change, however, strong differences continue to exist over the best methods for integrating science into environmental policymaking and for balancing scientific data with social and economic considerations.

Scientific complexity, convoluted channels of communication between scientists and policymakers, and political tendencies to focus on scientific uncertainties make it difficult for policymakers to chart a scientifically and politically acceptable course of action on climate change, with the result that 'science (which is the necessary underpinning for action) is too often employed

---

[60] Nicolas de Sadeleer, *Environmental Principles: From Political Slogans to Legal Rules* (OUP, Oxford, 2002) 174.
[61] Environmental Working Group, 'Luntz Memorandum on the Environment' <http://www.ewg.org/files/LuntzResearch_environment.pdf> accessed 3 April 2010.
[62] Eg, Michele Gilman, 'The President as Scientist-in-Chief' (2009) 45 *Willamette L Rev* 565, 571–73.
[63] Charles Kennedy, 'Full Text: Charles Kennedy's Speech on Climate Change' (19 December 2005) <http://www.guardian.co.uk/politics/2005/dec/19/speeches.liberaldemocrats> accessed 5 January 2010. 'You should be in no doubt here in America that the EU both regards climate change as one of the major priorities facing us and is very united on that issue'. Lord Whitty, 'Climate Change Policy: The View from Europe' (A Brookings Briefing Speech in Washington DC, 18 April 2005) <http://www.brookings.edu/events/2005/0418energy.aspx?rssid=environmental+regulation> accessed 5 January 2010.

in the cause of those who wish to take no action at all'.[64] Scientific consensus on anthropogenic forcing of the climate has slowly permeated political debate in the US. Yet scientific uncertainties continue to provide a popular tool for discrediting climate policy *per se* in the US and divergences continue to exist with regard to the extent to which a precautionary approach is warranted in regards to climate change, and what precisely a precautionary approach entails. That is, even agreeing that human-induced climate change is occurring, differing cultural perceptions of risk shape the manner in which US and EU policy-makers balance short- and long-term environmental, economic, and social costs.

Risk Perception

There is a growing body of literature discussing differing cultural perceptions of risk and the impact of risk perception on policymaking. Much of this literature suggests that, beginning in the 1990s, as compared to the US, the EU has become a more risk adverse society, especially in relation to consumer and environmental issues.[65]

The US helped launch the era of environmental law and policy and built a system of domestic environmental law centered on notions of precaution. One need not look beyond the US Clean Air Act, Clean Water Act, and Endangered Species Act to trace the US role in promoting a global regulatory approach to environmental protection based on principles of precaution and risk aversion. The US pioneered a precautionary approach to environmental law. As one commentator notes:

studies of public health, safety and environmental regulation published in the 1980s revealed striking differences between American and European practices for managing technological risks.... [T]hese studies showed that U.S. regulators were quicker to respond to new risks, more aggressive in pursuing old ones.[66]

The US's precautionary, risk averse approach, however, gradually gave way to more conservative approaches to environmental protection during the 1980s and 1990s. During the 1990s, the US slowly ceded its role as global environmental leader to the EU as the EU began taking steps to craft a comprehensive environmental law regime while the US struggled to maintain the rigor of the environmental framework it had taken such great pains to craft over the preceding two decades. Beginning in the early 1990s, the US domestic and international environmental leadership began to wane as European environmental leadership

---

[64] Sussman (n 14) 350–51, citing E Claussen, 'Global Environmental Governance: Issues for the New Administration' (2001) 43 *Environment* 29.
[65] Vogel, 'The Hare and the Tortoise Revisited' (n 7).
[66] Sheila Jasanoff, 'American Exceptionalism and the Political Acknowledgement of Risk' in Edward Burger Jr (ed), *Risk* (University of Michigan Press, Ann Arbor, 1993).

grew as a result of the rapid development of a comprehensive, risk-averse system of environmental protection.[67]

During this period, EU environmental policies increasingly took precaution as a guiding principle. The EU looked to the precautionary principle as a tool for responding to the inherent limitations in the certainty and availability of scientific data and for curtailing policymakers' ability to use scientific uncertainty to justify delays in adopting more rigorous environmental policies.[68] The EU's growing reliance on the precautionary principle reflected 'a series of regulatory failures and crises; broader citizen support for more risk-averse regulatory policies within Europe; and the growth of the regulatory competence of the EU'.[69] European environmental policies grew increasingly risk-averse as citizens responded to widely publicized regulatory failures by placing increased pressure on EU decision-makers to enact stringent protective measures.

The absence of similarly publicized regulatory failures in the US, together with the increased influence of the private sector on regulatory policy and growing public preferences for market-based responses to regulatory failures, pushed US environmental policy away from stringent, risk-averse regulatory structures towards more *laissez faire* regulatory approaches. The emergence of the EU system of environmental law during a period of regulatory failure, thus, gave rise to a more precautionary, risk-averse regulatory regime whose normative framework more closely resembled that of the US environmental regime of the 1970s, rather than its contemporary counterpart.

Public perceptions of regulatory failure created the conditions under which precaution came to dominate regulatory approaches to environmental protection in the EU. The debate over climate change emerged at the peak of this period, increasing the political salience of the issue and ensuring that the initial framing of the debate was characterized by public demand for timely and protective political response.

EU and US environmental regimes evolved down divergent regulatory pathways in the 1990s in large part due to differing perceptions of precaution and risk-aversion; these divergences played a critical role in shaping perceptions of and responses to climate change during a critical period for international climate change law. And, while many commentators argue that Europeans, in fact, are not more risk averse than Americans, but rather merely more averse to particular risks, the EU's vocal precautionary, risk-averse approach to trade in genetically modified organisms and global climate change positioned the EU as the global proponent of the precautionary principle and risk-averse approaches to environmental law.[70]

---

[67] Eg, Vogel, 'The Hare and the Tortoise Revisited' (n 7).
[68] Ibid 566–67; see also Ragnar Lofstedt and David Vogel, 'The Changing Character of Consumer and Environmental Regulation: A Comparison of Europe and the United States' (2001) 21 *Risk Analysis* 399.   [69] Vogel, 'The Hare and the Tortoise Revisited' (n 7) 568.
[70] Jonathan Wiener, 'Whose Precaution After All? A Comment on the Comparison and Evolution of Risk Regulatory Systems' (2003) 13 *Duke J Comparative & Intl L* 207; Jonathan

The EU embraced this characterization, with the effect that precaution became a dominate theme in European climate dialogue and in European representations to the international community. Consequently, as data about the extent, pace, and nature of climate change evolved, so too did European insistence for aggressive regional and international climate policies based on the precautionary principle and the need to avert predicted risks. In contrast, the climate change debate in the US arose during a period of regulatory recoil. Initial US responses to climate change were characterized by cautious skepticism that gradually evolved into open hostility towards costly legal measures in response to widespread characterization of the debate as riddled with scientific uncertainty and lack of consensus.

Science and political depictions of science have moved on since the initial framing of the debate. Yet, the initial framing of the debate along lines of sound science versus precaution indelibly influenced the development of climate change law and policy in the US and the EU, with continuing influences to this day.

Media and Civil Society

The sound science–precaution debate derived momentum from political framing. The extent to which questions of uncertainty, scientific consensus, and risk dominated US and EU climate debates, however, owed much to media coverage. The media has played a pivotal role in channeling US and EU climate policy down divergent paths.[71] With any issue of social and political relevance, the media plays a disproportionate role in determining how the public obtains information and, thus, in influencing public perceptions.[72] The prominent role the media plays has the secondary effect of prompting special interest groups to 'work extremely hard to promote public attention concerning particular risks'.[73]

Media

Media plays an especially important role in shaping public perceptions of environmental issues. The technical complexities associated with many environmental problems coupled with the 'lexicon of caution' and 'language of probability'

Wiener and Michael Rogers, 'Comparing Precaution in the United States and Europe' (2002) 5 *J Risk Research* 317, 319. Cf, Cass Sunstein, 'Precautions Against What? The Availability Heuristic and Cross-Cultural Risk Perception' (2005) 57 *Alabama L Rev* 75.

[71] See generally J Boykoff and M Boykoff, 'Journalistic Balance as Global Warming Bias: Creating Controversy Where Science Finds Consensus' [November/December 2004] *Extra! Magazine* <http://www.fair.org/index.php?page=1978> accessed 5 January 2010; J Sterman and L Sweeney, 'Understanding Public Complacency About Climate Change: Adults' Mental Models of Climate Change Violate Conservation of Matter' (2007) 80(3–4) *Climatic Change* 213; Mike De Souza, 'US media blamed for stall in climate plan' *The Gazette* (Montreal, 2007) A4 (stating the issue was too watered down, and British and Canadian newspaper coverage is more direct about gravity of global warming).

[72] J Barabas and J Jerit, 'Estimating the Causal Effects of Media Coverage on Policy Specific Knowledge (2009) 53) *American J Political Science* 73.   [73] Sunstein (n 70) 98.

that scientists use to discuss these problems requires media translation to make it intelligible to the general public, while also encouraging media exaggeration of underlying scientific uncertainties.[74] This has been especially true in the case of climate change, where the scientific, economic, and social complexity of the issue provides fertile fodder for media to pick and choose what aspect of the debate they will emphasize, and to what effect. Varying levels and styles of media coverage in the US and the EU over time have played a critical role in influencing public perceptions about climate change with concomitant effect on the types of legal and political responses the public will support or demand in response to climate change.[75]

The role the media plays in shaping public perceptions of climate coverage is influenced by:

- the extent to which reporters possess the scientific background to assess the reliability and importance of information on climate change;
- the willingness of scientists to speak with the media;
- the degree to which the broadsheet press and other mass media outlets engage with the complexities behind the sound bites; and
- the manner in which mainstream press coverage deals with questions of uncertainty, consensus, and risk.

In this regard, media coverage in the US and the EU has varied considerably.

In the US, during the formative years of international climate policy, detailed coverage of climate change occurred primarily in niche publications, while mainstream media coverage tended to be more sensationalized, less comprehensive, and weighted towards equal coverage of climate skeptics, emphasizing uncertainties generally without attempting to unpack the nuances of the debate.[76] In contrast, European broadsheet press (eg, *Guardian, Independent, Times*) and mass media (eg, BBC) have tended to engage in more in-depth analysis of the

---

[74] J Boykoff and M Boykoff, 'Climate Change and Journalistic Norms: A Case-Study of US Mass-Media Coverage' (2007) 38 *Geoforum* 1190.

[75] T Lowe and others, 'Does tomorrow ever come? Disaster narrative and public perceptions of climate change' (2005) Tyndall Centre Working Paper 72/2005; R Bord, R O'Connor, and A Fisher 'In what sense does the public need to understand global climate change?' (2000) 9 *Public Understanding of Science* 205.

[76] For example, compare headlines from mainstream media outlets in the US and the United Kingdom between January 1997 and January 1998: 'Global Warming: Globaloney?' (*The Boston Globe*), 'Holes in the Greenhouse Effect?' (*The Washington Post*), 'Not All Scientists Back Global Warming' (*The New York Times*), 'Don't Hold Your Breath' (*Newsweek*), 'Critics assail global-warming theory's "fudging", PC nature' (*The Washington Times*) versus 'Trouble is on its way; Four climate scientists warn that whatever happens at the Kyoto global summit, we must adapt to a future of bad weather with floods, hunger and drought' (*The Guardian*, London), 'Pollution puts Earth on course for hottest year' (*The Times*, London), 'Bad forecast for Scotland; Grim picture painted of floods and devastated wildlife through global warming' (*The Herald*, Glasgow). See also Bud Ward, 'A Higher Standard than 'Balance' in Journalism on Climate Change Science' (2008) 86 *Climatic Change* 13.

science, sociology, and economics of climate change, with less coverage focusing on questions of consensus, skepticism, and broadly conceptualized scientific uncertainties.

For example, much US media coverage has followed the traditional journalistic norms of presenting both sides of a debate—that is, the arguments for and against human induced global climate change—and then allowing the audience to decide on the issue for themselves.[77] As a result of this coverage, the arguments of a small minority of skeptics who deny the existence of climate change are disproportionately represented, wrongly misrepresenting to the public that they are equal in number and reputation to scientists who confirm the existence of anthropogenic forcing of the climate.[78] Media coverage, thus, has created an 'informational bias', suggesting that there is a high degree of scientific discord when, in fact, there is a high level of consensus on the question of anthropogenic climate change. Imbalanced media coverage, further, has created a feedback loop whereby climate skeptics have been empowered to 'emerge from conservative think tanks' to 'proliferate and amplify their "denial discourse"'.[79]

In discussing US media coverage of climate change, two prominent commentators note that '[t]he continuous juggling act [American] journalists engage in, often mitigates against meaningful, accurate, and urgent coverage of the issue of global warming' and that climate change has been characterized by 'informationally deficient mass media coverage'.[80] In contrast, media outlets in many European countries, such as the UK and Germany, have more consistently depicted climate change as an urgent scientific, social, and economic issue, focusing less on the claims of climate skeptics and allegations of underlying scientific uncertainties and more on the complexities of climate change, eg, climate justice, climate economics, adaptation, and mitigation alternatives.[81] Media coverage in both contexts has fluctuated over time and highlighted sensational aspects of climate change, but with different emphases and to different effect.

---

[77] J Boykoff and M Boykoff, 'Balance as bias: global warming and the U.S. prestige press' (2004) 14 *Global Envtl Change* 125 (finding that over a 15-year period, a majority (52.7%) of prestige-press articles featured balanced accounts that gave roughly equal attention to the views that humans were contributing to global warming and that exclusively natural fluctuations could explain the earth's temperature increase).

[78] Boykoff and Boykoff, 'Journalistic Balance as Global Warming Bias' (n 71) 22–25; B Ward (n 76).

[79] Ward (n 76), citing A McCright and R Dunlap, 'Challenging global warming as a social problem: an analysis of the conservative movement's counterclaims' (2000) 47 *Social Problems* 499; A McCright and R Dunlap, 'Defeating Kyoto: the conservative movement's impact on U.S. climate change policy' (2003) 50 *Social Problems* 348.

[80] Boykoff and Boykoff, 'Balance as bias' (n 77).

[81] Tammy Speers, 'A Picnic in March: Media Coverage of Climate Change and Public Opinion in the UK' in G Humphrys and M Williams (eds) *Presenting and Representing Environments* (Springer, Dordrecht, Netherlands, 2005); P Weingart, A Engels, and P Pansegrau, 'Risks of communication: discourses on climate change in science, politics, and the mass media' (2000) 9 *Public Understanding Science* 261.

Variations in media coverage during the 1990s and early 2000s played an important role in shaping the emerging public agenda on climate change, with uncertainty dominating US media coverage in contrast to the recurring theme of urgency dominating European media coverage.[82] The prevalence of the 'dueling scientist' debate in the US media fueled political efforts 'to demand more research before having to tinker with the status-quo reliance on fossil fuels'; whereas the prevalence of messages of social and scientific urgency in European media coverage fueled political efforts to demand more aggressive climate policies.[83] The role of the media in influencing legal and political strategies should not be downplayed; in both the US and the EU, the media has influenced the legal and political terrain of climate change with lasting effect.

As debate over climate change has progressed, the depth of media coverage has tended to deepen in both the US and the EU.[84] While mass media coverage has expanded, however, so too has distrust of the media as a reliable source on climate change. As the debate over climate change has intensified and moved beyond questions of 'is it or isn't it', citizens increasingly look beyond mass media to find reliable information on climate change. Yet, somewhat curbing this trend, in 2009, media coverage of IPCC errors in the 4th Assessment Report as well as the Climategate affair—involving a series of emails stolen from climate scientists—revealed the extent to which the media continues to be an extremely powerful force for shaping societal perceptions of the science and politics of climate change and the extent to which popular perceptions of climate change continue to be shaped by incomplete and sensationalist media coverage.

Civil Society

The influence of civil society on the climate change debate extends well beyond the media. Non-governmental organizations, businesses, and other organized interest groups play important parts in directing US and EU climate debates.[85]

NGOs have played a central role in the environmental sphere since the onset of the environmental movement and their role in climate politics has been equally central at the domestic and international levels. Civil society participation in international climate negotiations has received the majority of academic attention, but civil society participation at the domestic and regional levels is no less important. As discussed in Chapter 5, US environmental NGOs have aggressively advocated for federal climate change legislation and for the extension of existing environmental laws to cover climate change.

Persistent pressure for legal and political action from NGOs, however, was met early on with equally persistent pressure from business and special interest

---

[82] Eg, De Souza (n 71).
[83] Boykoff and Boykoff, 'Climate Change and Journalistic Norms' (n 74) 12.
[84] Eg, Mike Hulme, 'Mediated Messages about Climate Change: Reporting the IPCC Fourth Assessment in the UK Print Media' in T Boyce and J Lewis (eds) *Climate Change and the Media* (Peter Lang Publishing, New York, 2009).
[85] Eg, Chad Carpenter, 'Businesses, Green Groups and the Media: The Role of Non-Governmental Organizations in the Climate Change Debate' (2001) 77 *Intl Affairs* 313.

lobbies opposing greenhouse gas regulation. The push and pull effect of civil society advocacy is a common theme in environmental law and policymaking in the US. In the context of climate change, civil society advocacy has helped raise the profile of the issue while also breeding confusion and frustration due to the mixed messages different groups have spread through public campaigns, political lobbying and advertising.

EU politics have been similarly characterized by civil society advocacy efforts. In the EU, however, the advocacy message of environmental organizations played louder early on to a more receptive political system while industry advocates failed to invest in anti-climate change regulation campaigns to the same degree as interest groups in the US. In both Europe and America, while vociferous anti-climate campaigns have faded over time and been replaced by more subtle efforts to influence the scope of regulatory programs, environmental NGOs continue to lead sustained climate campaigns aimed at both citizens and policymakers.

Both EU and US political choices on climate change are marked by concessions and revisions due to the influence of environmental and industry interest groups. The direct effects of civil society activism are difficult to pinpoint, but the collective impact is apparent. In the US, for example, industry interest groups won early battles in shaping public and political opinion but have progressively lost ground to environmental NGOs that have gained wider audiences through outreach campaigns, political lobbying, and more direct efforts to influence policy, eg, litigation. The combined influence of civil society, business, and media on EU and US climate change policymaking has not been adequately explored, but is undoubtedly profound.

Environmental activism and industry lobbying together with media attention continue to shape the contours of public and political debate.[86] On-going debates in both the EU and the US over the proper role of nuclear energy in domestic climate change strategies reveal the continuing interplay between industry, environmental NGOs, and the media in shaping public and political responses to distinct dimensions of the climate debate.

Modes of Capitalism

Dissimilar modes of capitalism pervade EU and US approaches to climate change law and policy. Neoclassical economics dominate European and American decision-making, but the degree to which EU and US decision-making prioritize self-regulation and free-market theories differs. While US policymaking reflects the continuing hegemony of market fundamentalism, that is, 'a vastly exaggerated belief in the ability of self-regulating markets to solve problems',

---

[86] Mary Pettenger (ed), *The Social Construction of Climate Change* (Ashgate Publishing, Hampshire, England, 2007).

EU policymaking reflects a more tempered approach to capitalism, more aptly described as market socialism.[87]

In the US, dominant public philosophy favors 'free market' principles and reflects great skepticism towards governmental interference in the market. Responding to this trend, commentators have suggested that theories of market fundamentalism, that is, the notion that there is not society *per se* but 'just an enormous marketplace, peopled by rational actors pursuing their self-interest with the potential to create the highest gross domestic product in human history' have become hegemonic in the US.[88] As a result of the hegemony of market fundamentalism, rights, and protections that were hard fought for in the 1960s and 1970s, including labor, environmental, and welfare rights, have been gradually eroded by market forces.

It is within this free-market framework that the US climate change debate emerged. American economic prosperity and well-being is attributed, in large part, to fossil fuel intensive activities; as a result, increased regulation of these activities proved politically unpopular for the first two decades of the climate debate.[89] Greenhouse gas regulation was equated with domestic economic decline and global economic disadvantage. Emphasis on near-term prosperity and maintaining global economic primacy encouraged widespread public distrust of proposed governmental interference in the market for greenhouse gas emissions. The prevailing neoclassical paradigm weighed in favor of voluntary, self-regulatory approaches to addressing climate change.

Over time, however, emerging evidence of social, economic, and environmental externalities signaled the failure of free market approaches to climate change and the inevitability of governmental interference in greenhouse gas regulation. This shift is still on-going in the US; the public and politicians, alike, are engaged in on-going debate over the proper form and function of governmental involvement in controlling greenhouse gas emissions. Existing legislative proposals reflect continuing emphasis on market-based solutions, but tempered by governmental oversight and facilitation.

In the EU, by contrast, prevailing neoclassical paradigms manifest in a more moderate, socially-grounded style of capitalism. European capitalism has intensified with efforts by the EU to modernize and become a dominant global economic player. EU policy reflects increasing emphasis on deregulation and 'free' market functioning. Yet, European capitalism continues to emerge from and be directed by notions of social well-being. That is, social demands for welfare, labor, and environmental rights remain drivers of market choices, rather than the reverse. And, it is within this more tempered system of capitalism that the EU climate debate emerged.

---

[87] Fred Block, 'Confronting Market Fundamentalism: Doing "Public Economic Sociology"' (2007) 5 *Socio-Economic Rev* 319, 327. [88] Ibid 328.
[89] Karen Litfin, 'Environment, Wealth, and Authority: Global Climate Change and Emerging Modes of Legitimation' (2000) 2 *Intl Studies Rev* 119.

The EU mode of capitalism created room for policymakers to represent climate policy as both a social necessity and an economic opportunity. That is, prevailing US economic philosophy encouraged an image of greenhouse gas emissions regulation as unnecessary governmental interference in affluence-creating activities. Conversely, prevailing economic theory in the EU supported the depiction of climate change as an opportunity to use regulation to maximize social well-being with concomitant benefits for global economic competitiveness.[90] European capitalism contributed to EU efforts to frame climate policy as an opportunity for Community-wide efforts to improve energy security, incentivize innovation, improve global competitiveness, and revitalize European industry.

Shared adherence to neoclassical economic analysis unites US and EU climate policy around common goals of economic flexibility and global economic competitiveness. Variations in the dominant modes of capitalism, however, prove important points of departure in climate change law and policymaking. Thus, while economic rationales have proved powerful tools for influencing European and American climate policies, more often than not, the effect has been to encourage political action in the EU while discouraging similar actions in the US.

Ethics

Complex ethical questions often seem remote to the realities of modern environmental lawmaking. Ethical questions, however, play an important role in shaping political perspectives on climate change. Even when adopting a pragmatic perspective and even in the secular world of international climate change law, ethics do matter. Questions of climate ethics play an important part in shaping the language and contours of domestic and international climate change debates. While the often obscure and unapproachable debates that take place in philosophical circles matter little to policy negotiations, the principles that survive the vagaries of these debates and filter out into the public domain inform cultural perceptions and influence domestic political strategies and international negotiating positions. In particular, questions over how to balance the rights and responsibilities of the developed and developing world and how to assess our responsibilities towards future generations of world citizens bring ethics and climate change policymaking into direct contact. European and American political responses to climate change are intricately interlinked with these underlying ethical questions.

The UNFCCC firmly places ethics at the center of the international legal regime, providing that Parties to the Treaty 'should protect the climate system...on the basis of equity and in accordance with their common but dif-

---

[90] Eg, S Pfeifer and R Sullivan, 'Public Policy, Institutional Investors, and Climate Change: a UK Case Study' (2008) 89 *Climatic Change* 245.

ferentiated responsibilities and respective capabilities'.[91] This provision creates the parameters for one of the most salient ethics-based climate debates: what does common but differentiated responsibilities mean and how should it be actualized?[92]

The practice of differentiating responsibilities has existed in international law for years,[93] but the specific term of common but differentiated responsibilities was first formally used in 1992 by Principle 7 of the Rio Declaration on Environment and Development, stating that:

> In view of the different contributions to global environmental degradation, States have common but differentiated responsibilities. The developed countries acknowledge the responsibility they bear in the international pursuit of sustainable development in view of the pressures their societies place on the global environment and of the technologies and financial resources they command.[94]

That same year, the UNFCCC became the first multilateral environmental agreement to employ the term 'common but differentiated responsibilities' (CBDR). In the UNFCCC, the term is used to suggest that the international community shares a common responsibility for protecting the global atmosphere, but that the responsibility for addressing global climate change should be differentiated among the countries of the world based on past contribution to the problem as well as present capacity to respond.[95]

Inclusion of article 3 in the UNFCCC divided the international community. Many developed countries, including the US, opposed the inclusion of article 3 fearing that it created legal obligations additional to the primary obligations set out in article 4. When it became evident that inclusion of article 3 was vital to the continuing participation of developing countries, the US shifted course and

---

[91] United Nations Framework Convention on Climate Change (adopted 9 May 1992, entered into force 21 March 1994) 31 ILM 849 art 3(1) (UNFCCC).

[92] S Caney, 'Cosmopolitan Justice, Responsibility, and Global Climate Change' (2005) 18 *Leiden J Intl L* 747.

[93] See Christopher Stone, 'Common but Differentiated Responsibilities in International Law' (2004) 98 *AJIL* 276, 278. The 1972 Stockholm Declaration laid the foundations for CBDR in environmental law:

> Without prejudice to such criteria as may be agreed upon by the international community, or to standards which will have to be determined nationally, it will be essential in all cases to consider the systems of values prevailing in each country and the extent of the applicability of standards which are valid for the most advanced countries but which may be inappropriate and of unwarranted social cost for the developing countries.

Report of the United Nations Conference on the Human Environment (Stockholm, Sweden, 5–16 June 1972) (16 June 1972) UN Doc A/CONF.48/14 (Stockholm Declaration).

[94] Rio Declaration (n 50) principle 7; see also UNFCCC (n 91); see generally Philippe Sands, *Principles of International Environmental Law* (2nd edn, CUP, Cambridge, 2003) 225–28; Duncan French, 'Developing States and International Environmental Law: The Importance of Differentiated Responsibilities' (2000) 49 *ICLQ* 35.

[95] L Rajamani, 'The Principle of Common but Differentiated Responsibility and the Balance of Commitments under the Climate Regime' (2000) 9 *Rev Eur Community & Intl Envtl L* 120, 121.

sought to limit the scope of article 3. To limit the reach of article 3, the US introduced amendments:

- to add a Chapeau specifying that principles were to 'guide' parties;
- to replace the term 'State' with the term 'Parties';
- to add the term '*inter alia*' to the Chapeau to clarify that Parties are free to consider principles other than those specifically listed in article 3.[96]

The primary aim of these three changes was to pre-empt suggestions that the principles in article 3 constitute part of customary international law and, thus, take on greater normative status.[97] The US succeeded in curtailing the language of article 3. Despite these modifications, however, common but differentiated responsibility has emerged as the 'overall principle guiding future development of the regime'.[98]

Yet, entrenched disagreements continue to exist over the meaning of 'common but differentiated responsibilities'. The US's position is central to the debate. While the US has never openly rejected the notion of common but differentiated responsibilities, it advocates an interpretation of the term that differs from the interpretation suggested by the G77/China.

The US, along with Japan, Canada, Norway, Australia, and New Zealand, argues for an interpretation of CBDR that allows for differentiation of commitments between developed and developing countries while still expecting meaningful participation by developing countries in the climate regime. The US interpretation of CBDR is best exemplified by the 1997 Byrd–Hagel Resolution. The Byrd–Hagel Resolution, which was passed unanimously in the US Senate by a vote of 95–0, expressed the sense of the Senate that the US should not sign any protocol to the UNFCCC that did not include binding commitments for developing countries.

While the US ultimately rescinded any affiliation with the Kyoto Protocol in large part due to the fact that it failed to create emission reduction commitments from key developing countries, namely China and India, the resolution did not constitute an outright rejection of the Kyoto Protocol or the principle of CBDR. Rather, many Senators accepted that the US was 'more responsible' for causing anthropogenic climate change but rejected the notion that industrialized countries should bear the full burden of addressing climate change.

Instead, the Senators wanted a sharing of burdens and a clarification of commitments. The Byrd–Hagel Resolution expresses a desire for the respective responsibilities of developed and developing countries to be codified even if the

---

[96] Ibid 124; Daniel Bodansky, The U.N. Framework Convention on Climate Change: A Commentary, (1993) 18 *Yale J Intl L* 451.
[97] Rajamani (n 95) 124; Bodansky, 'The U.N. Framework Convention on Climate Change' (n 96).
[98] Rajamani (n 95) 124; Bodansky, 'The U.N. Framework Convention on Climate Change' (n 96).

resulting codification allows emissions increases.[99] Rather than rejecting the principle of CBDR, the Byrd–Hagel Resolution reframed the CBDR debate to include questions over the best way to meaningfully differentiate commitments under the climate change regime as opposed to interpreting CBDR as placing the full brunt of commitments on industrialized nations.[100] Since adopting the Byrd–Hagel Resolution in 1997, the US has consistently challenged any interpretation of CBDR that exempts developing countries from legally binding commitments under the international climate regime.

In contrast to the US position, the G77/China advocates an interpretation of CBDR that explicitly rejects the notion that developing countries should undertake any legally binding commitments and, instead, suggests that developing country responsibilities under the climate change regime are limited to efforts to promote sustainable development, adaptation, and domestic reporting.

The EU position falls somewhere between these two extremes, beginning closer to the position of the G77/China during early climate negotiations but inching closer to the US position as revealed during the 2009 Copenhagen negotiations. During early climate negotiations, the EU sought to mediate US and G77 by rejecting the possibility of developing country commitments in the short-term but stressing the importance of regulating developing country emissions in the long-term.[101]

The EU, however, is less ambiguous than the US in its acceptance of CBDR. EU climate policy is guided by the principle of CBDR and the notion that industrialized countries must lead global efforts to address climate change.[102] At the international level, the EU's ratification of the Kyoto Protocol signaled acceptance of the principles enunciated in article 3, including an interpretation of CBDR that omits developing countries from binding emissions reduction targets or timetables.

At the regional level, EU 'practice in a wide range of policy areas suggests an acceptance of common but differentiated responsibilities and of the idea of international equity more generally'.[103] Similarly, the decision to operate as a 'bubble' under the Kyoto Protocol reflected the EU's willingness to incorporate CBDR-style commitments into regional law through its burden-sharing agreement.

---

[99] Paul G Harris, 'Common but Differentiated Responsibility: The Kyoto Protocol and United States Policy' (1999) 7 *New York U Envtl L J* 27, 38–40.

[100] Ibid 41–42; Rajamani (n 95) 128.

[101] Farhana Yamin, 'The Role of the EU in Climate Negotiations' in J Gupta and M Grubb (eds), *Climate Change and European Leadership: A Sustainable Role for Europe?* (Kluwer Academic Publishers, Dordrecht, Netherlands, 2000) 62–63.

[102] Paul G Harris, 'The European Union and Environmental Change: Sharing the Burdens of Global Warming' (2006) 17 *Colorado J Intl Envtl L & Policy* 309, 338–40.

[103] Jutta Brunnee, 'Europe, the United States and the Global Climate Regime: All Together Now?' (2008) 24 *J Land Use & Envtl L* 1.

EU regional and international policymaking efforts reflect concerted attention to operationalizing the principle of CBDR in line with notions of equity.[104] For example, preceding the 2000 UNFCCC COP, the Commission recommended that European diplomats at the COP:

ensure that industrialized countries take real action at home to reduce emissions by securing a primary role for domestic policies and measures..., limit the extent to which 'sinks' that absorb carbon, such as forests, can be used by industrialized countries to offset emissions..., establish a comprehensive and tough regime to oversee Parties' compliance with the Protocol, including effective penalties with a clear economic impact for non-compliance by industrialized countries with their emission targets [and] help developing countries meet their Convention [reporting requirement] commitments and adapt to climate change through capacity building, transfer of clean technologies and financial assistance.[105]

The Commission also recognized the importance of bringing the large developing countries into the mix in the long-term, but it conditioned this expansion of commitments on industrialized countries ability to fulfill existing mitigation and resource and technology transfer commitments.[106] This firm stance on the responsibilities borne by industrialized countries reflects an almost bipolar position to that argued by US Senators in the 1997 Byrd–Hagel Resolution.

Leading up to and during the 2009 climate talks in Copenhagen, however, the EU stepped up efforts to increase the level of commitment extracted from rapidly developing countries in the post-Kyoto framework. In a talk given in July 2009, President of the European Commission, José Manuel Barroso stated:

Indeed, as the largest contributors to past emissions, developed countries have a special responsibility to take the lead. But this is not going to be enough. The emerging economies, for example, where growth in emissions is surging, must also join in the effort. We must all do our part, in line with the principle of common but differentiated responsibilities and respective capabilities.[107]

Barroso's statement reflects growing global concern about the collective effort necessary to effectively address climate change, but it does not clarify the extent to which the EU expects developing countries to participate in the post-Kyoto framework. Indeed, due to the sensitivity of this topic, most political negotiations have framed the topic in general terms that avoided polarizing the debate. Further, the EU continues to recognize the importance of equity considerations in establishing commitments, as reflected in the differentiated 2020 and 2050 emission goals proposed by the Europe Union.[108] Thus, while EU and US

---

[104] Ibid.
[105] Harris (n 102) 334–35, citing Commission (EC), 'Briefing Paper: The EU's Positions for COP6' (3 November 2000). [106] Ibid.
[107] Europa, 'José Manuel Durão Barroso, President of the European Commission, Press Conference Brussels, 6 July 2009' (Transcript of Speech, 6 July 2009) SPEECH/09/330.
[108] Brunnee (n 103) 38.

rhetoric now more closely align in calling for meaningful participation by the rapidly developing economies, the extent to which the EU and the US hinge their participation in a post-Kyoto regime on developing country participation is likely to differ significantly.

Questions over the most equitable way to differentiate commitments under the Kyoto Protocol split the US and the EU early on and have played a central role in directing US and EU climate strategies down different pathways. Economic concerns have consistently dominated US debate and decision-making, while questions of equity and differential treatment have largely been treated as secondary considerations to near-term domestic, economic, and social stability. In contrast, while economic concerns similarly pervade EU climate debate, questions of equity receive greater media coverage and political attention prompting the EU to construct a climate policy based upon principles of equity, central among which is the principle of CBDR. As a result, at the global level, the EU has earned greater moral authority to push for developing country participation in a post-Kyoto regime.

Ethical questions have been central to the development of divergent EU and US climate change strategies. The debate over CBDR is the most prominent ethical question influencing climate policies and on-going negotiations but it is by no means the only pressing ethical concern. Questions of inter-generational equity and climate justice, for example, raise pressing ethical dilemmas. Neither issue has played a determinative role in transatlantic policymaking, but both issues are receiving increased attention as the global community turns more attention towards adaptation.

Diverging ethical perspectives shaped early responses to climate change in the US and the EU in such a way as to embed cultural perspectives on the relative roles and responsibilities of developed and developing countries in addressing climate change. Ethical questions continue to be of profound importance in negotiating a post-Kyoto framework for international action on climate change. The political positions that the US and the EU adopt on two ethics-based questions—the responsibility of developing countries to undertake emissions reduction commitments and the appropriate balance between mitigation and adaptation measures—will indelibly influence negotiations for a post-Kyoto regime.

Conclusion

This chapter has examined how underlying socio-legal factors shape transatlantic climate change law and policymaking. This discussion is far from exhaustive but, rather, highlights the complexities involved in international climate negotiations among industrialized countries sharing many social and legal commonalities. What emerges from this analysis is the importance of placing climate change law and policymaking within its wider social context in order to understand the web of economic, political, cultural, and social interactions that shape domestic responses to climate change.

Examining how and why US and EU approaches to climate change law and policy vary has significance beyond a simple comparative study. The shape and success of the international climate regime hinges on the willingness and ability of influential countries, such as the US and the EU, to adopt leadership roles and to push international negotiations forward.[109] Absent the participation of powerful countries and absent the willingness of these countries to cajole widespread global participation, the international climate regime is unlikely to succeed. The EU's willingness to play an active leadership role has been fundamental to the continuing evolution of the climate regime. Participation and leadership by the US, however, is essential to ensuring the viability of the climate regime moving forward.

The centrality of US and EU participation to a functioning climate regime makes it imperative to unpack the factors responsible for driving European and American climate policies down different tracks. Recent political changes suggest that the US is poised to re-engage with international climate negotiations. Despite recent rhetoric and political activity, however, there remain real concerns about whether political leadership changes in the US are enough to compel change.[110] Given that political leadership is only one component of a network of factors influencing climate policy, the question becomes whether political leadership is the trigger factor that will compel complementary changes in public perception of risk and uncertainty, media coverage, economic philosophy, and ethical perspectives such as to spur US climate leadership.

The foregoing analysis suggests that while political leadership is a central factor in shaping modes of domestic policymaking and international engagement, entrenched social factors in both the US and EU continue to play an important role in impeding efforts to address climate change. Primary among these factors is an overwhelming focus on near-term economic well-being, which also plays a key part in shaping perceptions of risk, media coverage, and notions of ethics and equity. The intersection of political leadership changes and federal climate initiatives with the onset of economic recession creates additional hurdles to efforts to redirect US climate policy. EU climate policy is similarly affected by political leadership and economic considerations. As discussed throughout this chapter, however, the fragmented nature of EU political decision-making coupled with the EU's more moderate mode of capitalism mean that political, scientific, media, ethical, and economic factors are weighted dissimilarly and interact through more decentralized channels with the effect of enabling EU climate leadership to continue despite fluctuations in political leadership and economic well-being.

---

[109] Neil Carter, *The Politics of the Environment: Ideas, Activism, Policy* (Harvard University Press, Cambridge MA, 2001).

[110] Matthew Paterson, 'Post-Hegemonic Climate Politics?' (2009) 11 *British J Politics & Intl Relations* 140.

Political theorists increasingly question the premise that the greatest barriers to international environmental policymaking are 'differences between governments abroad, not differences within governments at home'.[111] In the context of the global climate regime, domestic social, political, and economic differences are decisive in determining international policymaking positions. The nuances of US and EU climate change law and policymaking are intricately interwoven with differences at home. To better understand these differences is to better understand the role the US and the EU have and will play in international climate change law and policymaking.

The socio-legal factors influencing EU and US climate change policymaking extend far beyond the elements considered here. Geographic location, for example, creates social and environmental differences that shape perspectives on climate change. Similarly, religion and religious values impact domestic climate debates. The cultural factors influencing US and EU climate change policymaking are as vast and complex as the scientific data underpinning the debate. They are also no less important. This chapter offers a glimpse into the socio-legal dimensions of transatlantic climate change policymaking and demonstrates the importance of examining socio-legal factors to the continuing process of international regime building.

---

[111] Chasek (n 20); see, eg, Sussman (n 14); DeSombre (n 20).

PART V

THE FUTURE OF INTERNATIONAL CLIMATE CHANGE POLITICS

9
Conclusions and the Way Forward

The Road Travelled

As 2010 neared and the 2009 UN Climate Change Conference in Copenhagen loomed, leaders from around the world looked to the European Union and the United States to discern the parameters for the global climate change debate. The ever pressing reality that participation by key developing countries, including China, India, and Brazil would be fundamental to global efforts to address climate change dominated pre-Copenhagen negotiations. Yet, even as the political and environmental roles of the rapidly developing economies grew in importance so did the roles of the US and the EU, albeit in very different ways. The US held the key to the ultimate survival of a global climate change agreement—without its full participation neither China nor India was likely to participate in a legally binding agreement. In essence, if the US opted out, once again global efforts to address climate change would fail. The US was perceived to be the indispensable political player; its participation could enable, if not guarantee, success. The EU did not possess a similar level of political 'do-or-die' power, but its role was equally important. The EU fueled international momentum on climate change. It stood virtually alone among industrialized countries in its persistent commitment to negotiating a new and improved global climate agreement. With the other vocal advocates for aggressive global action on climate change largely consisting of minor political players, such as the Alliance of Small Island States, the EU bore great responsibility for sustaining momentum for a meaningful and ambitious new global climate change agreement.

At a critical moment for climate change, divergent climate histories led the US and the EU to two very different, but fundamentally important, political positions in December 2009. Over the decades ensuing since global climate negotiations began in earnest in the 1980s, the US and the EU had ceded considerable political and economic power to the rapidly developing economies. Nevertheless, both the US and the EU continued to maintain decisive roles going into the 2009 Copenhagen Climate Change Conference.

The EU and the US brought the ups and downs of three decades of domestic climate politics to the global negotiating table with them in Copenhagen; hope hinged on the ability of the EU to maintain its progressive climate politics and the willingness of the US to abandon its long-standing history of political lethargy. Over the decades since climate change appeared on the global agenda, it has snaked its way from the margins of political debate to ascend to the highest level of social and political concern in both the US and the EU, but to very different effect. These tumultuous and varied political histories created the backdrop for post-Kyoto Protocol negotiations. As the international community gathered together to find a way forward in addressing global climate change, US and EU actions past and present served as indelible indicators of future possibilities.

The Escalation of Climate Change Politics in 2009

The 2009 Copenhagen Climate Change Conference marked the end of a critical decade for climate change politics. The first decade of the 21st century had proved pivotal to the physical and political analyses of climate change. Boosted by warnings from the IPCC that warming of the climate system is accelerating rapidly and that the global average temperature has increased by 0.74º Centigrade over the last century and is predicted to increase by between 1.8 to 6.4°C by the end of the 21st century and confirmation by the United Nations that climate change poses a 'massive threat to human development', social and political climate activism boomed during the first decade of the new millennium.[1] Climate change became a *cause célèbre* amongst not only traditional environmental activists but also amongst mainstream civil society, business, and political organizations. Guided by growing scientific consensus on the threats posed by anthropogenic climate change, activists and politicians worldwide began approaching climate change as a new type of environmental problem—one that threatens primary state interests. As the profile of climate change grew, however, the politics became ever more complex. State, regional, and multilateral efforts to address climate change proliferated, but initiatives differed widely in scope, focus, and style. Varying notions of uncertainty, risk, ethics, and equity presented daunting stumbling blocks to local, regional, and international efforts to develop consensus on the appropriate response to climate change.

Within this multifarious framework, two dominant themes emerged: economic wellbeing and security. Economics and security—the twin pillars of foreign policy—increasingly provided the foundations for the development of climate policy worldwide. One need only look to US President Barack Obama's handling of climate policy, which he intimately intertwines with energy policy, for evidence of these inter-linkages. The intertwining of climate change with

---

[1] Core Writing Team, R Pachauri and A Reisinger (eds), *Climate Change 2007: Synthesis Report* (Contribution of Working Groups I, II, and III to the Fourth Assessment Report of the Intergovernmental Panel on Climate Change) (IPCC, Geneva, 2007) 104; United Nations Development Programme, *Human Development Report 2007/2008: Fighting Climate Change—Human Solidarity in a Divided World* (5th edn, Palgrave Macmillan, New York, 2008) <http://hdr.undp.org/en/reports/global/hdr2007-2008/> accessed 3 April 2010, at v.

economic wellbeing and security simultaneously redefined climate change as an issue of primary concern to national sovereignty while also increasing the complexity of international negotiations since the manner in which these notions are defined and incorporated into law raises fundamental questions of equity.[2]

Even with the political boost given to climate politics via its marriage to more traditional state concerns and even accounting for the dramatic rise in social and political activism, the global community continued to struggle to find an effective, equitable, and sustainable solution to the problem of climate change. Climate change politics deepened only to expose growing divides—between the rich and the poor but also amongst those individuals and societies similarly situated.

During 2008, the opening year of the Kyoto Protocol's first compliance period, the exposed rifts appeared perilous. The US remained firmly outside the international arena, leaving little room to woo China and India into a new obligatory framework. Even the EU struggled to reach regional consensus on a way forward. In the ebb and flow of climate policy, 2008 brought about a new low only to then usher in one of the most dramatic and eventful years thus far in global climate change politics.

Beginning with the election of Barack Obama to the US Presidency and culminating with the UN Climate Change Conference in Copenhagen, 2009 proved to be a capstone year in efforts to elevate climate change to the forefront of international law and policy. In early 2009, President Obama took office on a platform that promised to prioritize domestic efforts to address climate change. Shortly thereafter, he began taking small steps to advance the creation of a US regulatory regime for greenhouse gases. President Obama's actions, in turn, prompted the US Congress to ramp up efforts to pass a piece of primary climate change legislation. As the world watched, the US began to take its first steps towards outlining a substantive climate change policy.

And the world was watching. Given the perceived indispensability of the US to the political process, small steps at the domestic level equated to significant indicators of progress for future international negotiations. The significance of the shift in US politics was made evident by the decision to award the 2009 Nobel Peace Prize to President Obama, in part due to his role in encouraging the US to 'pla[y] a more constructive role in meeting the great climatic challenges the world is confronting'.[3] The Nobel Peace Prize Committee's decision to award Obama the prize and, in so doing, to highlight his efforts to address climate change simultaneously rewarded Obama's early efforts while also exhorting him to see his efforts through. The awarding of the prize in December 2009 coincided with the UN Climate Change Conference in Copenhagen, creating additional incen-

---

[2] Carlarne, C, 'Risky Business: The Ups and Downs of Mixing Economics, Security and Climate Change' (2009) 10 *Melbourne Journal of International Law* 2.

[3] Nobelprize.org, 'Nobel Peace Prize for 2009' (Press Release, 9 October 2009) <http://nobelprize.org/nobel_prizes/peace/laureates/2009/press.html> accessed 3 January 2010.

tive for President Obama to attend the Climate Change Conference in person, something his predecessors had declined to do.

In late 2009, immediately preceding the December Climate Conference, the glow of Obama's Presidency and the momentum of the international community began to fade. The global economy continued to struggle, social and scientific debate on the certainty of human-induced climate change intensified, and key global negotiators warned that the international community would be unable to negotiate a new legally binding framework at the December Conference.[4] Partially combating this slump, on 24 November 2009, President Obama confirmed that he would attend the Copenhagen Conference in person and that the US was offering a goal of reducing domestic greenhouse gas emissions by 17% as against 2005 levels by 2020. Spurred on by this development, the following day, the Chinese Premier, Wen Jiabao, announced that he would also attend the Conference.

In the announcement, China for the first time offered the global community a clearer indication of what it would be willing to do to reduce its domestic emissions, stating that it would take voluntary measures to cut domestic emissions of carbon relative to economic growth by 40% to 45% by 2020 compared with 2005 levels.[5] This announcement did not indicate that China would be willing to commit to a legally binding emissions reduction framework, but it represented one of the most significant efforts by China to date to engage with international efforts to address climate change.

Compelled by the Chinese and US pledges and facing mounting international pressure, on 2 December 2009, India revealed its own voluntary target to cut domestic carbon intensity by 24% by 2020 as against 2005 levels. Rounding out these new commitments was an earlier declaration by Brazil that it aimed to reduce the anticipated level of its greenhouse gas emissions in 2020 by 36–39%.[6] The decision by the rapidly developing economies to use an emissions intensity standard, rather than committing to absolute emissions reductions continued a pattern long advocated and utilized by the US under George W. Bush's administration and came as no surprise given the reluctance of the developing nations to sacrifice the economic growth perceived as necessary to improve standards of living.

The collection of new national targets issued in late 2009 existed alongside the comparatively long-standing EU commitment to reduce greenhouse gas emissions 20% by 2020 using a 1990 baseline. Leading in to the Copenhagen Conference, the juxtaposition of the new US, Chinese, Indian, and Brazilian targets with the

---

[4] The so-called Climategate scandal caused social and scientific strife when the emails of numerous climate scientists were hacked and made public in an effort to undermine the reliability of scientific consensus on climate change. See Andrew Revkin, 'Hacked E-Mail is New Fodder for Climate Dispute' *New York Times* (New York, 20 November 2009) <http://www.nytimes.com/2009/11/21/science/earth/21climate.html> accessed 3 January 2010.

[5] Jonathan Watts, 'China Sets First Targets to Curb World's Largest Carbon Footprint' *Guardian* (26 November 2009) <http://www.guardian.co.uk/environment/2009/nov/26/china-targets-cut-carbon-footprint> accessed 3 January 2010.

[6] Gary Duffy, 'Brazil Proposes Carbon Cut Target' (BBC News Release, 14 November 2009) <http://news.bbc.co.uk/2/hi/8360072.stm> accessed 3 January 2010.

EU target created a bewildering patchwork of targets, none of which aligned. The EU target, using a 1990 baseline, was by far the most ambitious. The US target appeared attractive, but its reliance on a 2005 baseline meant that it was committing to reducing US emissions to roughly 3-4% below 1990 levels by 2020, a figure considerably below the EU target and below levels necessary to stabilize atmospheric greenhouse gases at a level deemed 'safe' by the IPCC. Meanwhile, the developing country targets were significant symbolic steps yet failed to offer mechanisms for curbing rapidly escalating emissions growth from three of the top ten global emitters. According to a UNFCCC study, even assuming that all of the big polluters met their stated targets, global temperatures would still rise by an average of 3ºC, a level that the IPCC predicted would mean disaster for millions of people.[7]

Following the outpouring of international expressions of intent, on 7 December 2009, the US EPA released its long-awaited endangerment finding, laying the groundwork for EPA to regulate greenhouse gases under the US CAA.[8] The publication of the endangerment finding marked a turning point for US climate policy by creating the first discrete legal mechanism for regulating domestic greenhouse gas emissions. This domestic measure, in turn, provided momentum to international negotiations by offering tangible evidence that the US was making concerted efforts to more fully engage with global climate politics. Three days later, on 10 December 2009, in his acceptance speech for the Nobel Peace Prize, President Obama reiterated his commitment to renew American leadership in international diplomacy, characterizing climate change as a threat to peace and security and calling upon 'the world to come together to confront climate change'.[9]

During the weeks immediately preceding the Copenhagen Climate Change Conference, news on the US, China, and India dominated the headlines, with stories about the EU occupying a secondary position. The outcome of Copenhagen appeared dubious, with pre-negotiations in Barcelona posing more questions than they answered and with the major polluters showing few signs of nearing common ground. At this juncture, the relative roles of the US and the EU became increasingly clear.

The EU occupied the role of symbolic leader, offering leadership through its actions and diplomacy. Lacking the political or economic clout to bully countries

---

[7] S Goldenberg, J Vidal, and J Watts, 'Leaked UN Report Shows Cuts Offered at Copenhagen Would Lead to 3C Rise' *Guardian* (17 December 2009) <http://www.guardian.co.uk/environment/2009/dec/17/un-leaked-report-copenhagen-3c> accessed 3 January 2010.

[8] US EPA, 'Endangerment and Cause or Contribute Findings for Greenhouse Gases under Section 202(a) of the Clean Air Act' (Description of Regulatory Climate Change Initiatives) <http://www.epa.gov/climatechange/endangerment.html> accessed 3 January 2010.

[9] The White House, Office of the Press Secretary, 'Remarks by the President at the Acceptance of the Nobel Peace Prize' (President Barack Obama's Acceptance Speech, Oslo, Norway, 10 December 2009) <http://www.whitehouse.gov/the-press-office/remarks-president-acceptance-nobel-peace-prize> accessed 3 January 2010.

such as China, India, or the US into adopting its more aggressive targets, the EU continued to push its established objectives through its framing of the debate, emphasizing the urgent economic, security, and human dimensions of climate change. Increasingly, however, what authority and influence Europe had appeared to be diminishing.

Conversely, the US charged into the debate with no firmly established diplomatic position, no pre-existing political legitimacy, and no history of domestic progress on climate change yet with assumed authority. Everything that the US had to offer rested on its promise of change. The Obama administration wielded its promise of change deftly, using it as a tool for recreating the image of the US as a worthy, moral, and indispensable actor and for eliciting promises of participation from its global counterparts. The US approached the Copenhagen Conference with confidence and authority belying its recent history of reticence and resistance. The EU, relying on its historical record to advance its agenda, was overshadowed by a bold US campaign to assert itself as one of, if not the essential player in international climate negotiations. Years of advocacy and action reserved the EU a crucial seat at the negotiating table, but the nuances of global politics and political personalities placed the US and the rapidly developing economies in the seats of power.

The 2009 Copenhagen Climate Change Conference

The Copenhagen Climate Change Conference opened on 7 December 2009 amidst growing social and political disorder. The US and the EU both maintained strong presences at the meeting. Yet, from the opening hours of the meeting it became evident that no one State, or bloc of States would be able to direct negotiations or ensure a unified outcome.

The Copenhagen Conference revealed the degree to which power—whether political, symbolic, or merely procedural—was dispersed among the State participants. The Alliance of Small Island States, the African bloc, and the least developed countries possessed virtually no economic or political power yet vocally and effectively wielded great moral authority, supported in great part by a very active civil society contingent. The rapidly developing economies, including China, India, Brazil, and South Africa diverged on many points but banded together to negotiate terms of an agreement favorable to the developing countries. China, individually, drew upon its status as the largest global greenhouse gas emitter and one of the most significant global economies to represent itself as the new indispensable nation holding the do-or-die political card. The larger G-77 initially presented a powerful negotiating bloc but gradually dissolved as irreconcilable differences emerged among the parties. The EU pushed for aggressive targets but appeared divided as the Danish hosts faltered in their Conference leadership abilities, French President Nicolas Sarkozy drifted from the party line, Germany failed to make a powerful showing, the UK dominated headlines with rhetoric, and the EU bloc, as a whole, was sidelined in major negotiations. The US assembled a powerful showing at the Conference and offered moments of dazzling diplomacy but was hampered by its rehabilitated-rogue status and its adoption of a political position that failed to align with any of the other major polluters.

The Copenhagen Conference negotiations demonstrated many of the difficulties inherent in international environmental law, including fundamental economic, ethical, and procedural barriers, as well as the dysfunctionality long characterizing the UNFCCC process. During the nearly 13 days of negotiations, the talks came to a standstill more than once as parties obstructed proceedings, delegates came to physical blows, and members of civil society staged round after round of protests. Long-standing concerns over the relative roles and responsibilities of developed and developing countries remained at the heart of deliberations over whether it was possible within the UNFCCC framework to agree upon a legal *or* political architecture that would equitably and effectively address climate change.

Traditional questions of roles and responsibilities took on new dimensions as parties from developed and developing nations called for a more nuanced understanding of the principle of common but differentiated responsibilities—one that would differentiate not just between developed and developing but among developing countries in reference to economic and environmental output. These calls came from the richest, the poorest, and the most vulnerable nations. The US and the EU called upon the rapidly developing economies to commit to meaningful actions to reduce greenhouse gas emissions, or at least the *growth* in greenhouse gas emissions. US and EU demands were motivated by science, but also by economic realities. In contrast, Tuvalu and other vulnerable nations demanded meaningful action on the part of all of the major polluters—regardless of economic standing—for reasons of basic survival. Mitigation commitments divided delegates along diverging lines with no clear bridging path. Questions of mitigation were further compounded by interlinked negotiations over adaptation funding.

On 17 December 2009, immediately preceding President Obama's planned arrival at the Conference, US Secretary of State Hillary Clinton made minor inroads into the increasing impasse by pledging that the US would help raise US$100 billion per year through 2020 to help poor and vulnerable nations cope with the effects of climate change. Secretary Clinton did not specify where the money would come from but suggested that it would come from public and private funds mobilized by the US and other nations. She cautioned, however, that the US offer of support was conditional on the major emitting developing countries agreeing to some form of binding, internationally verified emission targets. The US's conditional offer stirred the stalemate, simultaneously generating hope and criticism.

On the heels of Secretary Clinton's speech, on what was supposed to be the last day of the Conference, President Obama arrived to find negotiations deadlocked. Delegates were split not just on one or two key issues but on questions ranging from mitigation, to funding, to technology transfer, to carbon sinks. The Conference was mired in chaos. At this final hour, it appeared unlikely that the delegates would be able to reach any type of agreement. Speaking to a packed room full of exhausted and frustrated delegates, President Obama emphasized the urgency of the issue and America's commitment to international negotiations. Warning that 'there is not time to waste', he declared that:

America has made our choice. We have charted our course. We have made our commitments. We will do what we say. Now I believe it's the time for the nations and the people of the world to come together behind a common purpose. We are ready to get this done today—but there has to be movement on all sides to recognize that it is better for us to act than to talk; it's better for us to choose action over inaction; the future over the past—and with courage and faith, I believe that we can meet our responsibility to our people, and the future of our planet.[10]

Obama's eight-minute speech was greeted tepidly. The President's decision to address the Conference and, in so doing, to confirm the US's responsibility as one of the biggest polluters was an important symbolic gesture. Beyond its symbolic value, however, the speech offered no new advances to the negotiations. President Obama did not offer anything new in the way of mitigation or adaptation commitments, merely reiterating the previously stated target of cutting emissions 17% by 2020, and by 80% by 2050 and mobilizing financing per Secretary Clinton's announcement of the day before. The President's speech marked the return of the US to the center of high level climate politics but it failed to jump-start the stalled negotiations. At the start of the last formal day of the Conference, the delegates remained firmly deadlocked.

With the plenary sessions continually mired in conflict and the working groups hammering away at details in the absence of an overarching framework, there appeared to be little hope that the Copenhagen Climate Change Conference would produce anything more than an image of a world divided. At this point, however, the Conference diverged from the normal pattern of international environmental policymaking. Rather than leaving career diplomats and negotiators to finesse the terms of a deal to which the heads of state could give their stamp of approval, the heads of state present at the Conference engaged in a series of bilateral and multilateral meetings that proved pivotal to the ultimate ability of the Conference to produce any manner of agreement.

Throughout the Copenhagen Conference, delegates and heads of government from the rapidly developing economies of China, India, and Brazil engaged in private meetings over the terms of a possible climate change agreement. On the evening of 18 December 2009, following a full day of meetings with various

---

[10] The White House, Office of the Press Secretary, 'Remarks by the President at the Morning Plenary Session of the United Nations Climate Change Conference' (President Barack Obama, Copenhagen, 18 December 2009) <http://www.whitehouse.gov/the-press-office/remarks-president-morning-plenary-session-united-nations-climate-change-conference> accessed 3 January 2010.

heads of state, President Obama prepared to leave Copenhagen having made no discernible inroads into the impasse. Prior to his departure, President Obama was reportedly scheduled to meet with that the Chinese Premier, Wen Jiabao, only to learn that the Premier was in a private meeting with the Indian Prime Minister, Manmohan Singh, the Brazilian President, Luiz Inacio Lula da Silva, and South African President Jacob Zuma. Un-phased, President Obama entered the meeting.[11] His decision proved crucial. Walking into the room uninvited and unexpected, President Obama and his counterparts emerged after less than an hour with a non-binding political accord that eventually became the centerpiece document of the Copenhagen Conference. This three-page agreement, the 'Copenhagen Accord', contained nothing of the negotiating texts, yet offered a path forward amidst the collapse of formal negotiations.

In critical part, the Copenhagen Accord received the approval of the US and China—the two most important yet historically reticent political actors in global climate change politics. Noticeably absent from the multilateral negotiations that produced the Copenhagen Accord was the EU. The EU may not have been intentionally excluded from these critical last minute negotiations, but its absence revealed the extent to which symbolic and substantive power was shifting away from Europe towards the US, China, and, to a lesser degree, India. The EU's absence from the meeting was also significant in that it enabled the five participating nations to negotiate an agreement that in all likelihood would not have been possible if the EU had been present and pushing a more aggressive agenda.[12] The composition of the meeting was a historical coincidence but it reflected a reordering in the power of climate politics that had emerged during the Conference.

Following the announcement that the five heads of state from the US, China, India, Brazil, and South Africa had reached agreement, President Obama held a press conference heralding the agreement as 'a meaningful and unprecedented breakthrough', noting that '[f]or the first time in history all major economies have come together to accept their responsibility to take action to confront the threat of climate change'.[13] His optimism proved premature, however. President Obama and the other heads of state departed Copenhagen that evening leaving

---

[11] Philip Sherwell, 'Barack Obama denies accusations that he "crashed" secret Chinese climate change talks' *The Telegraph* (Dec. 19, 2010) <http://www.telegraph.co.uk/earth/copenhagen-climate-change-confe/6845952/Barack-Obama-denies-accusations-that-he-crashed-secret-Chinese-climate-change-talks.html> accessed 18 June 2010. For a slightly different account of the incident, see Jake Tapper, 'High Drama in Copenhagen (per Administration Officials)' (ABC News Blog, 18 December 2009) <http://blogs.abcnews.com/politicalpunch/2009/12/high-drama-in-copenhagen-per-administration-officials.html> accessed 3 January 2010.

[12] While the European Union did not formally object to the Copenhagen Accord, it criticized it as weak and lacking in ambition. President of the European Commission, José Manuel Barroso, 'Joint statement of the EU Presidency and the European Commission—Copenhagen Climate Accord' (News Release, 19 December 2009) <http://ec.europa.eu/commission_barroso/president/index_en.htm> accessed 3 January 2010.

[13] The White House, Office of the Press Secretary, 'Remarks by the President During Press Availability in Copenhagen' (President Barack Obama, Copenhagen, 18 December 2009) <http://www.whitehouse.gov/the-press-office/remarks-president-during-press-availability-copenhagen> accessed 3 January 2010.

an exhausted group of delegates to decide what to do with the new agreement, which on its own had no recognized status within the framework of the UNFCCC. During a debate that lasted throughout the night and into the next morning, the UNFCCC COP was unable to secure the consensus vote necessary to adopt the Copenhagen Accord as an official UNFCCC decision. Overt objections on the part of five of the 193 countries represented, including Bolivia, Cuba, Nicaragua, Sudan, and Venezuela, blocked formal adoption of the Accord. In a bid to recognize the Accord and to allow for implementation of certain of its provisions, the COP voted to 'take note' of the Accord and thereby to include it as part of a UNFCCC COP-15 decision.[14]

Yet, because the agreement itself was not formally adopted, its precise status remains ambiguous. By taking note of the document, the COP included it in the annals of UNFCCC COP-15/CMP-5 'decisions', but the Copenhagen Accord remains a political document only. It does not have any recognized legal status and it remains questionable whether it can even be described as a 'soft-law' instrument since it was neither signed nor formally adopted by the parties to the Conference at the meeting.[15] Following the decision to 'take note' of the Accord in a COP decision, however, Parties to the Convention will be able to sign up to the Accord. Parties who choose to opt-in to the Accord will be listed in a Chapeau and will be eligible to record quantified economy-wide emissions targets for 2020 or nationally appropriate mitigation actions and to become eligible to receive funding.

The general, three-page Copenhagen Accord stands in direct contrast to the legally binding, detailed 20-page Kyoto Protocol that had been negotiated a dozen years earlier. The agreement reflects deepening collective action problems that impeded the COP and prevented parties from being able to agree upon any type of framework for legally binding emission reduction targets. The Copenhagen Accord was equally lauded as a solid framework for moving forward and as a total failure.[16]

Supporters note that the Copenhagen Accord includes key elements of progress, including provisions mandating that developed countries identify by 31 January 2010 new commitments to be made by them with respect to emissions reductions by 2020 and that certain developing countries identify nationally appropriate

---

[14] UNFCCC, COP-15, 'Draft decision -/CP.15: Outcome of the work of the Ad Hoc Working Group on Long-term Cooperative Action under the Convention' (18 December 2009) <http://unfccc.int/files/meetings/cop_15/application/pdf/cop15_lca_auv.pdf> accessed 3 January 2010. Delegates argued that it was necessary for the COP to recognize the agreement in some manner in order to realize certain of the financing provisions, which the Accord describes as being under the guidance of the UNFCCC COP. Ibid.

[15] See Radoslav S Dimitrov, 'Inside Copenhagen: The State of Climate Governance' (2010) 10[2] *Global Environmental Politics* 18, 21.

[16] David Doniger, 'The Copenhagen Accord: A Big Step Forward' (NRDC Climate Center Blog, 21 December 2009) <http://switchboard.nrdc.org/blogs/ddoniger/the_copenhagen_accord_a_big_st.html> accessed 3 January 2010. But see Bill McKibben, 'Copenhagen: Things Fall Apart and an Uncertain Future Looms' (Yale Environment 360 Opinion, 21 December 2009) <http://www.e360.yale.edu/content/feature.msp?id=2225> accessed 3 January 2010.

mitigation actions at first instance by 31 January 2010. The discretion built into the mitigation component of the agreement, however, undermines the value of this section. It is doubtful whether shifting away from a mitigation strategy based on binding targets and timetables to one based on voluntary pledges offers a sound framework for the climate regime. The funding provisions of the agreement, thus, offer the most significant development.

The Copenhagen Accord commits parties to providing '[s]called up, new and additional, predictable and adequate funding', including a commitment by developed countries to contribute funding approaching US$30 billion between 2010–2012 to support mitigation and adaptation activities in developing countries, with adaptation funding to be prioritized for the most vulnerable developing countries; a commitment by developed countries to a goal of jointly mobilizing US$100 billion per year by 2020 to help meet the needs of developing nations, conditional upon transparency with respect to the implementation of meaningful mitigation actions by developing countries; the establishment of first, a High Level Panel 'under the guidance of and accountable to the Conference of the Parties' to explore potential sources of revenue and second, the Copenhagen Green Climate Fund as the operational entity of the financial mechanism of the Convention.[17]

Beyond the provisions of the Copenhagen Accord, the Copenhagen Climate Change Conference produced a series of decisions, including two further decisions of note extending the mandates of the Ad Hoc Working Group on Long-Term Cooperative Action under the Convention (AWG-LCA) and the Ad Hoc Working Group on Further Commitments for Annex I Parties under the Kyoto Protocol (AWG-KP)—the two working groups created in 2007 as part of the Bali Action Plan to enhance implementation of the UNFCCC.[18] The extension of the two Ad Hoc Working Groups permits the continuation of formal efforts to develop a consensus-based plan for future implementation of the UNFCCC and the Kyoto Protocol.

Even with new funding provisions on the table and the continuation of the Ad Hoc Working Groups, it is questionable whether the Copenhagen Accord managed to salvage the UNFCCC process or whether it, in fact, ceded power to a much smaller group of political actors. Leading into the Copenhagen Conference, there was great doubt over the ability of the parties to overcome the collective action problems inherent in the UN process. With a record 115 heads of state present, there was a brief window of opportunity for using the UNFCCC forum to reach global consensus, but this proved impossible. Instead, global consensus proved more elusive than ever while the probability of a small group of actors

---

[17] 'Draft decision -/CP.15' (n 14) para 8; see Dimitrov (n 15)21.
[18] 'Draft decision -/CP.15 (n 14); UNFCCC, CMP-5, 'Draft decision -/CMP.5: Outcome of the work of the Ad Hoc Working Group on Further Commitments for Annex I Parties under the Kyoto Protocol' (2009) <http://unfccc.int/files/meetings/cop_15/application/pdf/cmp5_awg_auv.pdf> accessed 3 January 2010.

determining the global course of action over the long-term became increasingly likely.

The inability of the UNFCCC COP to reach consensus on almost any of the core issues identified in 2007 by the Bali Action Plan created conditions conducive for a smaller group of actors to determine the framework for future negotiations. The decision by a small group of states to act on behalf of the remaining 188 parties with the expectation that their agreement would be accepted reveals the extent to which the UNFCCC process has become dysfunctional, the degree to which power is becoming concentrated with a few powerful political actors, and the challenges that negotiators face between 2010–2012 in pursuing legally binding emission reduction commitments from any or all members of the global community. The momentum for policymaking shifted away from consensus decision-making, creating and uncertain future for the UNFCCC. In this way, the Copenhagen Climate Change Conference marked a turning point in global climate negotiations. The parties were unable to free themselves from the 'fault lines' that had 'imprisoned... them for years' but they did manage to agree upon a skeletal framework for moving forward.[19] Although the Copenhagen Accord failed to offer a solid foundation for combating climate change, it succeeded in keeping negotiations alive. For this reason, the US hailed the agreement as a success. This interpretation, however, has been question by European leaders who view the agreement as diluting the legal foundations of the climate regime and undermining accepted modes of decision-making.[20]

The divergent responses of the US and the EU to the Copenhagen Conference reflect long-standing differences in approach. While President Obama heralded the Copenhagen Accord as a 'success', 'an important *first* step, and 'a meaningful and unprecedented breakthrough',[21] José Manuel Durão Barroso, the President of the European Commission, characterized the agreement as 'better than no accord... but [not] a huge step',[22] bemoaned that fact that the EU would 'not get all we had hoped for', and reiterated the region's commitment to 'continue to lead the world on *legally* binding emissions targets and climate finance for the poorest people on the planet'.[23]

The EU's tepid response to the Copenhagen Accord reflected disappointment not only in the agreement itself but in the outcome of US re-engagement in international climate politics. Led by President Obama, the US actively

---

[19] 'Remarks by the President During Press Availability in Copenhagen' (n 13).
[20] Eg, Christian Egenhofer and Anton Georgiev, 'The Copenhagen Accord—a first stab at deciphering the implications for the EU' (Climate Change Centre for European Policy Change Commentaries, 28 December 2009) <http://www.ceps.eu/book/copenhagen-accord-first-stab-deciphering-implications-eu> accessed 30 December 2009.   [21] Ibid (emphasis added).
[22] 'Joint statement of the EU Presidency and the European Commission—Copenhagen Climate Accord' (n 12).
[23] Europa, 'José Manuel Durão Barroso, President of the European Commission, Statement of President Barroso to the Plenary of the Copenhagen conference on climate change COP 15 Copenhagen' (Press Release, 18 December 2009) SPEECH/09/587 (emphasis added).

re-engaged in international climate negotiations and, in so doing, helped broker the Copenhagen Accord. The EU had long called for US re-engagement. The results of US efforts to renew its international leadership, however, resulted in a reorientation of climate politics that sidelined the EU, threatening not only the EU's symbolic leadership but also its efforts to advance a collective, legally binding framework for climate change. A new political order emerged during the Copenhagen Climate Conference. The US and China are at the center of that political order.[24] It remains to be seen how the EU fits within this new framework and whether the re-ordering advances EU-led efforts to create a global framework for addressing climate change or, in fact, prompts devolution of power to a smaller group of political players.

Reassessing the Roles of the European Union and the United States

The often portrayed picture of the EU as the flag bearer for progressive climate policy and the US as the deplorable political laggard masks policy successes and failures in both contexts and ignores key points of convergence at the macro level. As the analysis in Chapter 7 reveals, this depiction is blunt and unhelpful. It masks turbulent EU member state relations on questions of climate governance and overlooks significant flaws in European climate laws and policies. Equally, it ignores meaningful political effort at the state and regional level in the US and fails to recognize the inroads made into federal resistance via litigation and regulation. Yet, the statement is not untrue at a macro level.

Despite existing weaknesses, the EU is establishing the parameters for a measurable, reviewable, and improvable climate regime. In so doing, it has embedded the importance of progressive efforts to address climate change in state, regional, and international legal and political systems. While the EU-ETS remains a work in progress and great strides in reducing emissions made by individual member states, such as Germany and the UK, are partially attributed to historic circumstances, the EU-ETS and complementary legal mechanisms offer increasingly effective tools for regional emissions reductions while key member states, including Germany, have made real and significant emissions reductions. For all of its shortcomings, the evolving EU climate regime is establishing the parameters for a measurable, reviewable, and improvable climate regime.

The US, on the other hand, continues to lack clearly defined objectives, legal parameters, and institutional capacity for addressing climate change at the federal level. Regional and state measures offer opportunities for progress and review while federal litigation and regulation offers genuine mandates for change. Yet, for all this, as of early 2010—midway through the first Kyoto Protocol

---

[24] Eg, Robin Lustig, 'Copenhagen: The Dawn of a New Political Reality' (BBC Radio Blog, 21 December 2009) <http://www.bbc.co.uk/blogs/worldtonight/2009/12/copenhagen_the_dawn_of_a_new_p.html> accessed 3 January 2010.

compliance period—the US continued to lack crucial measurable or enforceable initiatives for addressing climate change at the domestic level, and public and political debate continued to rage over the validity of constraining greenhouse gas emissions.

Further, by 2009, EU reports indicated that the EU-15 and the EU-27 were on track to meeting Kyoto commitments and that measures put in place to meet emissions reduction targets ensured continuing emissions reductions and improved abilities to decouple economic growth from emissions growth. In contrast, in the US, absolute emissions continue to increase at approximately 1.2% annually with very few measures in place to enable near-term emissions reductions. Thus, at a critical point in global climate politics, the US and the EU continued to diverge in their ability to lead by example.

The historical evolution in US and EU climate policy exemplifies how climate change has risen to the top of political agendas in divergent contexts while the spans separating US and EU climate policy to date epitomize the struggles inherent in on-going global efforts to address climate change. Neither the EU nor the US offer unqualified lessons in success, but both offer many lessons, some of which reveal successes but all of which offer opportunities to learn from social, political, and regulatory experiments. In the case of the US, efforts to date have resulted in few macro-level accomplishments but numerous micro-level successes that create replicable models. The EU, by comparison, has undertaken a progressive campaign to develop a comprehensive region-wide climate strategy that blends macro and micro-level initiatives. As a result, EU climate policy offers a richer and more advanced body of policy initiatives for analysis, replication, and rebuttal.

As Chapter 8 explores in some detail, however, when looking to either the US or the EU for lessons on developing a climate policy regime, it is critical to look beyond the mere expressions of law and policy and even beyond the success or failure of certain programs of action to determine why certain courses of action were chosen and to locate both the choice of the course of action and the success of the course of action within the larger socio-legal context. As the twists and turns of US and EU climate politics in 2009 demonstrate, reference to political leadership or primary domestic legal frameworks alone reveals very little about the substance or meaningfulness of US and EU climate policies or their likely evolution. Social, political, and economic differences remain determinative in framing international policymaking positions. From governance systems, to perceptions of risk, to media and civil society involvement, to economic norms, to perceptions of fairness and equity, the present and future of US and EU climate change law and policymaking are embedded within and influenced by distinctive cultural contexts. Understanding the cultural coordinates that frame regional climate change policymaking is a prerequisite to improving understanding of the roles that the US and the EU have and will play in the global climate change regime, in terms of their ability to offer either programmatic models of success or global political leadership.

The tumultuous nature of global climate change politics in 2009, for example, highlighted at the international level what had become evident in US and EU politics during the later years of the decade, that is, how an overriding focus on near-term economic well-being and the concomitant shaping of public perceptions of the 'costs' of climate change is hampering efforts not only to redirect US climate policy but also to sustain EU leadership and to expand the global climate regime to include the rapidly developing economies in a more meaningful way.

The pathways and the perils evident in US and EU climate policy are microcosms of larger international efforts to address climate change. The pushes, pulls, successes, and failures of transatlantic climate policy reflect the sheer difficulties inherent first, in international policymaking generally and second, within the specific context of climate change. Global climate change presents one of the most difficult problems the international community has ever faced. Recent events suggest that the United Nations is not yet equipped to address the issue, and national politics reveal that, in most cases, domestic politicians have neither the political will nor the regulatory tools at their disposal to structure effective policy regimes. Against this daunting backdrop, US and EU efforts to develop climate policies over the last two decades offer instructive lessons to developed and developing countries alike as they seek to avoid political and regulatory pitfalls in structuring domestic climate change regimes. Just as the parties to the UNFCCC look to economic flexibility mechanisms, technology transfer, and development aid to encourage developing economies to bypass the environmental errors of the industrialized nations, so must the parties look to the first generation of national climate policy regimes to bypass early climate policy failures and move forward with more effective legal and political strategies.

The Path Ahead

The era of the EU as the symbolic sentry and the US as the rogue of the global climate movement is ending and a new, more complex political alignment is beginning. The international community has long pressed the US to re-engage in international climate policy and to implement progressive domestic action on climate change. The US was viewed as 'the indispensable nation' whose presence or absence from international climate negotiations controlled the ability of the international community to build a meaningful global climate regime. The rapid re-engagement of the US in international climate politics in 2009, however, failed to offer the panacea needed to facilitate global consensus and action on climate change. US efforts to renew global climate leadership, in fact, revealed the extent to which global power is spread among nations. US re-engagement came at a time when China is acquiring great political and economic power. Hegemonic power is no longer clearly vested with the US and the alliances in climate politics are more complex than ever, creating a political sphere where no one player has the political or economic clout to compel others to adopt their approach to climate change.[25]

Within this new framework, the likelihood of global consensus and concerted global action on climate change diminishes and the importance of effective regional and national action increases. The reality that the roles of the EU and the US within this new framework are in flux makes their domestic and international actions more important than ever. If the EU and the US hope to play a meaningful role in global climate politics moving forward, they will have to adopt a dual track approach. First, they must lead by action. Second, they must endeavour to once again place themselves at the centre of the circles of power. The EU has laid the substantive and symbolic groundwork for pursuing this approach; the US has not. The US dominated the Copenhagen Conference with evocative rhetoric and with the promise of action, but its failure to put its own house in order renders its rhetoric hollow and its hold on power tenuous. As a result of its continuing reluctance to prioritize climate policy, the US now finds itself grasping for a smaller wedge of the power and influence that it once so confidently wielded.

The global climate change regime faces an uncertain future. US and EU leadership cannot guarantee success moving forward; yet, it is an essential component to success. Similarly, while neither the US nor the EU can offer the global community a comprehensive climate policy roadmap, their efforts past and present offer critical lessons for crafting more effective climate policies at every level of governance.

---

[25] Eg, Ian Clark, 'Bringing Hegemony Back In: the United States and International Order' (2009) 85 *Intl Affairs*.

Bibliography

Agrawala, S., 'Context and Early Origins of the Intergovernmental Panel on Climate Change' (1998) 39 *Climactic Change* 605, 606.
—— and Andresen, S., 'US Climate Policy: Evolution and Future Prospects' (2001) 12[2-3] *Energy and Environment* 117.
—— 'Leaders, pushers and laggards in the making of the climate regime' (2002) 12 *Global Environmental Change* 41.
Altman, R. C., 'The Great Crash, 2008: A Geopolitical Setback for the West' (2009) 88[1] *Foreign Affairs* 2.
Aminzadeh, S. C., Note, 'A Moral Imperative: The Human Rights Implications of Climate Change' (2007) 30 *Hastings International and Comparative Law Review* 231.
Associated Press, 'California Regulators Vote to Ban Utilities from Buying "Dirty" Power' *International Herald Tribune* (25 January 2007).
Bäckstrand, K. and Lövbrand, E., 'Climate Governance Beyond 2012: Competing Discourses of Green Governmentality, Ecological Modernization and Civic Environmentalism' in M Pettenger (ed), *The Social Construction of Climate Change* (Ashgate Publishing, Hampshire, 2007) 124.
Baker, P. and Cohen, Jon, 'Bush To Face Skeptical Congress: Iraq Overshadows Domestic Outreach' *Washington Post* (Washington DC, 23 January 2007).
Barabas, J. and Jerit, J., 'Estimating the Causal Effects of Media Coverage on Policy Specific Knowledge (2009) 53 *American Journal of Political Science* 73.
Barringer, F. and Revkin, Andrew C., 'Bills on Climate Move to Spotlight in the New Congress' *New York Times* (New York, 18 January 2007) A-24.
Bayer, J., 'Re-balancing State and Federal Power: Toward A Political Principle of Subsidiarity in the United States' (2004) 53 *American University Law Review* 1421, 1424
Benedick, R. E., *Ozone Diplomacy* (2nd edn, Harvard University Press, Cambridge, 1998) 11.
Block, F., 'Confronting Market Fundamentalism: Doing "Public Economic Sociology"' (2007) 5 *Socio-Economic Review* 319, 327.
Bodansky, D., 'Targets and Timetables: Good Policy But Bad Politics?' in J. Aldy and R. Stavins (eds), *Architectures for Agreement: Addressing Global Climate Change in the Post-Kyoto World* (CUP, New York, 2007).
—— The U.N. Framework Convention on Climate Change: A Commentary (1993) 18 *Yale Journal of International Law* 451.
Bokwa, A., 'Climactic Issues in Polish Foreign Policy' in Paul G. Harris (ed), *Europe and Global Climate Change: Politics, Foreign Policy and Regional Cooperation* (Edward Elgar Publishing, Northampton MA, 2007) 113.
Bord, R., O'Connor, R., and Fisher, A., 'In what sense does the public need to understand global climate change?' (2000) 9 *Public Understanding of Science* 205.
Boykoff J. and Boykoff, M., 'Balance as bias: global warming and the U.S. prestige press' (2004) 14 *Global Environmental Change* 125.

—— 'Climate Change and Journalistic Norms: A Case-Study of US Mass-Media Coverage' (2007) 38 *Geoforum* 1190.
—— 'Journalistic Balance as Global Warming Bias: Creating Controversy Where Science Finds Consensus' [November/December 2004] *Extra! Magazine*.
Brunnée, J., 'Europe, the United States, and Global Climate Regime: All Together Now?' (2008) 24 *Journal of Land Use & Environmental Law* 1.
Bundestag, G. (ed), *Protecting the Earth's Atmosphere: An International Challenge* (German Bundestag, Bonn, 1989) 439–41.
Bursik, M., 'Response: The Czech president's climate change denial is irrelevant' *Guardian* (25 March 2009).
Buzan, B., 'A leader without followers? The United States in world politics after Bush' (2008) 45 *International Politics* 554.
Caney, S., 'Cosmopolitan Justice, Responsibility, and Global Climate Change' (2005) 18 *Leiden Journal of International Law* 747.
Cannon, J. and Riehl, Jonathan, 'Presidential Greenspeak: How Presidents Talk about the Environment and What it Means' (2004) 23 *Stanford Environmental Law Journal* 195, 210–11.
Carlarne, C., 'Climate Change—The New "Superwhale" in the Room: International Whaling and Climate Change Politics—Too Much in Common?' (2007) 80 *Southern California Law Review* 101.
—— 'From the USA with Love: Sharing Home-Grown Hormones, GMOs, and Clones with a Reluctant Europe' (2007) 37 *Environmental Law* 2.
—— 'Good Climate Governance: Only a Fragmented System of International Law Away?' (2008) 30 *Law & Policy* 4.
—— and others, 'Maturity and Methodology: Starting a Debate about Environmental Law Scholarship' (2009) 21 *Journal of Environmental Law* 213.
Carlarne, C., 'Risky Business: The Ups and Downs of Mixing Economics, Security and Climate Change' (2009) 10 *Melbourne Journal of International Law* 2.
Carpenter, C., 'Businesses, Green Groups and the Media: The Role of Non-Governmental Organizations in the Climate Change Debate' (2001) 77 *International Affairs* 313.
Carson, R., *Silent Spring* (Houghton Mifflin, Boston, 1962).
Carter, N., *The Politics of the Environment: Ideas, Activism, Policy* (Harvard University Press, Cambridge MA, 2001).
Cass, L., *The Failures of American and European Climate Policy: International Norms, Domestic Politics, and Unachievable Commitments* (SUNY Press, New York, 2007) 105.
Cavender, J., and Jäger, J., 'The History of Germany's Response to Climate Change' (1993) 5 *International Environmental Affairs* 3, 6–9.
Chanin, R. L., 'California's Authority to Regulate Mobile Source Greenhouse Gas Emissions' (2003) 58 *New York University Annual Survey of American Law* 699.
Chasek, P. S., 'The Global Environment in the Twenty-First Century: Prospects for International Cooperation' in P. S. Chasek (ed), *The Global Environment in the Twenty-First Century* (United Nations University Press, New York, 2000).
Clark, I., 'Bringing Hegemony Back In: The United States and International Order' (2009) 85 *International Affairs* 23
Clarke, K. C. and Hemphill, Jeffrey J., 'The Santa Barbabra Oil Spill: A Retrospective' (2002) 64 *Yearbook of the Association of Pacific Coast Geographers* 157–62.

Coglianese, C. and Lazer, D., 'Management-Based Regulation: Prescribing Private Management to Achieve Public Goals' (2003) 37 *Law & Society Review* 691.

Coglianese, C. and Nash, J., *Leveraging the Private Sector: Management-Based Strategies for Improving Environmental Performance* (Resources for the Future Press, Washington DC, 2006).

Curl, A., 'Government Scientists Claim Censorship on Global Warming' *Federal Times* (Springfield, 30 January 2007).

Dabelko, G. and Simmons, P., 'Environment and Security: Core Ideas and US Government Initiatives' (1997) 17 *School of Advanced International Studies Review* 127.

Dernbach, J. C., 'Facing Climate Change: Opportunities and Tools for States' (2004) 4 *Widener LJ* 1, 2.

de Sadeleer, N., *Environmental Principles: From Political Slogans to Legal Rules* (OUP, Oxford, 2002) 174.

De Souza, M., 'US media blamed for stall in climate plan' *The Gazette* (Montreal, 2007) A4.

DeSombre, E. R., *Domestic Sources of International Environmental Policy: Industry, Environmentalists, and U.S. Power* (MIT Press, Cambridge MA, 2000).

Dessler, A. and Parson, E., *The Science and Politics of Global Climate Change: A Guide to the Debate* (CUP, New York, 2006) 15.

Dimitrov, R. S., 'Inside Copenhagen: The State of Climate Governance' (2010) 10[2] *Global Environmental Politics* 18, 21.

Dinnell, A. M. and Russ, A. J., 'The Legal Hurdles to Developing Wind Power as an Alternative Energy Source in the United States: Creative and Comparative Solutions' (2007) 27 *Northwestern Journal of International Law & Business* 535, 569.

Drezner, D. W., 'The New World Order' (2007) 86[2] *Foreign Affairs* 34.

Driesen, D., 'Thirty Years of International Environmental Law: A Retrospective and Plea for Reinvigoration' (2003) 30 *Syracuse Journal of International Law and Commerce* 353, 362.

Duffy, G., 'Brazil Proposes Carbon Cut Target' (BBC News Release, 14 November 2009).

Durant, R. F., *The Administrative Presidency Revisited: Public Lands, the BLM, and the Reagan Revolution* (SUNY Press, Albany NY, 1992).

Egenhofer, C. and Georgiev, A., *The Copenhagen Accord—a first stab at deciphering the implications for the EU* (Climate Change Centre for European Policy Change Commentaries, 28 December 2009).

Engel, K. H., 'Mitigating Global Climate Change in the United States: A Regional Approach' (2005) 14 *New York University Environmental Law Journal* 54, 64.

—— 'State Environmental Standard-Setting: Is There a "Race" and is it "to the Bottom"?' (1997) 48 *Hastings Law Journal* 271.

—— and Saleska, S. R., 'Facts are Stubborn Things: An Empirical Reality Check in the Theoretical Debate over the Race-to-the Bottom in State Environmental Standard Setting' (1998) 8 *Cornell Journal of Law and Public Policy* 55.

—— 'Subglobal Regulation of the Global Commons: The Case of Climate Change' (2005) 32 *Ecology Law Quarterly* 183, 215.

Esty, D. C., 'Revitalizing Environmental Federalism' (1996) 95 *Michigan Law Review* 570, 600–03.

—— 'Good Governance at the Supranational Scale: Globalizing Administrative Law' (2006) 115 *Yale Law Journal* 1497.

Falkner, R., 'The Political Economy of "Normative Power" Europe: EU Environmental Leadership in International Biotechnology Regulation' (2007) 14 *Journal of European Public Policy* 507.

Faure, M. G. and Nollkaemper, A., 'International liability as an instrument to prevent and compensate for climate change' (2007) 23A *Stanford Environmental Law Journal* 123.

Fiorino, D., *The New Environmental Regulation* (MIT Press, Cambridge, 2006).

Fisher, D. R., *National Governance and the Global Climate Change Regime* (Rowman & Littlefield Publishers Inc, Lanham, 2004).

Fisher, E., *Risk Regulation and Administrative Constitutionalism* (Hart Publishing, Oxford, 2007) 669.

French, D., 'Developing States and International Environmental Law: The Importance of Differentiated Responsibilities' (2000) 49 *International & Comparative Law Quarterly* 35.

Friedman, T., *Hot, Flat and Crowded: Why We Need a Green Revolution—and How it Can Renew America* (Farrar, Straus and Giroux, New York, 2008) 17.

Gerrard, M. B., and Howe, J. C., 'Climate Change Litigation in the U.S.' (Chart) <http://www.arnoldporter.com/resources/documents/ClimateChangeLitigationChart.pdf>.

Ghaleigh, N. S., 'Emissions Trading Before the European Court of Justice: Market Making in Luxembourg', 29, in D. Freestone and C. Streck (eds), *Legal Aspects of Carbon Trading: Kyoto, Copenhagen and Beyond*, (OUP, Oxford, 2009).

Gilbreath, C., 'Federalism in the Context of Yucca Mountain: *Nevada v Department of Energy*' (2000) 27 *Ecology Law Quarterly* 577, 592.

Gilman, M., 'The President as Scientist-in-Chief' (2009) 45 *Willamette Law Review* 565, 571–73.

Giovinazzo, C. T., 'Defending Overstatement: The Symbolic Clean Air Act and Carbon Dioxide' (2006) 30 *Harvard Environmental Law Review* 99.

Goldenberg, S., Vidal, J., and Watts, J., 'Leaked UN Report Shows Cuts Offered at Copenhagen Would Lead to 3C Rise' *Guardian* (17 December 2009).

Grubb M. and Brack, D., *The Kyoto Protocol: A Guide and Assessment* (Brookings Institution Press, Washington DC, 1999) 89–114.

Gupta, J. and Grubb, M. (eds), *Climate Change and European Leadership: A Sustainable Role for Europe?* (Kluwer Academic Publishers, Dordrecht, Netherlands, 2000).

Hansjurgens, B., 'Introduction' in Bernd Hansjurgens (ed), *Emissions Trading for Climate Policy: US and European Perspectives* (CUP, New York, 2005).

Harris, P. G. (ed), *Europe and Global Climate Change: Politics, Foreign Policy and Regional Cooperation* (Edward Elgar Publishing, Northampton MA, 2007) 74.

—— 'Common but Differentiated Responsibility: The Kyoto Protocol and United States Policy' (1999) 7 *New York University Environmental Law Journal* 27, 38–40.

—— 'The European Union and Environmental Change: Sharing the Burdens of Global Warming' (2006) 17 *Colorado Journal of International Environmental Law & Policy* 309, 338–40.

Harswick, T. L., 'Developments in Climate Change' (2002) *Colorado Journal of International Environmental Law & Policy* 25, 31.

Hatch, M., *The Europeanization of German Climate Change Policy* (Conference paper for the European Union Studies Association Biennial Conference in Montreal 2007).

Hawkins, K., *Environment and Enforcement* (OUP, Oxford, 1984).

Heller, T., 'The Path to EU Climate Change Policy' in Jonathan Golub (ed), *Global Competition and EU Environmental Policy* (Routledge, New York, 1998).

Herbert, Y., 'President Bush, See You in Court Judging the Cost of Climate Change' (25 August 2004) *The Dominion*.

Holtrup, P., 'The Lack of U.S. Leadership in Climate Change Diplomacy' in B. May and M. H. Moore (eds), *The Uncertain Superpower: Domestic Dimensions of US Foreign Policy after the Cold War* (Leske & Budrich, Opladen, 2003).

Hulme, M., 'Mediated Messages about Climate Change: Reporting the IPCC Fourth Assessment in the UK Print Media' in T. Boyce and J. Lewis (eds), *Climate Change and the Media* (Peter Lang Publishing, New York, 2009).

Hutter, B., 'Socio-legal Perspectives on Environmental Law: An Overview' in B. Hutter (ed), *A Reader in Environmental Law* (OUP, Oxford, 1999); K. Hawkins, *Environment and Enforcement* (OUP, Oxford, 1984).

Jaggard, L., 'The Reflexivity of Ideas in Climate Change Policy: German, European and International Politics' in Paul G. Harris (ed), *Europe and Global Climate Change: Politics, Foreign Policy and Regional Cooperation* (Edward Elgar Publishing, Northampton MA, 2007) 323.

—— *Climate Change Politics in Europe: Germany and the International Relations of the Environment* (Tauris Academic Studies, New York, 2007) 20.

Jasanoff, S., 'American Exceptionalism and the Political Acknowledgement of Risk' in Edward Burger Jr (ed), *Risk* (University of Michigan Press, Ann Arbor, 1993).

John, M., 'Obama Climate Pledge gets Cautious EU Welcome' *Reuters* (5 April 2009).

Joyner, C. and others, 'Common But Differentiated Responsibility' (2002) 96 *American Society of International Law Proceedings* 358.

Kagan, R. A., *Adversarial Legalism: The American Way of Law* (Harvard University Press, Cambridge MA, 2001).

—— 'Power and Weakness: Why the United States and Europe see the World Differently' (2002) 113 *Policy Review* 1.

—— 'Conclusions on Iraq' (Council of Ministers' Conclusions, 2003).

—— 'Globalization and legal change: The "Americanization" of European Law?' (2007) 1 *Regulation & Governance* 99.

—— 'Questioning the "Americanization" of European Law' (2007) 6(1) *Newsletter of European Politics & Society Section of the American Political Science Association* 5.

Kagan, R. A. and Axelrad, L. (eds), *Regulatory Encounters: Multinational Corporations and American Adversarial Legalism* (University of California Press, Berkeley CA, 2000).

Kahn, G., 'The Fate of the Kyoto Protocol Under the Bush Administration' (2003) 21 *Berkeley Journal of International Law* 548.

Kay, K., '"Toxic Texan" Has Poor Green Record' *Times* (London, 23 August 2002) 19.

Keeth, D., 'The California Climate Law: A State's Cutting-Edge Efforts to Achieve Clean Air' (2003) 30 *Ecology Law Quarterly* 715, 719.

Keiser, A., 'Carbon Sequestration Options Under the Clean Development Mechanism to Address Land Degradation' (2000) 92 *World Soil Resources Reports* 7, 7–11.

Kimber, C., 'Environmental Federalism: A Comparison of Environmental Federalism in the United States and the European Union' (1995) 54 *Maryland Law Review* 1658, 1659.

King, K. F., 'The Death Penalty, Extradition, and the War Against Terrorism: US Responses to European Opinion about Capital Punishment' (2003) 9 *Buffalo Human Rights Law Review* 161, 162.

Komurcu, M., 'Cultural Heritage Endangered by Large Dams and Its Protection under International Law' (2002) 20 *Wisconsin International Law Journal* 233, 284.

Lacasta, N. S., Dessai, S., and Powroslo, E., 'Rio's Decade: Reassessing the 1992 Earth Summit: Reassessing the 1992 Climate Change Agreement: Consensus Among Many Voices: Articulating the European Union's Position on Climate Change' (2002) 32 *Golden Gate University Law Review* 351, 352.

Lahsen, M., 'Technocracy, Democracy, and U.S. Climate Politics: The Need for Demarcations' Special (2005) 30 *Science Technology & Human Values* 137.

Laidlaw, A., 'Is it Better to be Safe than Sorry?: The Cartagena Protocol Versus the World Trade Organisation' (2005) 36 *Victoria University of Wellington Law Review* 427.

Lee, M., *EU Environmental Law: Challenges, Change and Decision-Making* (Hart Publishing, Oxford, 2005) 21.

Lisowski, M., 'Playing the Two-Level Game: US President Bush's Decision to Repudiate the Kyoto Protocol' (2002) 11[4] *Environmental Politics* 101–19.

Litfin, K., 'Environment, Wealth, and Authority: Global Climate Change and Emerging Modes of Legitimation' (2000) 2 *International Studies Review* 119.

Lofstedt, R. and Vogel, D., 'The Changing Character of Consumer and Environmental Regulation: A Comparison of Europe and the United States' (2001) 21 *Risk Analysis* 399.

Lohen, T., 'Climate Change Heats up Washington' (13 February 2007) <http://www.alternet.org/environment/47891>.

Longworth, R. C., '"Bush Doctrine" Arises From the Ashes of Sept. 11' *Chicago Tribune* (Chicago, 7 March 2002) 4.

Mank, B. C., 'Standing and Global Warming: Is Injury to All Injury to None?' (2005) 35 *Environmental Law* 8–9.

Manners, I., 'Normative Power Europe: A Contradiction in Terms?' (2002) 40 *Journal of Common Market Studies* 235.

Marshall, C., Fialka, J., and Radick, L., 'Renewable industry cheers Obama budget while coal and nuclear jeer' *New York Times* (New York, 8 May 2009).

Mason, J., 'U.S. to Take Reins in Global Climate Talks' *Reuters* (Washington, 24 April 2009).

Matthew, R. A., 'The Environment as a National Security Issue' (2000) 12 *Journal of Political History* 101, 109.

McCright, A. and Dunlap, R., 'Defeating Kyoto: the conservative movement's impact on U.S. climate change policy' (2003) 50 *Social Problems* 348.

McEldowney, J. and McEldowney, S., *Environmental Law and Regulation* (Blackstone Press, London, 2001) 89.

McKeown, T., 'Climate Change, Population Movements, and Conflict' in Carolyn Pumphrey (ed), *Global Climate Change: National Security Implications* (Strategic Studies Institute, Carlisle PA, 2008) 99.

McKibben, B., 'Copenhagen: Things Fall Apart and an Uncertain Future Looms' (Yale Environment 360 Opinion, 21 December 2009).

McKibbin, W. J. and Wilcoxen, P. J., *Climate Change Policy After Kyoto: Blueprint for a Realistic Approach* (Brookings Institution Press, Washington DC, 2002) 43–44.

McKinstry Jr, R. B., 'Laboratories for Local Solutions for Global Problems: State, Local and Private Leadership in Developing Strategies to Mitigate the Causes and Effects of Climate Change' (2004) 12 *Penn State Environmental Law Review* 15, 16.

Mooney, C., 'Beware "Sound Science." It's Doublespeak for Trouble' *Washington Post* (Washington DC, 29 February 2004) B2.

Murphy, J., Levidow, L., and Carr, S., 'Regulatory Standards for Environmental Risks: Understanding the US–European Union Conflict over Genetically Modified Crops' (2006) 36 *Social Studies of Science* 1.

Neuhoff, K., Cust, J., and Keats, K., 'Implications of Intermittency and Transmission Constraints for Renewables Deployment' in M. Grubb, T. Jamasb, and M. Pollitt (eds), *Delivering a Low-Carbon Electricity System: Technology, Economics and Policy* (CUP, New York, 2008).

Niiler, E., 'House Probes Global Warming Censorship' (Marketplace broadcast, 30 January 2007).

Nordhaus, R. R. and Danish, K. W., 'Assessing the Options for Designing a Mandatory U.S. Greenhouse Gas Reduction Program' (2005) 32 *Boston College Environmental Affairs Law Review* 97, 106.

Nye Jr, J. S., 'The American National Interest and Global Public Goods' (2002) 78 *International Affairs* 233.

O'Riordan, T. and Jager, J. (eds), *Politics of Climate Change: A European Perspective* (Routledge, London, 1996).

Oberthur, S. and Kelly, C. R., 'EU Leadership in International Climate Policy: Achievements and Challenges' (2008) 43(3) *The International Spectator* 35.

Oberthür, S. and Ott, H., *The Kyoto Protocol: International Climate Policy for the 21st Century* (Springer, New York, 1999).

Ochs, A. and Sprinz, D., 'Europa riding the hegemon? Transatlantic climate policy relations' in D. B. Bobrow and W. Keller (eds), *Hegemony Constraint: Evasion, Modification, and Resistance to American Foreign Policy* (University of Pittsburgh Press, Pittsburgh, 2008) 144.

Paarlberg, R., 'Lapsed Leadership: U.S. International Environmental Policy Since Rio' in N. J. Vig and R. S. Axelrod (eds), *The Global Environment: Institutions, Law, and Policy* (CQ Press, Washington DC, 1999).

Parenteau, P., 'Anything Industry Wants, Environmental Policy Under Bush II' (2004) 14 *Duke Environmental Law & Policy Forum* 363.

Paterson, M., 'Post-Hegemonic Climate Politics?' (2009) 11 *British Journal of Politics and International Relations* 140.

Percival, R. V., 'Environmental Federalism: Historical Roots and Contemporary Models' (1995) 54 *Maryland Law Review* 1141, 1157–78.

Petersmann, E. U. and Pollack, Mark A. (eds), *Transatlantic Economic Disputes The EU, the US, and the WTO* (OUP, Oxford, 2004) 66.

Petsonk, A., 'The Kyoto Protocol and the WTO: Integrating Greenhouse Gas Emissions Allowance Trading Into the Global Marketplace' (1999) 10 *Duke Environmental Law & Policy Forum* 185.

Pettenger, M. (ed), *The Social Construction of Climate Change* (Ashgate Publishing, Hampshire, England, 2007).

Pfeifer, S. and Sullivan, R., 'Public Policy, Institutional Investors, and Climate Change: a UK Case Study' (2008) 89 *Climatic Change* 245.

Pidot, J. R., *Global Warming in the Courts: An Overview of Current Litigation and Common Legal Issues* (2006) Georgetown Environmental Law & Policy Institute.

Poirier, M. R., 'Property, Environment, Community' (1997) 12 *Journal of Environmental Law and Litigation* 43, 64.
Porter, G., Brown, J. W., and Chasek, P. S., *Global Environmental Policy* (Westview Press, Boulder CO, 2000).
Posner, E. A. and Vermeule, A., 'The Credible Executive' (2007) 74 *University of Chicago Law Review* 865, 871.
Rabe, B. G., 'Greenhouse & Statehouse: The Evolving State Government Role in Climate Change [2002] *Pew Center on Global Climate Change* 36–9, 152.
—— *Statehouse and Greenhouse: The Emerging Politics of American Climate Change Policy* (Brookings Institution Press, Washington DC, 2004).
—— 'North American Federalism and Climate Change Policy: American State and Canadian Provincial Policy Development' (2004) 14 *Widener Law Journal* 121.
—— 'State Competition as a Source Driving Climate Change Mitigation' (2005) 14 *New York University Environmental Law Journal* 1, 17.
Rajamani, L., 'The Principle of Common but Differentiated Responsibility and the Balance of Commitments under the Climate Regime' (2000) 9 *Review of European Community & International Environmental Law* 120, 121.
—— 'From Berlin to Bali and Beyond: Killing Kyoto Softly?' (2008) 57 *International & Comparative Law Quarterly* 909.
Ramseur, J. L., *CRS Report for Congress: Climate Change: Action by States To Address Greenhouse Gas Emissions*, 9–10 (18 January 2007).
Rees, M. and Evers, Rainer, 'Proposals for Emissions Trading in the United Kingdom' (2000) 9 *Review of European Community & International Environmental Law* 232.
Reimann, M., 'The Progress and Failure of Comparative Law in the Second Half of the Twentieth Century'(2002) 50 *American Journal of Comparative Law* 671, 685.
Reitze Jr, A. W., 'Federal Control of Carbon Dioxide Emissions: What are the Options?' (2009) 36 *Boston College Environmental Affairs Law Review* 1.
Reus-Smit, C. (ed), *The Politics of International Law* (CUP, New York, 2004) 86.
Revkin, A., 'Hacked E-Mail is New Fodder for Climate Dispute' *New York Times* (New York, 20 November 2009).
Richardson, B. J. and Chanwai, K. L., 'The UK's Climate Change Levy: Is It Working?' (2003) 15 *Journal of Environmental Law* 39.
Rodi, M., 'Public Environmental Law in Germany' in J Rene and others (eds), *Public Environmental Law in the European Union and US: A Comparative Analysis* (Kluwer Law International, The Hague, 2002).
Rogers, K. E., 'Germany's Efforts to Reduce Carbon Dioxide Emissions from Cars: Anticipating a New Regulatory Framework and its Significance for Environmental Policy' (2008) 38 *Environmental Law Reporter News & Analysis* 10214, 10214–15.
Rose-Ackerman, S., *Controlling Environmental Policy: The Limits of Public Law in Germany and the United States* (Yale University Press, New Haven CT, 1995).
Rugh, D. W., 'Clearer, But Still Toxic Skies: A Comparison of the Clear Skies Act, Congressional Bills and the Proposed Rule to Control Mercury Emissions from Coal-Fired Power Plants' (2003) 28 *Vermont Law Review* 201.
Samans–Dunn, J., 'The City of Philadelphia—The Government and Community Work Together to Reduce Greenhouse Gas Emissions' (2004) 12 *Penn State Environmental Law Review* 207.
Sands, P., *Transnational Environmental Law* (Kluwer Law International, London, 1999).

—— *Principles of International Environmental Law* (2nd edn, CUP, Cambridge, 2003).
—— *Lawless World: America and the Making and Breaking of Global Rules* (Penguin Books Ltd, London, 2005).
Sarnoff, J. D., 'The Continuing Imperative (But Only from a National Perspective) For Federal Environmental Protection' (1997) 7 *Duke Environmental Law & Policy Forum* 225, 227.
Sathaye, J. and Walsh, M., 'Transportation in Developing Nations: Managing the Institutional and Technological Transition to a Low-Emissions Future' in Irving Mintzer (ed), *Confronting Climate Change: Risks, Implications and Responses* (CUP, Cambridge, 1992).
Scherer, R. and Marks, A., 'New Environmental Cops: State Attorneys General' [2004] *Christian Science Monitor* (22 July 2004)
Schreurs, M. A., *Environmental Politics in Japan, Germany, and the United States* (CUP, Cambridge, 2002).
—— 'The Climate Change Divide: The European Union, the United States, and the Future of the Kyoto Protocol' in Norman J. Vig and Michael G. Faure (eds), *Green Giants: Environmental Policies of the United States and the European Union* (MIT Press, Cambridge, 2004).
Schreurs, M. A. and Tiberghien, Y., 'Multi-Level Reinforcement: Explaining European Union Leadership in Climate Change Mitigation' (2007) 7(4) *Global Environmental Politics* 19.
Schreurs, M. A., Selin, H., and VanDeveer, S. D., (eds), *Transatlantic Environment and Energy Politics* (Ashgate, 2009).
Schubert, R., Schellnhuber, H. J., and Buchmann, N., *Climate Change as a Security Risk* (Earthscan Publications, London, 2007).
Scott, J. and Holder, J., 'Law and New Environmental Governance in the EU' in G. de Burca and J. Scott (eds), *Law and Governance in the EU and US* (Hart Publishing, Oxford, 2006).
Shaffner, E., 'Repudiation and Regret: Is the United States Sitting Out the Kyoto Protocol to its Economic Detriment?' (Comment) (2007) 37 *Environmental Law* 441, 443.
Smith, K. E., *European Union Foreign Policy in a Changing World* (Polity, Cambridge, 2003)
Smith, M. E., *Europe's Foreign and Security Policy: The Institutionalization of Cooperation* (CUP, Cambridge, 2004).
Speers, T., 'A Picnic in March: Media Coverage of Climate Change and Public Opinion in the UK' in G. Humphrys and M. Williams (eds), *Presenting and Representing Environments* (Springer, Dordrecht, Netherlands, 2005).
Sprinz, D. F. and Weifrs, M., 'Domestic Politics and Global Climate Policy' in U. Luerbacher and D. F. Sprinz (eds), *International Relations and Global Climate Change* (MIT Press, Cambridge MA, 2001).
Sterio, M., 'The Evolution of International Law' (2008) 31 *Boston College International and Comparative Law Review* 213, 245.
Sterman, J. and Sweeney, L., 'Understanding Public Complacency About Climate Change: Adults' Mental Models of Climate Change Violate Conservation of Matter' (2007) 80 [3–4] *Climatic Change* 213.
Stewart, R., 'A New Generation of Environmental Regulation?' (2001) 29 *Capital University Law Review* 21.

Stone, C., 'Common but Differentiated Responsibilities in International Law' (2004) 98 *American Journal of International Law* 276, 278.
Sunstein, C., 'Precautions Against What? The Availability Heuristic and Cross-Cultural Risk Perception' (2005) 57 *Alabama Law Review* 75.
Sunstein, C. R., 'Of Montreal and Kyoto: A Tale of Two Protocols' (2007) 31 *Harvard Environmental Law Review* 1, 27.
Sussman, E., *Climate Change Litigation: Past, Present and Future: ABA Renewable Energy Resources Committee Program* (21 June 2006).
Sussman, G., 'The USA and Global Environmental Policy: Domestic Constraints on Effective Leadership' (2004) 25 *International Political Science Review* 349.
Teske, P., *Regulation in the States* (The Brookings Institution, Washington DC, 2004) 8.
Thoms, L., 'A Comparative Analysis of International Regimes on Ozone and Climate Change with Implications for Regime Design' (2003) 41 *Columbia Journal of Transnational Law* 795, 797.
Titus, J. G., 'Does the U.S. Government Realize that the Sea is Rising? How to Restructure Federal Programs so that Wetlands and Beaches Survive' (2000) 30 *Golden Gate University Law Review* 717.
Urry, J., *Mobilities* (Polity Press, Cambridge, 2007).
Van Lenten, C., *New York Tackles Climate Change: Promoting Renewable Energy and Capping Greenhouse Gas Emissions* (Report by New York Academy of Sciences, 20 October 2005).
Veno, J., 'Flying the Unfriendly Skies: The European Union's Proposal to Include Aviation in Their Emissions Trading Scheme' (2007) 72 *Journal of Air Law & Commerce* 659, 673–5.
Vierecke, A., 'Complex Problem Solving in the Co-operation Between Science and Politics? The workings of the Enquête Commissions of the German Bundestag in the Fields of Technological Development and Environmental Policy' in Elmar Stuhler and Dorien DeTombe (eds) *Complex Problem Solving: Cognitive Psychological Issues and Environment Policy Applications* (Hampp, Mering, Germany, 1999) 110–18.
Vig, N. J. and Faure, M. G. (eds), *Green Giants: Environmental Policies of the United States and the European Union* (MIT Press, Cambridge, 2004).
Vig, N. J. and Kraft, M. E., (eds), Environmental Policy: New Directions for the Twenty-First Century (6th edn, CQ Press, Washington DC, 2006) 12.
Vogel, D., *National Styles of Regulation: Environmental Policy in Great Britain and the United States* (Cornell University Press, New York, 1986).
—— 'The Hare and the Tortoise Revisited: The New Political of Consumer and Environmental Regulation in Europe' (2003) 33 *British Journal of Political Science* 557.
Vogler, J. and Bretherton, C., 'The European Union as a Protagonist to the United States on Climate Change' (2006) 9 *International Studies Perspectives* 1.
Vogler, J. and Stephan, H., 'The European Union in Global Environmental Governance: Leadership in the Making?' (2007) 7 *International Environmental Agreements* 389.
von Moltke, K., 'Three Reports on German Environmental Policy' (September 1991) 33(7) *Environment* 1.
Warbrick, C., McGoldrick, D., and Davies, P., 'Global Warming and the Kyoto Protocol' (1998) 47 *International & Comparative Law Quarterly* 446, 458.
Ward, B., 'A Higher Standard than "Balance" in Journalism on Climate Change Science' (2008) 86 *Climatic Change* 13.

Ward, H., 'Game Theory and the Politics of Global Warming: the State of Play and Beyond' (2006) 44 *Political Studies* 850, 861.

Watts, J., 'China Sets First Targets to Curb World's Largest Carbon Footprint' *Guardian* (26 November 2009).

Weiland, P. S., 'Federal and State Preemption of Environmental Law: A Critical Analysis' (2000) 24 *Harvard Environmental Law Review* 237.

Weingart, P., Engels, A., and Pansegrau, P., 'Risks of communication: discourses on climate change in science, politics, and the mass media' (2000) 9 *Public Understanding of Science* 261.

Wiener, J. B., 'Protecting the Global Environment' in John Graham and Jonathan Wiener (eds), *Risk vs Risk: Tradeoffs in Protecting Health and the Environment* (Harvard University Press, Cambridge, MA, 1995) 193, 214.

—— 'On the Political Economy of Global Environmental Regulation' (1999) 87 *Georgetown Law Journal* 749.

—— 'Whose Precaution after All? A Comment on the Comparison and Evolution of Risk Regulatory Systems' (2003) 13 *Duke Journal of Comparative & International Law* 207.

—— 'Better Regulation in Europe' in J. Holder and C. O'Cinneide (eds), *Current Legal Problems 2006: Vol 59* (OUP, New York, 2006).

—— 'Precaution' in Daniel Bodansky, Jutta Brunnée, and Ellen Hey (eds), *Oxford Handbook of International Environmental Law* (OUP, Oxford, 2007) 597, 599.

—— and Rogers, M., 'Comparing Precaution in the United States and Europe' (2002) 5 *Journal of Risk Research* 317.

Winkler, H., Brouns, B., and Kartha, S., 'Future Mitigation Commitments: Differentiating among Non-Annex I Countries' (2006) 5 *Climate Policy* 469.

Yamin, F., 'The Role of the EU in Climate Negotiations', in Joyeeta Gupta and Michael Grubb (eds), *Climate Change and European Leadership: A Sustainable Role for Europe?* (Kluwer Academic Publishers, Dordrecht, Netherlands, 2000).

Index

absolute emissions reductions 37, 38, 41, 177, 245, 246
Acquis 152
actions and omissions 12, 16, 23, 239, 248, 323
activists 293, 348
Ad Hoc Working Group on Further Commitments for Annex I Parties under the Kyoto Protocol (AWG-KP) 357
Ad Hoc Working Group on Long-Term Cooperative Action under the Convention (AWG-LCA) 357
adaptation 3, 17, 34, 54, 65, 76, 116, 153, 166–7, 179–80, 212–13, 222, 258, 287–8, 291, 301–6, 332, 339, 341, 353, 354, 357
see also European Union; United States
Adaptation Fund 54, 305
additionality 222
administrative agencies 22, 29, 39, 82, 196, 198, 249–52, 268
administrative law 22, 248–52, 320, 323, 325
adversarial legalism 248, 249–50, 322–4
agenda setting 244
agriculture 10, 13, 43, 64, 74–5, 163–8, 175–6, 180, 188, 190, 205, 211, 218, 220–1, 228, 257, 274, 287, 290, 292, 302–3
Air Quality Directive 167
Alabama 269, 330, 371
Alaska 67, 119, 126, 267, 269
Alliance of Automobile Manufacturers 82–4, 103, 267
Alliance of Small Island States (AOSIS) 347, 352
alternative energy 75, 85, 202, 295
alternative fuel 69, 85, 87
American Clean Energy and Security Act (ACES) 54, 55
American Recovery and Reinvestment Act (ARRA) 52, 53, 92
Annex B party 228
Annex I countries 6, 9, 237
anthropogenic *see* climate change, anthropogenic
Arizona 67, 69, 70, 75, 77
Arkansas 269
auctioning 175, 188, 206, 213, 230, 273, 274, 275, 281, 282, 304
Australia 9, 90, 128–9, 133, 178, 237–8, 281, 338, 356

Austria 159, 161, 183, 185, 187, 230, 273
auto manufacturers 81–5, 103–4, 295–300
aviation 166, 175, 178, 222, 258, 294, 296

Bali Action Plan 306, 357–8
Bali Roadmap 241, 244
Barroso, José Manuel 237, 272, 308, 340, 355, 358
Beckett, Margaret 289
Belgium 159, 161, 183
Berlin Mandate 7, 8
best available control technologies (BACT) *see* Clean Air Act
bilateral consultations, 13
biofuels 42, 85–6, 163, 166, 169–70, 184, 186, 188, 191, 210–12, 245, 274
biofuels directive 191
Biosafety Protocol 325
Bolivia 356
Brazil 59, 237, 240, 264, 279, 301, 306, 347, 350–55, 365
Britain *see* United Kingdom
Brown, Gordon 224, 315, 369
Browner, Carol 52, 106
bubble, EU 159–60, 162, 183, 213, 215–16, 228, 280, 339
Bulgaria 160, 184, 187, 230
burden sharing agreement 160, 161
Bush, George H.W. 7, 30
Bush, George W. 9, 15, 17, 27, 29, 30, 31, 35–8, 41, 45–9, 58, 59, 60, 78, 119, 124, 179, 237–8, 240, 244, 260, 278–9, 298, 314, 315, 327
business *see* industry
Byrd-Hagel Resolution 8, 240, 338

California 25–6, 40, 46, 53, 55, 63, 64–9, 71, 73, 75, 77–82, 83–88, 96, 100–104, 110, 113, 119, 120, 122, 139, 140, 178, 232, 255, 259, 266–7, 299, 310, 363, 364, 367
California Air Resources Board (CARB) 78, 80, 86–7
Canada 59, 90, 131, 338
Premiers 65, 66, 67
Provinces 72
cap-and-trade 49–50, 54, 64, 73, 76, 124, 247, 282
carbon sequestration 36, 42, 43, 71
capture and storage 166–7, 193, 224, 240, 258

carbon sinks 134, 167, 185, 191, 354
carbon taxes 270
  see also energy taxes
Carbon Trust 217–19
causation 100, 113–14
Center for Biological Diversity
  (CBD) 114–15, 119–20, 315, 325
Center for International Environmental Law
  (CIEL) 125–9
Center for Strategic and International Studies
  (CSIS) 285, 286, 288
certified emissions reduction 10
Chernobyl 197, 200
China 6, 12, 35, 56, 59–60, 230, 237, 240,
  264, 272, 279, 282, 289, 293, 297,
  301, 306–7, 338–9, 347, 349–55,
  359, 361, 372
chlorofluorocarbons (CFCs) 5
Christian Democratic Union (CDU) 195,
  200
Christian Socialist Union (CSU) 195, 200
Cities for Climate Protection Campaign
  (CCPC) 89
civil society *see* European Union; United
  States
Clean Air Act (CAA) 79, 80, 83–4, 94, 104–
  10, 112, 114, 119–24, 167, 247–8,
  254, 277, 351
  best available control technology
    (BACT) 122
  endangerment finding 55, 108, 109–10,
    121, 123, 277, 300, 351
  new source review and Title V 123, 124
  prevention of significant deterioration
    (PSD) 122, 123
  regulation of motor vehicle greenhouse
    gases, 81
  waiver 53, 55, 79, 80–4, 104, 122
Clean Development Mechanism (CDM) 10,
  198, 277, 280
Clean Water Act (CWA) 120, 121
Clear Skies Act 124, 125, 370
Climate Change Act 65–6, 70–1, 125, 221–5
climate change
  anthropogenic impact on 4–6, 109, 113,
    115, 118, 120, 125, 135, 214, 264, 328,
    332, 338, 348
climate change levy 217–9
Climate Change Programme (UKCCP) 217,
  221
Climate Change Technology Plan
  (CCTP) 45–7
climate justice 332, 341
  see also environmental justice
Climate Protection Initiative 212
climate science 46, 52, 113, 322
Clinton Climate Initiative 93

Clinton, Hillary 353–4
Clinton, William J. 7, 30, 52, 93, 106, 315
CNA Corporation 285, 286, 288
CO2 *see* emissions, types
coal 55, 75, 87, 103, 112, 113, 114, 117, 122,
  124, 165, 199, 200, 206, 209, 214,
  227, 229, 230, 241, 267, 268, 271,
  272, 273, 275, 368
  see also fossil fuels; natural gas; oil
coal-fired 117, 122, 124, 200, 209, 229
co-decision 145
collective action 64, 126, 160, 267, 356–7
combined heat and power (CHP) 209, 212,
  219
comitology 251
command-and-control 29, 30, 62, 193, 323
Commerce Clause 24
Commission *see* European Commission
Committee on Climate Change
  (CCC) 222, 223
common but differentiated responsibility
  (CBDR) 337–41
common coordinated policies and measures
  (CCPMs) 164, 245
Common Foreign and Security Policy
  (CFSP) 302
compensation 231, 273
competence 32, 144, 146–7, 149, 152–8,
  190, 270, 319, 329
competitiveness 30, 41, 54, 155, 218, 280–1,
  324, 336
compliance 3, 10, 17, 29, 33, 38, 53, 73, 101,
  106, 136, 138, 148, 151, 159–61,
  171, 176, 183, 206, 211, 223, 225,
  228, 229, 252, 256, 257, 264, 320,
  340, 349, 360
Conference of the Parties to the
  UNFCCC (COP)
  COP-1 7, 8, 356
  COP-2 8
  COP-15 356
Congress 7, 8, 21–2, 25–31, 36, 38, 40, 47,
  49–56, 73, 76, 81–2, 86, 92, 116,
  118, 123, 125, 198, 244, 248,
  249–50, 254, 266–8, 275, 286–8,
  298, 305, 319–21, 349, 363, 370
Connecticut 65, 72–4, 101–3, 113–14, 267
consensus 11, 18, 26, 29, 50, 56, 82, 135,
  195, 197, 232, 242, 244, 256,
  263–67, 270–5, 301, 318, 327, 328,
  330–2, 348–50, 356–62
Constitution 21–4, 66, 148, 194–6, 249
Convention on Biological Diversity
  (CBD) 315
cooperation 3–5, 11–17, 29, 44, 60, 70,
  76–7, 89, 92, 137, 140, 144, 156,
  166, 179, 181, 192, 196–7, 199–20,

Index

232, 250, 256–7, 261, 272, 275, 292, 302, 303, 304
Copenhagen Accord 179, 355–59, 365
Copenhagen Climate Change Conference 12, 45, 160, 307, 347–55, 357–8, 362
corporate average fuel economy (CAFE) 53, 111
see also motor vehicle emissions
Council of Ministers 12, 189, 226, 251, 367
Council of the European Union (the Council) 71, 134, 144–9, 150, 151–4, 159, 163, 167–8, 171, 173, 175, 177, 187, 190, 231, 243, 249, 251–3, 257, 272–3, 289, 295–6, 317, 318, 319
Council on Environmental Quality (CEQ) 112, 249
Court of Justice of the European Communities 98–9, 144, 146, 174, 186, 312, 366
Croatia 184
Cuba 356
Cyprus 160
Czech Republic 160, 184, 187, 226, 271, 273, 308

da Silva, Luiz Inacio Lula 355
dash for gas 241
decentralized 143, 206, 251, 277, 301, 304, 320, 323–4, 342
democracy 13, 17, 192, 194, 215, 226, 253
Denmark 159, 161, 183, 185, 187, 275
Department of Defense (DOD) 286
Department of Energy and Climate Change (DECC) 224
Department of the Interior (DOI) 22, 118–19, 121, 249, 252
developed countries 7, 44, 57, 134, 184, 254, 271, 298, 302, 307, 337, 340, 356, 357
developing countries 6–11, 36, 44, 56–60, 133, 179, 207, 229, 232, 263, 264, 284, 287, 302–7, 337–41, 347, 351–7, 361
dialogue 14, 16, 18, 75, 133, 139, 140, 180–1, 191, 195, 246, 257, 292, 303, 307, 309, 316, 319, 330, 333
Dimas, Stavros 283
diplomacy 33, 351, 353
directorates-general (DGs) 145–6, 161, 190
discourse 3, 47, 195, 332
dispute settlement 13, 312, 325, 326
distribution of allowances 300
District of Columbia 14, 26, 47–8, 62, 72, 85, 89, 104–5, 107, 110, 173, 242, 262, 266, 268, 276, 281, 283, 285, 314, 321, 327, 363, 365–72

Diversified Energy Initiative (DEI) 65
domestic sector 44, 164, 217, 218, 220, 247

Eastern Climate Registry (ECR) 65
Eastern European 230, 232, 271, 273
ECCP *see* European Climate Change Program; European Climate Change Program II
economic development 17, 25, 27, 46, 136, 152, 156, 230–2, 246, 268, 306–7
economic recession 52, 87, 177, 189, 213–4, 227–8, 281, 283, 342
economics 3, 12, 47, 58, 250, 279, 282, 288, 292, 307, 315, 332, 334
economies in transition (EIT) 229, 232
electricity 40–1, 53, 64, 65, 87, 90, 102, 163, 166, 169–70, 184, 191, 206–13, 217, 219, 220, 247, 298
emissions
 reduction obligation 6, 56, 162, 183
 targets 184, 247, 356, 358
 trading directive 187, 206
 types
  CO_2 5, 9, 36, 41–2, 46, 61, 63–4, 72–3, 78, 83–4, 87, 94, 96, 100–108, 115, 120–25, 134–5, 140, 161–5, 168, 171–8, 186, 188, 191, 199, 201, 203–5, 207, 211–12, 216, 219, 221–30, 241, 247, 257, 258, 266–7, 269, 272, 274, 277, 281, 292–8, 307, 321
  nitrous oxide 108, 115, 165, 171
  sulfur dioxide 63, 124, 173
 see also historical emissions
emissions reduction obligation, 11, 15, 35–6, 38, 56, 67, 72, 160, 183, 193, 215, 218, 224, 228, 238, 240, 242, 244, 245–6, 253, 262–3, 279–80, 306–7
Emissions Trading (ET) 10, 276
Emissions Trading Scheme (ETS) 151–2, 164–8, 171–8, 184–5, 187–8, 191, 206, 213–8, 228, 230, 245–6, 257–8, 263, 273–5, 280–2, 296, 304, 359
 Phase I 73, 173, 230, 259, 273, 281, 282, 303
 Phase II 73, 173, 230, 259, 273, 281, 282
 Phase III 230, 273, 281, 282
Endangered Species Act (ESA) 114–21, 138, 253
endangerment finding 55, 108, 109, 110, 121, 123, 277, 300, 351
energy *see* energy taxes; nuclear energy; oil; renewable energy
energy efficiency 15, 38, 42, 54, 59, 65–71, 76, 90, 93, 95, 163–8, 178–9, 181, 185–6, 202, 206–7, 210–21, 225, 229, 246–7, 254, 272, 281

Energy Independence and Security Act of 2007 53, 247, 298
Energy Policy Act of 2005 39, 53
Energy Policy and Conservation Act of 1975 (EPCA) 111
energy taxes 40, 202, 203, 217, 294, 296
see also carbon taxes
enforcement 22–3, 29, 98, 139, 147–8, 152, 175, 186, 190, 196, 223, 250, 251–2, 259, 319–24
Enquete Commission 200–1, 372
Environmental Action Programme (EAP) 158
Environmental Appeals Board (EAB) 122, 123
environmental impact assessment (EIA) 302
environmental impact statement (EIS) 111
environmental justice 55
see also climate justice
Environmental Protection Agency (EPA) 22–3, 28–9, 32, 39, 42, 52, 53, 55–6, 60, 79–84, 104–12, 119–25, 247–55, 259, 267–8, 277–8, 292, 300, 320–1, 351
equity 68, 222, 250, 306, 307, 310, 336, 339–42, 348, 349, 360
Estonia 160, 174, 184, 187, 230
EU-12 184–5, 228
EU-15 159–69, 172, 182–5, 191–4, 213, 216, 225, 241–6, 253, 280, 293–4, 360
EU-27 160, 182–5, 193, 228, 253, 294, 360
European Climate Change Program (ECCP) 159, 162–7, 182–3, 193, 245, 257, 293–7, 301
see also ECCP II
European Climate Change Program II (ECCP II) 66–7, 297, 301
European Commission (the Commission) 126–7, 144–55, 159, 163–70, 172–79, 181, 186–8, 190, 200, 231, 251–3, 257, 270, 272–3, 295, 301, 303, 317–19, 325–6, 340
European Community (EC) 13, 145, 148, 150–56, 158–9, 162–4, 166–77, 182–3, 186, 188, 192, 226–8, 241, 246–7, 253, 257, 270, 280, 288, 295–6, 301, 303, 311, 316, 324–5, 340
European Economic and Social Committee 175, 180–1
European Economic and Social Committee (EESC) 179
European Economic Community (EEC) 144, 157, 161, 163–4, 220
European Energy Exchange (EEX) 206
European Environment Agency (EEA) 157–9, 163, 183–7, 252, 257
European Parliament (EP) 144–6

European Union (EU)
civil society 18, 34, 114, 123, 128, 130, 140, 179, 180–1, 190, 200, 203, 226–7, 254–5, 310–1, 322–3, 333–4, 348, 352–3, 360
ethics 336–7, 341–2, 348
evolution of environmental law 23, 25
governance 12, 16, 18, 21–3, 32, 34, 96, 143–50, 154, 158, 162, 189, 190–5, 207–8, 214–15, 225–7, 232, 241, 256, 289, 310–12, 316–8, 321, 323–4, 359, 360, 362
media 3, 26, 34, 132, 201, 296, 310–1, 313, 315, 330–4, 341–2, 360, 365, 373
modes of capitalism 310, 334, 336
national security 51, 54, 102, 278, 283, 285, 286, 287, 288, 290, 291, 292
policy evolution 96
policy mix 248
political structure 255, 266
precautionary principle, 156, 202, 261, 325–7, 329, 330
risk perception, 310, 328
Executive Order 30, 70, 71, 72, 85, 86, 87
Export-Import Bank (Ex-Im) 99, 100–101

Federal Ministry for Economic Cooperation and Development 207, 291
Federal Ministry for the Environment, Nature Conservation and Nuclear Safety (BMU) 197–8, 202–9, 211–4, 216
Federal Ministry of Economics and Technology 136
federalism 21, 67, 95, 137, 138, 148, 196, 256
financial assistance 99, 231–2, 300, 340
financial crisis 12
Finland 14, 159, 161, 183
Fish and Wildlife Service (FWS) 114
flexibility mechanisms 10, 162, 166, 172, 183–6, 225, 243, 270, 276–7, 280, 361
foreign policy 102, 260, 261, 287, 292, 313, 318, 348
forestry 64, 175, 180, 183, 191, 205, 303
fossil fuel 16, 85, 88, 91, 99, 101, 124, 170, 206, 208–9, 229, 259, 268, 286, 307, 333, 335
see also coal; natural gas; oil
framing 52–3, 276, 301, 327, 329–30, 352, 360
France 161, 183–4, 187, 194, 260, 271, 275
fuel economy 41, 81–3, 104, 110–11, 247, 298–300

G77 6, 338–9
G8 58–9, 179, 202, 276, 285
General Agreement on Tariffs and Trade (GATT) 13

genetically modified organism (GMO) 326, 364
Georgia 269
German Advisory Council on Climate Change (WGBU) 201
Germanwatch 135–6
Germany 56, 58–9, 135–6, 159, 161, 182–87, 192–209, 212–6, 225, 230, 232–3, 239, 241–2, 253, 271–3, 281, 289–90, 297, 321, 324, 332, 352, 359, 364, 367, 370, 371, 372
Gerrard, Michael 112, 248, 366
glacier 131
Global Change Research Program 304
Global Climate Change Initiative 37, 44, 47, 48
Goldberg, Donald 126–7
Gore, Al 8, 30, 34, 240, 243–4, 273, 292
governance 12, 16, 18, 21–3, 32, 34, 96, 143–50, 154, 158, 162, 189–92, 195, 207–8, 214–15, 225–7, 232, 241, 256, 289, 310–12, 316–18, 321, 323–4, 359, 360, 362
  see also European Union; United States
grassroots 89, 137, 140
Greece 161, 174, 183, 184, 187, 226, 311
Green Party 195, 197, 200
greenhouse gas (GHG) 10, 37, 46, 71, 73–80, 91, 94, 101, 108, 110, 112, 123, 165, 171, 177–9, 184–5, 205, 209, 224, 228–9, 237–41, 245–6, 254, 257, 259, 262–70, 276–7, 279, 280–3, 292–302, 304, 307, 320, 326
  commitments 7–10, 15, 34, 36, 39, 41, 50, 58, 59, 66–7, 69–70, 91, 96, 136, 158–62, 165, 171–2, 176, 178–9, 183, 185, 189, 191, 194, 205, 209, 214, 216, 219, 225, 228–9, 231, 238–40, 242, 246–7, 253, 255, 257, 264, 294–5, 301, 312, 315, 338–9, 340–1, 350–60
  emissions 5–11, 15–16, 23, 35, 37–9, 41–59, 62–6, 68–96, 100–140, 151–2, 158–9, 160–79, 183–91, 193–4, 197–229, 237, 240–2, 245–7, 253–62, 265–9, 271–2, 276–7, 279–80, 282, 286, 292–301, 304, 307, 315, 326, 335–6, 339–40, 350–60
  obligations 6–11, 22, 35–6, 38, 40, 42–4, 66, 69, 88, 91, 96, 110, 112, 116, 127–8, 131, 136, 138, 150–62, 164, 167, 182–89, 193–4, 203, 214, 217, 221–2, 225, 231, 240, 244–5, 248, 250, 253, 256–8, 272, 275, 277, 280, 306, 307, 337
  reduction 7–8, 10, 15, 27, 35–9, 46, 50, 54, 56, 58, 65–7, 71, 74–80, 86, 90–1, 93, 96, 104, 115, 124, 160–66, 174–85, 188, 193, 196, 202, 204–5, 209, 212, 214, 216–7, 219, 221–9, 238, 240, 242, 244, 245–7, 253–62, 264, 266, 274, 276, 278, 280, 290, 294, 301, 303, 307, 320, 326, 338, 339, 341, 350, 356, 358, 360
  registry 39, 65, 66, 73–7, 163
  reporting 57, 61, 70, 73–6, 85, 151–2, 171, 223, 257–60, 269, 339, 340
  targets 56, 58–9, 61, 63, 67–8, 85, 136, 150, 161–70, 173, 176, 181, 182–88, 204–5, 211, 214–15, 217–18, 220–5, 231–2, 246–7, 253, 266, 269–78, 290, 295, 339, 340, 350–60
Gross Domestic Product (GDP) 194, 231, 273, 282, 307
guidance 112, 121, 223, 356, 357

harmonization 24, 143–9, 151, 155, 175, 193, 227, 297
Heads of State 59, 179, 254
hegemony 260, 334–5, 364
hot air 228
House of Representatives 21, 27, 36–7, 39, 48–9, 51, 53–5, 60, 72, 78, 83, 109, 112, 124–5, 240, 268, 277–9, 283, 287–8, 299–300, 305, 308, 319, 351, 354–5, 369
human health 13, 25, 109, 125, 154, 156, 303, 321
human rights 17, 126–8, 192, 315
Hungary 160, 184, 187, 226, 230, 271, 311
hydrofluorocarbons 108

Idaho 67, 267, 269
Illinois 64, 267
India 12, 56, 59, 237, 240, 264, 279, 289, 301, 306, 338, 347, 349, 350–55
indispensable party 237, 301, 308, 347, 352, 361
industrialization 13, 25, 310
industrialized countries 6–9, 11, 17, 240, 276, 338–41, 347
industry 24, 27, 30–1, 33, 38, 40–2, 44, 50, 54–5, 65, 76, 79, 80, 83–4, 101–2, 104, 135–6, 162, 164–5, 167, 177, 181, 184, 188, 190, 198, 200, 202–9, 212, 214, 218–19, 225–6, 230, 242, 248, 275, 281–3, 292–8, 303, 334, 336, 368
  see also business
insurance 99, 103, 114, 303
Integrated Pollution and Control Directive (IPCC) 284, 288, 293, 301, 307, 348, 351
Integrated Pollution Prevention and Control Directive (IPCC) 148, 168

intensity 37, 38, 41, 44, 46, 86, 99, 113, 165, 174, 208, 225, 228, 246, 300, 350
  carbon 41, 44, 46, 86, 165, 174, 207, 225, 228, 247, 298, 350
  energy 37, 38, 40–1, 45–6, 113, 165–6, 208, 225, 228, 246, 300, 348
  greenhouse gas 37–9, 41, 44–6, 86, 113, 165, 174, 225, 228, 245, 247, 300, 350, 351
Inter-American Commission on Human Rights (IACHR) 125–9
interest groups 27, 33–4, 197, 200, 248, 313, 319, 321, 330, 333–4
Intergovernmental Panel on Climate Change 5, 34, 47, 109, 132–3
Interministerial Working Group (IMA) 201
International Council for Local Environmental Initiatives (ICLEI) 90–2
International Court of Justice (ICJ) 128–9
International Energy Agency 282
international environmental agreements 32, 133, 315, 318
International Renewable Energies Agency (IRENA) 207
international trade 12–13, 222, 326
Inuit Circumpolar Conference 126
Ireland 159, 161, 174, 183, 220, 226, 311
isolationist 260–1
Italy 161, 182–7, 230, 271–3

Jackson, Lisa P. 52, 80–1, 84, 105–10, 119, 172, 249–50, 259, 300
Japan 9, 59, 90, 201, 237, 240, 280, 338, 371
Jiabao, Wen 350, 355
Johnson, Stephen 80
joint implementation (JI) 10, 198, 276, 280

Kansas 67, 267
Kentucky 269
Klausc, Vaclav 187, 273, 318
Kyoto Protocol 3–11, 14–17, 30–1, 34–8, 41–2, 44, 57, 63, 67, 89, 91, 93, 95, 106, 133–4, 158–66, 171–5, 179, 182–86, 191, 194, 199, 203–4, 207, 213, 216, 221, 225, 228–9, 232, 238–45, 257, 260–3, 270, 276, 278, 280, 296, 306, 312, 314–5, 338–9, 341, 349, 356–59, 366–72
  commitments 7–10, 15, 34, 36, 41, 67, 91, 158–60, 162, 165, 171–2, 179, 183, 185, 191, 194, 216, 225, 228–9, 238–40, 242, 257, 295, 312, 315, 338–9, 341, 356, 358
  flexibility mechanisms 10, 162, 166, 172, 183–6, 225, 243, 270, 276–7, 280, 361

implementation 10, 18, 29, 32, 40, 45, 86, 108, 110, 138, 155, 158, 162–6, 169, 172–3, 183–6, 190, 198, 202, 223, 227, 229, 231, 239, 244, 256, 259, 272, 300, 312, 318–24, 356–7
  negotiations 5–12, 15, 31, 47, 52, 53, 56–9, 96–7, 125, 135, 156, 158, 178–9, 187, 190, 193, 199, 203, 211–15, 228–9, 232, 237–44, 253, 255, 260–63, 270, 271, 275–82, 307–8, 313–19, 333, 336, 339–42, 347–8, 349–59, 361
  ratification 7, 9, 34, 159–60, 229, 238, 240–44, 315, 318–19, 339
  see also post-Kyoto, negotiations

laggard 232, 241, 243, 308, 359
Latvia 160, 174, 184, 230, 251
leakage 222, 281, 282
least-developed countries (LDCs) 179, 264, 284, 352
legal culture 151, 192, 227, 244, 245, 309
legally binding obligations 6, 8, 9, 34, 56, 158
linking directive 172, 173
Lithuania 160, 174, 184, 187, 225, 230
litigation 34, 60, 83, 84, 97–8, 99–104, 112–4, 121–2, 133–40, 147, 248–50, 253, 255–6, 267, 277, 299, 322–3, 334, 359
lobbying 334
local climate change initiatives 73, 138, 256
  cities 40, 63, 89, 90–5, 99, 105, 114, 139, 239, 289
  mayors 89, 91, 92
Louisiana 267–9
Luxembourg 98, 154, 159, 161, 174–5, 183, 186–7, 366

Maastricht Treaty 145, 148, 152–6, 316
Maine 50, 65–7, 72–4, 82–3, 267
Malta 160, 170, 174, 187, 275
manufacturing 10, 54, 84, 112, 230, 247, 273, 282, 300
market-based mechanism 15, 30, 39, 49, 54, 69, 75, 85, 87, 124, 246, 247, 277, 280, 282, 304, 321, 323, 329, 335
Massachusetts 39, 41, 55, 63–6, 72–3, 83, 87, 104–10, 114, 121–2, 248, 253, 255, 266–7
media 3, 26, 34, 132, 201, 296, 310–15, 330–34, 341–2, 360, 365, 373
Merkel, Angela 59, 179, 195, 196, 201, 231, 271–2, 290, 318, 319
Michigan 267–8, 328, 365, 367
Midwest Greenhouse Registry (MGR) 65
Midwestern Greenhouse Gas Reduction Accord (MGGRA) 64
Missouri 269

mitigation 3, 6, 17, 38, 54, 65, 70, 116, 222, 238, 257, 258, 259, 276, 281, 288–93, 301–6, 332, 340, 341, 353–7
Montreal Protocol on Substances that Deplete the Ozone Layer 5, 7, 9, 35, 159, 160, 203, 229, 240, 243–4, 261, 262, 276–7, 306, 340
Moon, Ban-Ki 289
motor vehicle emissions 81
 see also corporate average fuel economy (CAFE)
multilateralism 313, 314

national allocation plan (NAP) 171–2
National Climate Change Technology Initiative 45–6
National Climate Protection Programme (NCPP) 204–6
national communications 57
National Environmental Policy Act (NEPA) 100–1, 111–12, 114, 119, 138
National Highway Traffic Safety Administration (NHTSA) 109, 110, 111, 298, 300
National Oceanic and Atmospheric Administration 37, 52, 262, 304, 306
National Security Strategy, 290
natural gas 87, 90, 165, 199, 210, 223, 229, 241
Natural Resources Defense Council (NRDC) 115, 117, 356
Nebraska 67, 267, 269
Netherlands 159, 161, 171, 183, 191, 239, 270, 332, 339, 366, 371, 373
New England Governor's Conference Climate Change Action Plan 2001 (NEG/ECP) 65–7
New Jersey 52, 63, 72–3, 102, 266–7
New Mexico 67, 69–72, 75, 82–4, 267
new source review *see* Clean Air Act
New York 50, 63–4, 72–3, 93, 95, 100, 102, 110, 112, 144, 152, 191, 195, 260, 266, 267, 270, 275–82, 293, 298, 306, 313–14, 320, 323, 331, 333, 339, 348, 350, 363, 364–70, 372–3
New York City 95, 102, 110
New Zealand 90, 134, 136, 338
Nicaragua 356
Nigeria 135
Nobel Peace Prize 292, 349, 351
non-Annex I countries 237
nongovernmental organizations (NGOs) 27, 102, 114, 137, 248, 299, 312, 333, 334
nonjusticiable 107, 113
normative status 57, 338
norms 22, 128, 244, 261–3, 267, 309, 312, 321–2, 324, 332, 360
North Carolina 64, 269

North Dakota 67, 75, 267, 269
Norway 171, 338, 351
nuclear energy 195, 197, 199, 200, 208, 209, 271, 334
nuisance 103, 104, 113, 153
 public nuisance 84, 102

Obama, Barrack H. 38, 47, 49, 51–8, 80–3, 91, 104, 119, 179, 240–1, 246, 261–2, 275–9, 283, 287–8, 299–300, 307–8, 327, 348–55, 358, 367–8
obligations 6–11, 22, 35–6, 38, 40, 42–4, 66, 69, 88, 91, 96, 110, 112, 116, 127–8, 131, 136, 138, 150, 152, 156–64, 167, 182–9, 193–4, 203, 214, 217, 221–2, 225, 231, 240–8, 250, 253–8, 272, 275, 277, 280, 306–7, 337
ocean acidification 120, 121
Office for Climate Change (OCC) 224
Office of Management and Budget (OMB) 277, 278
Ohio 25, 267
oil 25–26, 29, 47–55, 99, 103, 113–14, 135, 165, 171, 211, 223, 229, 241, 267, 278, 279, 319
 see also fossil fuels
Oklahoma 67, 267, 269
Oregon 52, 63–9, 75–8, 87, 94, 112, 266, 267
Organisation for Economic Co-operation and Development (OECD) 136, 282
Overseas Private Investment Corporation (OPIC) 99, 100, 101
ozone depletion 5

Pachauri, Rajendra 284, 348
Pennsylvania 25, 72, 73, 92, 125
pH 120, 121
Philadelphia, Pennsylvania 92
photovoltaic 203
 see also solar
Poland 160, 174, 182, 184, 187, 192–4, 215, 225–33, 271–2, 275, 311
 Communist rule in 193, 226–31
polar bear 114–21, 252
policies and measures 90, 164, 166, 183–7, 217–8, 221, 225, 229, 239, 245, 276, 280, 340
political question 102–3, 113
politics 3–18, 34, 46, 85, 100, 194–5, 198, 200–2, 226, 238, 241–2, 251, 260, 268, 276, 278, 293, 314, 318, 323–6, 342, 348, 350–60, 362–72
 global/international 11, 12, 144, 187, 194, 215, 260–1, 313, 316, 317, 352
 global/international climate 56, 192, 194, 308, 317, 351, 358, 360, 361, 362

politics—*continued*
  global/international environmental 3–9, 13, 31–4, 37, 62, 128, 133, 152, 190, 244, 261, 263, 265, 286, 311–18, 325, 328, 337, 343, 353–4
pollutant 27, 29, 36, 83, 90, 105, 106, 121–4, 167, 211, 277
polluter pays 156, 199, 227
portfolio 41, 46, 72, 76, 85, 88, 95, 181, 217, 247
portfolio standard 41, 72, 95
  see also renewable energy portfolio
Portland, Oregon 63, 94, 95
Portugal 161, 183, 187, 226, 311
post-Kyoto
  agreement 11, 56–7, 59, 191, 237, 246, 254, 263–4, 301, 306–7, 340
  negotiations 12, 56–9, 190, 193, 237, 244, 253, 255, 258, 260–4, 270, 271, 275–280, 281, 307, 308, 340, 341, 342, world 12, 13, 58, 59, 60, 191, 237, 264, 302, 307, 308, 337, 347, 349, 351, 354, 358, 364, 367
poverty 10, 218, 220, 287, 290, 303
Powering the Plains Initiative (PPI) 65, 74, 75
power-sharing 24, 193, 319
precautionary principle 156, 202, 261, 325–30
  see also European Union; United States
pre-emption 24, 80–3, 99, 113–14, 138, 256
Presidential memorandum 53, 55, 104, 106
Prevention of Significant Deterioration (PSD) *see* Clean Air Act, prevention of significant deterioration
primary legislation 82, 198, 251, 252
private industry *see* industry
proportionality 146, 149–50, 251
public sector 40, 44, 75, 93, 218–19, 221, 303
public utilities 54, 65, 72, 85, 223, 300

rapidly developing economies 11, 237–8, 240–1, 347, 350–4, 361
Reagan, Ronald 26, 29, 30–1, 365
re-engagement 11, 51–9, 241, 279, 342, 358–9, 361
regime-building 9, 11, 15
regional climate change initiatives 62, 64, 269
  role of 11, 26–7, 34, 44, 47, 52, 55, 58, 64, 78, 97, 98, 101, 118, 137, 150, 168, 180–1, 237, 240, 251, 272, 303, 316, 320–5, 333–4, 347, 351
  see also specific initiatives
Regional Greenhouse Gas Initiative (RGGI) 64–5, 72–3
registry 39, 65–6, 73–5, 77, 163

regulation 13, 24, 30–6, 53, 60–3, 66, 78–9, 82–7, 97–8, 103, 111, 122–3, 137–8, 145, 149, 153, 188–90, 193, 217, 247–56, 266–8, 277, 295, 299, 311, 320–3, 327–8, 334–6, 359
  see also Clean Air Act; Clean Water Act; California
regulatory framework 10, 15–16, 56, 110, 169, 188, 207, 219, 221, 252, 256, 274
renewable energy 39–41, 52–3, 61, 64–5, 68–76, 86, 88, 91, 93, 95, 101, 134, 163, 166–70, 175, 178, 185–6, 191, 202–9, 231, 245–7, 267–9, 271, 275, 279
Renewable Energy Sources Act (EEG) 207–9
renewable obligation commitment (ROC) 40–1
requirement 67, 209, 272, 340
research 5, 14–15, 17, 35–46, 51, 54, 69, 75, 109, 130, 135, 138, 146, 166, 178–9, 189, 198, 285, 286, 290, 298, 302–3, 333
Reunification 194–6, 213–14, 242
Rhode Island 65, 72–3, 82–4, 102, 267
Rio Declaration on Environment and Development 261, 337
Rio Earth Summit 239, 314, 316, 367
risk 14, 30, 107, 127, 131, 223, 244, 281, 289, 302, 310–11, 325–31, 342, 348, 360
  assessment 34, 53, 138–9, 150, 159, 167, 221, 241, 258–9, 301, 325
  role of 11, 26–7, 34, 44, 47, 52, 55, 58, 64, 78, 97–8, 101, 118, 137, 150, 168, 180, 181, 237, 240, 251, 272, 303, 316, 320–5, 333–4, 347, 351
roadmap 16, 74, 101, 201, 362
roles and responsibilities 6, 143, 147, 248, 262, 319, 341, 353
Romania 160, 184, 230
rulemaking 29, 78, 80, 107, 111, 150, 250
Russia 9, 225, 238

Sarkozy, Nicolas 231, 352
Schwarzenegger, Arnold 71, 79, 85, 86, 87
sea level 128, 286, 305
secondary rules 22, 251, 252, 322, 323
secretariat 57
security 3, 12, 17, 32, 46–51, 55, 57, 62, 67, 127, 144–5, 157, 168, 199, 211, 215, 223–4, 260–1, 278–9, 283–92, 298, 301–6, 313, 336, 348, 349, 351–2
  environmental 3, 12, 32, 33, 51, 55, 62, 68, 127, 144, 145, 157, 168, 211, 214, 260–1, 278–9, 283–93, 296, 314–15, 336, 348, 353, 361

human 13, 127, 156, 258, 284, 302–3, 348, 350, 352
international 3, 12, 13, 17, 31, 46–51, 57, 62–4, 127, 158, 199, 215, 260–1, 279, 283–92, 301, 305–315, 336–337, 348-352, 355,
Senate 7, 8, 21, 49, 50–6, 60, 66, 72, 77–8, 86, 124, 226, 240, 268, 277, 287, 315, 318, 338
Singh, Manmohan 98, 174, 355
Single European Act of 1986 (SEA) 154
sinks *see* carbon sinks
Slovak Republic 160, 174, 184, 226, 230, 271
Slovenia 160, 185, 187
Social Democratic Party (SPD) 195, 200
social movements 27, 321
society, role of *see* European Union; United States
socio-legal 3, 16, 17, 309, 341, 343, 360
soft law 151
solar 41, 86, 169, 203, 206, 212, 228, 266, 272
see also photovoltaic
sound science 39, 262, 321, 326, 327, 330
South Africa 59, 90, 352, 355
South Carolina 268, 269
South Dakota 67, 74–5, 267, 269
Southwest Climate Change Initiative (SCCI) 65
sovereignty 18, 58, 143, 149, 192–3, 314–15, 317, 349
Spain 159, 161, 183–5, 187
specific initiatives 263
stakeholder 271
standard 50, 80–1, 86–7, 94, 114, 136, 155, 177, 191, 246, 266, 298, 312, 350
standing 56, 63, 81–2, 100–7, 113–14, 138, 215, 260, 348, 350, 353, 358
state implementation plans (SIPs) 29, 32
state initiatives 66, 304
impetus 55, 62, 199, 265
laboratories of democracy 61
see also Arizona, California, New Mexico, New Jersey, Oregon
stationary sources 110, 171, 197
see also Clean Air Act
Stern Review 47
Stern, Todd 47, 52, 56, 240
subsidiarity 24, 143, 149–50, 251, 253, 275, 319, 320
subsidies 212, 214
Sudan 356
supplementarity 239, 276–7
Supreme Court 23–4, 39, 55, 83, 104–8, 117, 121, 253
sustainability 90, 95, 157, 168, 223
sustainable development 14, 16–17, 154–6, 198, 201, 206, 227, 337, 339

Sweden 14, 159, 161, 170, 174, 183–4, 337
symbolic 60, 77, 85, 96, 137–40, 167, 191, 215, 244, 255–6, 264, 308, 351–5, 359–62

tax 39–42, 92, 205, 211, 270, 296, 298, 324
credit 70, 75, 136, 163
incentive 40, 169, 217, 247, 349
technology transfer 44, 166, 256, 340, 354, 361
Tennessee 102, 269
Tennessee Valley Authority (TVA) 102, 117, 118
Texas 41, 64, 67, 88, 232, 267, 268
tort 103, 113, 114, 140, 255, 324
transparency 57, 173–4, 226, 232, 258, 357
transport 41–2, 44, 82–8, 90, 96, 108, 112, 163–70, 175–6, 178, 180–8, 191, 204, 205–6, 212, 217–22, 245, 257, 274, 292–302
Treaty of Lisbon (Lisbon Treaty) 143–48, 156, 157
Treaty of Rome 144, 156
Treaty on European Union (TEU) 144–5, 152–6, 316
Treaty on the Functioning of the European Union (TFEU) 148, 155–7
Tuvalu 128–9, 353

uncertainty, scientific 330
unilateral 12, 139, 260–2, 281, 313–14
Union of Concerned Scientists (UCS) 51
United Kingdom (UK) 40–1, 161–2, 164, 174, 182–4, 192–4, 214–25, 232–3, 241–3, 271, 282, 289–90, 296, 319, 332–3, 336, 352, 359, 367, 369, 370–1
United Nations (UN) 10, 12, 45, 59, 129, 161, 246, 261–3, 273, 276, 284, 288–9, 303, 312, 314, 316, 325, 337, 347–51, 357, 366
United Nations Conferences on Environment and Development 6
United Nations Convention on the Law of the Sea 314–15
United Nations Framework Convention on Climate Change (UNFCCC) 3–11, 15, 34–9, 44, 46, 55–9, 93–4, 133–4, 158–60, 167, 182–3, 191, 194, 199, 200–7, 216, 225, 228, 230, 239, 240–4, 258, 260, 262, 264, 270, 280, 303, 306, 314–15, 336–40, 351, 353–8, 361
history 3, 4, 9, 12, 21, 26, 35, 77, 96, 140, 177, 192–5, 215, 233, 266, 275, 335, 348, 352, 355
objectives 4, 39, 40, 43, 48, 54, 63, 70, 90, 94, 147, 153, 156, 158, 160, 166,

United Nations Framework Convention on Climate Change (UNFCCC)—*continued*
170, 179, 207, 212–13, 216, 221, 231, 247–52, 262–3, 275, 297, 305, 315, 320, 352, 359
United Nations Security Council 289, 290, 312
United States
  civil society 18, 34, 114, 123, 128, 130, 140, 179–81, 190, 200, 203, 226–7, 254–5, 310–11, 322–3, 333–4, 348, 352–3, 360
  ethics 336–7, 341–2, 348
  evolution of environmental law 23, 25
  governance 12, 16, 18, 21–3, 32, 34, 96, 143–50, 154, 158, 162, 189–92, 195, 207–8, 214–15, 225–7, 232, 241, 255–6, 289, 310–18, 321, 323–4, 359–60, 362
  media 3, 26, 34, 132, 201, 296, 310–11, 313, 315, 330–34, 341–2, 360, 365, 373
  modes of capitalism 310, 334, 336
  national security 51, 54, 102, 278, 283, 285–92
  policy evolution 96
  policy mix 248, 260
  political structure 255
  precautionary principle 156, 202, 261, 325–7, 329, 330
  risk perception 310, 328
US Conference of Mayors (USCM) 90–2
Utah 67, 75, 122, 125, 267

values 132, 138, 168, 192, 337, 343
Venezuela 356
Vermont 49, 65, 72–3, 82–4, 102, 267, 370
Vienna Convention for the Protection of the Ozone Layer 5, 6, 35, 315

Virginia 125
voluntary agreements 78, 167, 188, 247, 295
vulnerability 48, 284, 302, 305

Wagner, Martin 126, 127, 128
Waldsterben 197
waiver *see* Clean Air Act
Washington 14, 26, 47–53, 62, 67–9, 75, 78, 85, 87, 121, 128, 173, 240, 242, 262, 266–8, 276, 281, 283, 285, 299, 314, 321, 327, 331, 363–72
waste 13, 29, 94, 156, 163, 165, 176, 190, 197, 205, 222–3, 227, 354
Waterton-Glacier International Peace Park 131
Watt-Cloutier, Sheila 126–7, 129
welfare 11, 24, 54, 105, 108–9, 138, 244, 283, 287, 323, 324, 335
West Coast Governors' Global Warming Initiative (WCG) 65
West Virginia, 267, 268, 269
Westen Climate Initiative (WCI) 65, 75, 76
Western Governor's Association (WGA) 65, 67, 77
wind energy 169, 207, 208, 212
World Heritage Committee, 131, 132, 133
World Heritage Convention (WHC) 125, 129, 130, 131, 132
World Mayors Council on Climate Change 92
World Trade Organization (WTO) 10, 12, 13, 325, 326, 369
World War II 12, 144, 195, 215, 226, 238, 260
Wyoming 67, 267, 269

Zuma, Jacob 355